Urban Ecosystem Services III

Urban Ecosystem Services III

Editors

Alessio Russo
Giuseppe T. Cirella

MDPI
Basel • Beijing • Wuhan • Barcelona • Belgrade • Novi Sad • Cluj • Manchester

Editors

Alessio Russo
School of Arts
University of Gloucestershire
Cheltenham
United Kingdom

Giuseppe T. Cirella
Faculty of Economics
University of Gdansk
Sopot
Poland

Editorial Office
MDPI
St. Alban-Anlage 66
4052 Basel, Switzerland

This is a reprint of articles from the Special Issue published online in the open access journal *Land* (ISSN 2073-445X) (available at: www.mdpi.com/journal/land/special_issues/3rd_Urban_Ecosystem_Services).

For citation purposes, cite each article independently as indicated on the article page online and as indicated below:

Lastname, A.A.; Lastname, B.B. Article Title. *Journal Name* **Year**, *Volume Number*, Page Range.

ISBN 978-3-0365-8747-9 (Hbk)
ISBN 978-3-0365-8746-2 (PDF)
doi.org/10.3390/books978-3-0365-8746-2

Cover image courtesy of Alessio Russo

Contents

About the Editors

Alessio Russo

Alessio Russo is a senior lecturer and academic course leader for the Master of Landscape Architecture at the University of Gloucestershire, Cheltenham, United Kingdom. Before joining the University of Gloucestershire, he worked in Russia as an associate professor at RUDN University in Moscow and as a professor and head of the Laboratory of Urban and Landscape Design at Far Eastern Federal University in Vladivostok. He holds a Bachelor in Science in Plant Production from the University of Naples, a Post-Graduate Specialization in Healing Garden Design from the University of Milan, and a Master in Science in Landscape Design and Planning from the University of Pisa. He received his Ph.D. in Urban Forestry from the University of Bologna. Outside of academia, Dr. Russo has worked as a landscape architect in the United Kingdom, Italy, and the United Arab Emirates, dealing with sustainable design and planning.

Giuseppe T. Cirella

Giuseppe T. Cirella is a professor of Human Geography at the Faculty of Economics, University of Gdansk, Sopot, Poland. He received his Ph.D. in Environmental Engineering (specialization: Sustainability) from Griffith University, Australia. He is the founder of the Polo Centre of Sustainability and is the director and head of research. He has acted as a principal investigator and coordinator in a number of international projects and is a reviewer and member of the editorial board of several reputed international journals on sustainability and the environment. He has extensive interdisciplinary and cross-cultural experience in socioeconomics as well as expertise in landscape architecture, urban planning, and societal development.

Preface

In today's urbanised world, where cities thrive, the importance of ecosystem services takes the spotlight. These services are crucial for the health and well-being of people. As cities face challenges, it is vital to integrate the idea of ecosystem services into urban planning and design. This fusion is key, and it is essential for city planners and decision-makers to grasp the benefits that nature can bring to the built environment.

This fusion hinges on scientific knowledge and advanced research. This foundation is what drives the creation of sustainable cities. This brings us to our third reprint–a collection of twelve chapters. These chapters are reprints from the Special Issue on Urban Ecosystem Services III, originally published by *Land*. This reprint is a testament to the ongoing teamwork between Dr Russo and Dr Cirella, both deeply dedicated to uncovering the importance of this topic.

Inside these chapters, the authors shed light on the details of ecosystem services and highlight their huge importance in different countries. We firmly believe that this reprint isn't only for students. It is just as valuable for others like policymakers, architects, ecologists, landscape architects, and urban designers and planners.

Within these pages, we provide the tools needed to bridge the gap between bustling city life and the offerings of the natural world. This reprint shows how cities and ecosystem services can coexist and offers a roadmap as our world quickly progresses towards urban green development.

Alessio Russo and Giuseppe T. Cirella
Editors

land

MDPI

Editorial

Urban Ecosystem Services: Advancements in Urban Green Development

Alessio Russo [1,*] and Giuseppe T. Cirella [2]

[1] School of Arts, Francis Close Hall Campus, University of Gloucestershire, Cheltenham GL50 4AZ, UK
[2] Faculty of Economics, University of Gdansk, ul. Armii Krajowej 119/121, 81-824 Sopot, Poland
* Correspondence: arusso@glos.ac.uk

Citation: Russo, A.; Cirella, G.T. Urban Ecosystem Services: Advancements in Urban Green Development. *Land* **2023**, *12*, 522. https://doi.org/10.3390/land12030522

Received: 12 February 2023
Accepted: 15 February 2023
Published: 21 February 2023

Urban ecosystems are under pressure as a result of rapid urbanization [1]. When compared to more profitable residential, commercial, and industrial land uses, such ecosystems are rarely economically competitive [1]. The research shows that the multifaceted wellbeing that people gain from nature has decreased as a result of the changing human–nature relationships in urban areas [2]. Thus, there is a need to incorporate nature in our urban environments to deliver several ecosystem services [3]. Addressing current urban challenges with nature-based approaches has the potential to enhance and restore ecosystem services [4]. The literature has defined "Urban Ecosystem Services" (UES) as those ecosystem services that are offered in urban and peri-urban regions by green infrastructure (GI), such as trees, wetlands, parks, lakes, street vegetation, allotment gardens, historical and botanical gardens, and green roofs [5,6]. Nowadays, UES is a key concept that is influencing how urban landscapes are planned, designed, and managed in the direction of sustainable and livable cities [7]. They are the benefits that people derive from the natural environment within urban areas. These services play a crucial role in promoting sustainability and wellbeing in cities, and their importance has been increasingly recognized in recent years [8]. Urban green development, which refers to the integration of green spaces into urban environments, has been identified as a key means of providing UES and promoting sustainability in cities [9]. For this reason, an emphasis of studying the urban ecosystem, particularly UES, is needed. After two previous Special Issues (SIs) on UES [10,11], we decided that there was a need for new studies that could contribute to the literature on sustainable urban green development. Therefore, this SI contains ten original articles and one review.

Lee et al. [12] developed a framework to diagnose and prioritize vulnerable areas of urban ecosystem regulation services that can be utilized in urban planning. In China, He et al. [13] investigated the internal scale law of the urban system at the municipal and district scales. Additionally, they carried out a spatial autocorrelation analysis using the landscape expansion index. They evaluated the connection between the urban scaling law and the compactness of urban morphological development in this way. China's urbanization process is unique, with uneven regional resource endowment and distribution impeding integrated city development [13]. Understanding the relationship between urban indicators and the size of an urban population is critical for understanding the size of a city, its state of urban economic development, and promoting the development of inter-regional macroeconomic co-urbanization [13]. Furthermore, the urban scaling law reflects the outcomes of numerous indicators such as nature, economy, and policy. It also depicts the process of urban evolution, reflects the future trend of the city's development, and broadens people's understanding of China's urbanization on a spatial scale. He et al.'s [13] findings indicate that the temporal scaling law on the city scale has a more significant linear law than the single-year scaling law. In the Global South, Brill et al. [14] evaluated how residents in the city of Cape Town understand and value the many cultural ecosystem services related to freshwater ecosystems that are provided by the various landscape features originating in an urban protected area [14]. The results show that both built infrastructure and blue infrastructure, which serves the demands of various user groups, are equally important in producing cultural services [14]. In Italy, [15] conducted a pilot study in the municipality of Lecce to analyze the amount of urban green areas at urban and suburban levels (i.e., at the district scale) using an Urban Green Index. The joint analysis of the spatial composition and configuration of urban green spaces was carried out through the integration of three landscape metrics and the Urban Landscape Services Index was estimated and mapped at the urban and suburban scale as a support measure [15]. Another study in Italy, by Oliveira et al. [16], evaluated the potential benefits of increasing tree coverage within the boundaries of the Metropolitan City of Naples. The integration of the i-Tree Canopy online tool and the life cycle assessment method were used for the quantification of ecosystem functions, including an economic perspective and pollution sequestration, to support future policies for projects and investments aimed at expanding urban green spaces [16]. The researchers evaluated the scenarios for current and potential tree coverage scenarios as well as the benefits related to them [16]. The findings showed that an additional 2.4 million trees might be planted, resulting in 51% more benefits for pollutant removal, carbon sequestration, and stormwater management [16]. In Poland, Durlak et al. [17] combined several non-invasive methods to assess the resistance of city trees' trunks to fractures as well as damage and cavities inside the trunks. This study also included a tree valuation method that is based on internal evaluations as well as comparative analyses of methodologies used abroad, notably in EU countries. The advantage of determining a tree's value is that it may be used to convey to city residents the economic benefits of monumental trees [17]. Moreover, Chen et al. [18], using physiological eye movement and heart rate variability data, looked at psychological information based on positive and negative emotional adjectives and virtual reality and photoplethysmographic technology to evaluate subject satisfaction with regard to urban green space plant community landscape scenes. The study showed that the impact of various plant community structures on people's satisfaction with urban green spaces varies. In particular, the single-layer grassland had the most visual appeal [18]. The research of people's satisfaction with urban green spaces is helpful for the sustainable development of their design and can raise people's awareness of environmental protection, according to the findings [18]. High vegetation cover in urban green spaces increased environmental satisfaction, and single-layer grassland and tree-shrub-grass composite woodland community areas improved people's physiological and psychological responses [18]. In Africa, Kabanyegeye et al. [19] assessed anthropogenic disturbances on green spaces along a Bujumbura urban–rural gradient. The research shows that Bujumbura's green spaces are primarily concentrated in urban areas, with cemeteries in peri-urban areas and sports green spaces visible all along the urbanization gradient. These green spaces are more vulnerable to trampling, which is more prevalent in administrative entities with a peri-urban morphological status than in administrative entities with an urban status. Finally, statistically significant pairwise associations of anthropogenic disturbances were found. The findings highlight the importance of protecting these green spaces from all types of anthropogenic disturbances by increasing the population's and municipal authorities' eco-responsible awareness [19].

In the midwestern United States of America, the study by Davis and Stoyko [20] aimed to identify the barriers of planting various pollinator-friendly plants in private yards, as well as to determine which incentives to plant these native plants might be the most persuasive. It also sought to ascertain whether there are any procedural knowledge gaps regarding how to plant, care for, or where to buy these pollinator friendly plants [20]. The findings indicated that the respondents do not have a strong intention to plant these native species, particularly *Asclepias syriaca*. Surprisingly, the intention to plant these does not statistically differ when assistance with prices, labor, or the availability of online resources is offered [20].

Amani-Beni et al. [21] conducted qualitative field studies in six recognized Persian gardens in four provinces of Iran using socio-cultural guidelines derived from a literature review. The findings suggested that combining elements of formal landscape design, non-edible decorative plants, and traditional artwork would increase the attractiveness of Persian gardens [21]. The study concluded that there would be an increase in ecosystem services including provisioning and cultural services if urban agriculture were implemented in Persian gardens [21]. Finally, Mokhtari et al. [22] developed a theoretical framework to better understand the interactions between social–biophysical patterns and processes that contribute to the Urban Heat Island (UHI). They conducted a theoretical review to define UHI complexity by employing the concept of dynamic heterogeneity of pattern, process, and function in the UHI phenomenon [22]. Furthermore, as a model template, a hypothetical heterogeneity spiral (i.e., driver–outcome spiral) related to the UHI was conceived. The adopted theoretical framework can provide a comprehensive view of the UHI, aiding in the understanding of UHI spatial variations in long-term studies [22].

In conclusion, it should be stressed that urban green spaces, such as parks, gardens, and green roofs, provide numerous UES, including air and water purification, climate regulation, and recreation. Urban green spaces can help to remove air pollutants and improve air quality, reducing the health impacts associated with exposure to pollutants [23,24]. They also play an important role in reducing the UHI effect, which can raise temperatures in cities to levels that are uncomfortable and even hazardous for human health [25,26].

Moreover, urban green spaces provide habitats for wildlife and can contribute to biodiversity within cities [8]. One of the key advancements in urban green development is the integration of GI into the built environment. GI refers to the network of green spaces and natural features that provide UES within urban areas. This approach to urban planning and design aims to connect and enhance existing green spaces, as well as to create new ones. The integration of GI into the built environment can adopt many forms, such as green roofs, permeable pavements, and living walls [27]. In all, urban green spaces provide a wide range of UES that are essential for promoting sustainability and wellbeing in cities. The integration of GI into the built environment and the use of sustainable design techniques represent important advancements in urban green development. These developments demonstrate a growing recognition of the importance of urban green spaces in promoting sustainability and provide a roadmap for the creation of more livable and sustainable cities in the future.

Author Contributions: Conceptualization, A.R. writing—original draft preparation, A.R.; writing—review and editing, A.R. and G.T.C. All authors have read and agreed to the published version of the manuscript.

Funding: This research received no external funding.

Data Availability Statement: All data pertaining to the research is contained within the article.

Acknowledgments: We would like to thank the team at *Land* for helping and coordinating the first three SIs of the Ecosystem Services series.

Conflicts of Interest: The authors declare no conflict of interest.

References

1. Richards, D.R.; Thompson, B.S. Urban Ecosystems: A New Frontier for Payments for Ecosystem Services. *People Nat.* **2019**, *1*, 249–261. [CrossRef]
2. Lapointe, M.; Gurney, G.G.; Coulthard, S.; Cumming, G.S. Ecosystem Services, Well-being Benefits and Urbanization Associations in a Small Island Developing State. *People Nat.* **2021**, *3*, 391–404. [CrossRef]
3. Russo, A.; Cirella, G.T. Urban Sustainability: Integrating Ecology in City Design and Planning. In *Sustainable Human-Nature Relations: Environmental Scholarship, Economic Evaluation, Urban Strategies*; Cirella, G.T., Ed.; Springer: Singapore, 2020; pp. 187–204. ISBN 978-981-15-3049-4.
4. Pan, H.; Page, J.; Cong, C.; Barthel, S.; Kalantari, Z. How Ecosystems Services Drive Urban Growth: Integrating Nature-Based Solutions. *Anthropocene* **2021**, *35*, 100297. [CrossRef]
5. Lourdes, K.T.; Hamel, P.; Gibbins, C.N.; Sanusi, R.; Azhar, B.; Lechner, A.M. Planning for Green Infrastructure Using Multiple Urban Ecosystem Service Models and Multicriteria Analysis. *Landsc. Urban Plan.* **2022**, *226*, 104500. [CrossRef]

6. Russo, A.; Escobedo, F.J.; Cirella, G.T.; Zerbe, S. Edible Green Infrastructure: An Approach and Review of Provisioning Ecosystem Services and Disservices in Urban Environments. *Agric. Ecosyst. Environ.* **2017**, *242*, 53–66. [CrossRef]
7. Tan, P.Y.; Zhang, J.; Masoudi, M.; Alemu, J.B.; Edwards, P.J.; Grêt-Regamey, A.; Richards, D.R.; Saunders, J.; Song, X.P.; Wong, L.W. A Conceptual Framework to Untangle the Concept of Urban Ecosystem Services. *Landsc. Urban Plan.* **2020**, *200*, 103837. [CrossRef] [PubMed]
8. Grimm, N.B.; Faeth, S.H.; Golubiewski, N.E.; Redman, C.L.; Wu, J.; Bai, X.; Briggs, J.M. Global Change and the Ecology of Cities. *Science* **2008**, *319*, 756–760. [CrossRef]
9. Maas, J.; Verheij, R.A.; Groenewegen, P.P.; de Vries, S.; Spreeuwenberg, P. Green Space, Urbanity, and Health: How Strong Is the Relation? *J. Epidemiol. Community Health* **2006**, *60*, 587–592. [CrossRef]
10. Russo, A.; Cirella, G.T. Urban Ecosystem Services: Current Knowledge, Gaps, and Future Research. *Land* **2021**, *10*, 811. [CrossRef]
11. Russo, A.; Cirella, G.T. Urban Ecosystem Services: New Findings for Landscape Architects, Urban Planners, and Policymakers. *Land* **2021**, *10*, 88. [CrossRef]
12. Lee, D.; Oh, K.; Suh, J. Diagnosis and Prioritization of Vulnerable Areas of Urban Ecosystem Regulation Services. *Land* **2022**, *11*, 1804. [CrossRef]
13. He, Q.; Huang, L.; Li, J. Rediscovering the Scaling Law of Urban Land from a Multi-Scale Perspective—A Case Study of Wuhan. *Land* **2022**, *11*, 914. [CrossRef]
14. Brill, G.C.; Anderson, P.M.L.; O'Farrell, P. Relational Values of Cultural Ecosystem Services in an Urban Conservation Area: The Case of Table Mountain National Park, South Africa. *Land* **2022**, *11*, 603. [CrossRef]
15. Valente, D.; Marinelli, M.V.; Lovello, E.M.; Giannuzzi, C.G.; Petrosillo, I. Fostering the Resiliency of Urban Landscape through the Sustainable Spatial Planning of Green Spaces. *Land* **2022**, *11*, 367. [CrossRef]
16. Oliveira, M.; Santagata, R.; Kaiser, S.; Liu, Y.; Vassillo, C.; Ghisellini, P.; Liu, G.; Ulgiati, S. Socioeconomic and Environmental Benefits of Expanding Urban Green Areas: A Joint Application of i-Tree and LCA Approaches. *Land* **2022**, *11*, 2106. [CrossRef]
17. Durlak, W.; Dudkiewicz, M.; Milecka, M. A Combined Methods of Senile Trees Inventory in Sustainable Urban Greenery Management on the Example of the City of Sandomierz (Poland). *Land* **2022**, *11*, 1914. [CrossRef]
18. Chen, X.; Wang, Y.; Huang, T.; Lin, Z. Research on Digital Experience and Satisfaction Preference of Plant Community Design in Urban Green Space. *Land* **2022**, *11*, 1411. [CrossRef]
19. Kabanyegeye, H.; Sikuzani, Y.U.; Sambieni, K.R.; Mbarushimana, D.; Masharabu, T.; Bogaert, J. Analysis of Anthropogenic Disturbances of Green Spaces along an Urban–Rural Gradient of the City of Bujumbura (Burundi). *Land* **2023**, *12*, 465. [CrossRef]
20. Davis, A.; Stoyko, J. Barriers to Native Plantings in Private Residential Yards. *Land* **2022**, *12*, 114. [CrossRef]
21. Amani-Beni, M.; Xie, G.; Yang, Q.; Russo, A.; Khalilnezhad, M.R. Socio-Cultural Appropriateness of the Use of Historic Persian Gardens for Modern Urban Edible Gardens. *Land* **2021**, *11*, 38. [CrossRef]
22. Mokhtari, Z.; Barghjelveh, S.; Sayahnia, R.; Qureshi, S.; Russo, A. Dynamic and Heterogeneity of Urban Heat Island: A Theoretical Framework in the Context of Urban Ecology. *Land* **2022**, *11*, 1155. [CrossRef]
23. Junior, D.P.M.; Bueno, C.; da Silva, C.M. The Effect of Urban Green Spaces on Reduction of Particulate Matter Concentration. *Bull. Environ. Contam. Toxicol.* **2022**, *108*, 1104–1110. [CrossRef] [PubMed]
24. Diener, A.; Mudu, P. How Can Vegetation Protect Us from Air Pollution? A Critical Review on Green Spaces' Mitigation Abilities for Air-Borne Particles from a Public Health Perspective—With Implications for Urban Planning. *Sci. Total Environ.* **2021**, *796*, 148605. [CrossRef] [PubMed]
25. Kruize, H.; van der Vliet, N.; Staatsen, B.; Bell, R.; Chiabai, A.; Muiños, G.; Higgins, S.; Quiroga, S.; Martinez-Juarez, P.; Aberg Yngwe, M.; et al. Urban Green Space: Creating a Triple Win for Environmental Sustainability, Health, and Health Equity through Behavior Change. *Int. J. Environ. Res. Public Health* **2019**, *16*, 4403. [CrossRef]
26. Wang, C.; Ren, Z.; Dong, Y.; Zhang, P.; Guo, Y.; Wang, W.; Bao, G. Efficient Cooling of Cities at Global Scale Using Urban Green Space to Mitigate Urban Heat Island Effects in Different Climatic Regions. *Urban For. Urban Green.* **2022**, *74*, 127635. [CrossRef]
27. Langemeyer, J.; Wedgwood, D.; McPhearson, T.; Baró, F.; Madsen, A.L.; Barton, D.N. Creating Urban Green Infrastructure Where It Is Needed—A Spatial Ecosystem Service-Based Decision Analysis of Green Roofs in Barcelona. *Sci. Total Environ.* **2020**, *707*, 135487. [CrossRef]

land

Article

Diagnosis and Prioritization of Vulnerable Areas of Urban Ecosystem Regulation Services

Dongwoo Lee [1], Kyushik Oh [2,*] and Jungeun Suh [2]

1 Research Institute of Spatial Planning & Policy, Hanyang University, Seoul 04763, Korea
2 Department of Urban Planning and Engineering, Hanyang University, Seoul 04763, Korea
* Correspondence: ksoh@hanyang.ac.kr

Abstract: Rapid urbanization and population growth have led to drastic degradation of urban ecosystem regulation services (ERS). Urgently needed is the identification of vulnerable areas where ERS are being intensively deteriorated, and preparation of measures to respond to them. This study developed a framework to diagnose and prioritize vulnerable areas of urban ERS. The vulnerability of urban ERS that include carbon storage capacity, flood-risk mitigation capacity, and heat stress reduction capacity was diagnosed with a resolution of 100 m × 100 m grid. Priority areas to improve urban ERS were delineated using hot spot analysis, and the diagnosed results of the urban ERS were categorized by eight combination types including exposure, sensitivity, and adaptability. The spatial and societal problems included in the priority areas were further investigated by overlaying hot spot areas with eight combination maps. Finally, spatial management measures for the priority areas were suggested based on the analysis results. From the detailed diagnosis results of the vulnerable ERS areas, this study provides a framework to link the concept of ERS vulnerability with urban planning. Furthermore, effective spatial planning guidelines can be prepared to improve urban ERS by spatially delineating priority areas to improve urban ERS vulnerability.

Keywords: urban ecosystem regulation services; vulnerability diagnosis; hotspot analysis; spatial combinations; urban planning

Citation: Lee, D.; Oh, K.; Suh, J. Diagnosis and Prioritization of Vulnerable Areas of Urban Ecosystem Regulation Services. *Land* **2022**, *11*, 1804. https://doi.org/10.3390/land11101804

Academic Editors: Alessio Russo and Giuseppe T. Cirella

Received: 5 September 2022
Accepted: 12 October 2022
Published: 14 October 2022

1. Introduction

The idea of ecosystem service (ES) emerged in the 1970s, and definitions and meanings of ES have been developed through literature research [1,2] to the 1990s. One of the most widely used definitions was established by millennium ecosystem assessment [3], which states that ES includes the benefits that humans receive from nature for human health and wellbeing [4,5]. The functions of ES are categorized as "provisioning, regulating, cultural, and supporting". Based on this conceptual framework, research has been mainly conducted by focusing on quantifying the social and economic value of ES, trade-offs among functions of ES, and mapping methods [6]. Based on scientific research results, there have been continuous attempts to apply the concept of ES as a planning and design tool to achieve sustainability [7–9].

In particular, ecosystem regulating services (ERS), including purification of air and water, climate regulation, carbon sequestration, and runoff mitigation and flood control, serve to promote safe environments for citizens [10–12]. ERS are often useful in improving environmental quality of urban spaces while mitigating unwanted negative impacts due to human development. Therefore, efforts are increasing worldwide to improve the multiple functions of ERS using urban planning [13,14]. Despite these efforts however, ERS are often ignored or undervalued in the planning process because most results of studies related to urban ERS assessment have been mainly focused on quantitative "explanations" of current urban ERS conditions rather than practical "applications" in urban planning [14,15].

In order to enhance ERS through urban planning, areas that have high vulnerability should be spatially delineated [16,17]. Vulnerability assessment studies in the field of ecology have been relatively recently launched compared with other social science studies [18]. Moreover, the definition of ecosystem vulnerability is applied differently according to the purpose of each study and generally, it is defined as the degree of experiencing harm due to exposure to hazards [19]. The conceptual ecosystem vulnerability diagnosis model consisting of exposure, sensitivity, and capacity to recover was developed by Van Straalen [20]. This model has also been applied as a basic analysis framework for the International Panel of Climate Change (IPCC) vulnerability assessment. Applying this conceptual model, studies on ecosystem vulnerability have been mainly diagnosed on the regional scale for specific systems such as river basins, coal fields, watersheds, wetlands, and mountain regions [18]. Through such studies, vulnerable areas of ecosystem degradation [21,22] and the potential impact on ecosystems by land use changes including urban development, restoration projects, etc. [23] have been analyzed. Recently, attempts to involve social and economic factors in ecosystem vulnerability diagnosis have been increasing. However, vulnerability assessments have still mainly focused on natural and physical factors, while population and social infrastructure have received little consideration [24,25]. Furthermore, most studies have primarily focused on the vulnerability of natural ecosystems at the regional scale (low resolution), and thus it remains difficult to specify as to which spaces are at risk in entire urban spaces [26]. Such limitations make it less effective to apply vulnerability assessment results directly to urban planning.

On the other hand, identifying priority areas to improve environmental quality in urban planning has been an important research topic for decades. The spatial occurrences and frequency of environmental problems, or accessibility to environmental goods and services, have been mainly analyzed to identify priority areas. Traditionally, ES research focused on the investigation of priority areas for conservation [27,28]. In recent years, the provision of green infrastructure (GI) to improve existing ERS has been widely conducted in the urban planning process [14,29,30], and research to identify priority areas for GI supply to mitigate flood risk and urban heat islands is also being actively conducted [31–34]. Such efforts stem from the fact that environmental inequality for citizens has a direct negative effect on the health as well as safety of citizens [35,36]. Thus, more efforts are being made to supply environmental goods and services to the identified priority areas. In this regard, spatial and societal problems in priority areas should be investigated to ensure the effectiveness of the improvements [37].

As mentioned above, the importance of ERS is increasing as a planning element to achieve the sustainability of cities. Due to the drastic degradation of urban ERS by rapid urbanization, urgently needed is the identification of priority areas where ERS are being intensively deteriorated. Systematic and concrete investigation of spatial and societal problems in priority areas is also necessary in order to establish effective improvements. However, studies that diagnose and prioritize vulnerable areas of ERS in detail for entire urban areas are still insufficient.

The purpose of this study was to develop a framework to diagnose and prioritize vulnerable areas of urban ERS that can be utilized in urban planning. To achieve this, a framework of vulnerability diagnosis of urban ERS was established, and the priority areas needing improvement were delineated by hot spot analysis. Spatial and societal problems were also investigated by combining the urban ERS vulnerability diagnosis results to determine why the priority areas have high vulnerability. Finally, this study suggests future spatial management measures to reduce ERS vulnerability in priority areas.

2. Materials and Methods

This study consists of four parts: (1) definition of urban ERS vulnerability concepts and establishment of an urban ERS vulnerability diagnosis framework, (2) diagnosis of urban ERS vulnerability, (3) prioritization of areas of high vulnerability, and (4) suggestions for spatial management measures for these areas. The detailed process is presented in

Figure 1. Among the urban ERS, this study selected carbon storage capacity (CSC), flood mitigation capacity (FMC), and heat stress reduction capacity (HSRC) as urban ERS indices because they have been emphasized through national environmental research under the direction of the Korea Ministry of Environment [38].

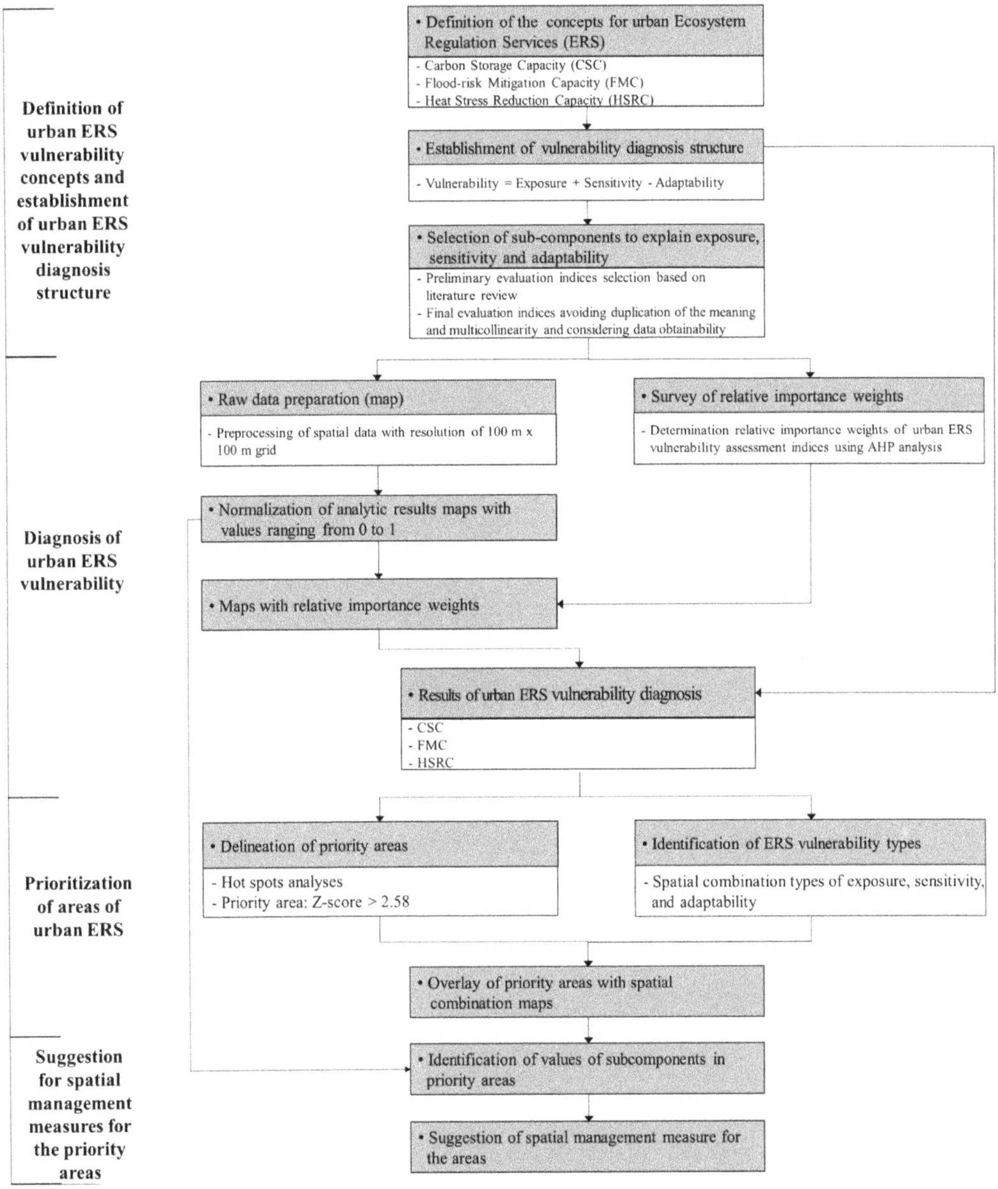

Figure 1. Framework to diagnose and prioritize vulnerable areas of urban ERS.

2.1. *The Study Area*

This study was conducted in the city of Suwon located in South Korea. Suwon is a basin-type city with a population of 1.19 million and a total area of 115 km^2 (Figure 2). Urbanization has resulted in a population increase of about 100,000 people over the last 10 years. At the same time, the natural land cover, including forests and agroforestry, has decreased by about 9 km^2, while the urbanized area has increased by about 6 km^2. A decrease in natural land cover has resulted in a reduction in carbon storage. Moreover, the continuous expansion of the urbanized area has contributed to the exacerbation of heat islands and extreme temperatures. Furthermore, over the past decade, immense damage has been caused by torrential rains that have been occurring year after year, and thus active measures to restore ERSs are urgently needed. The military facility areas located in the southern area were excluded from the analysis due to legal limitations prohibiting access.

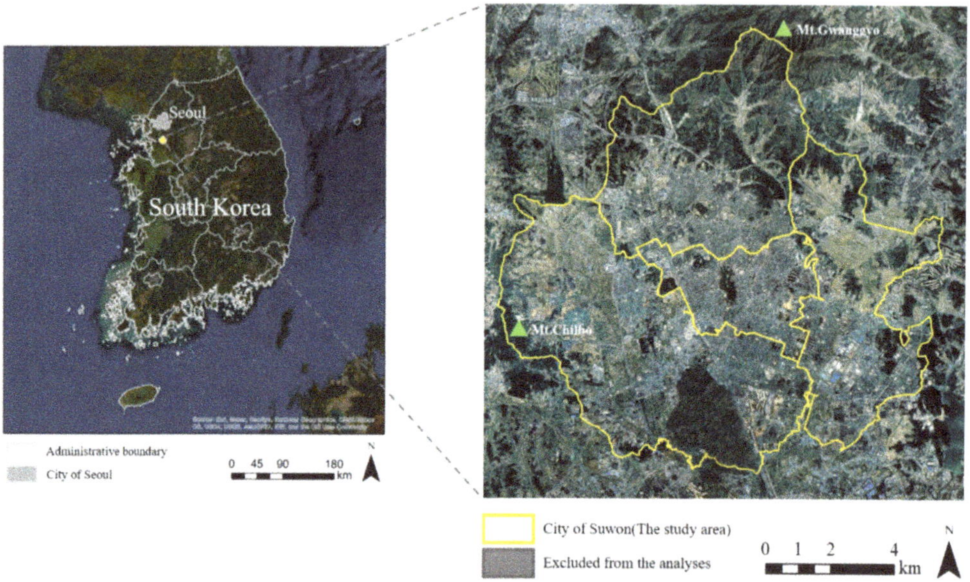

Figure 2. The study area: Suwon, South Korea.

2.2. *Definition of Urban ERS Vulnerability Concepts and Establishment of the Urban ERS Vulnerability Diagnose Framework*

Vulnerability of urban ecosystems has been defined as the "potential for loss or harm" [19], "the degree to which system, subsystem, or system components are likely to experience harm due to exposure to hazards" [39,40], "the possibility that a particular space will be damaged by external pressures" [21,22], "the possibility that external pressures will cause negative changes in natural, human, economic and social factors" [18,41], and "the side effects of natural disasters and environmental changes" [42]. Based on these ideas of vulnerability assessment, this study defined the vulnerability of ERS in the urban area as "the potential for negative changes in urban space components due to exposure". As mentioned, the three ERS indices that are currently emphasized by the Korea Ministry of Environment were selected for this study. Based on the definition of ERS vulnerability, the vulnerability of CSC, FMC, and HSRC are defined in Table 1.

Table 1. Vulnerability concepts for CSC, FMC, and HSRC.

	CSC	FMC	HSRC
Definitions	Areas with high potential to reduce carbon storage capacity in urban spaces	Areas where flooding in urban space may generate casualties and property damage	Areas where the deterioration of urban thermal environment can cause human health problems

In line with the IPCC assessment system for evaluating vulnerability assessment [43], this study employed a basic analysis structure based on the function of exposure, sensitivity, and adaptability, as shown in the following Equation (1).

$$\text{Vulnerability} = \text{Exposure} + \text{Sensitivity} - \text{Adaptability} \quad (1)$$

Exposure is a term that describes the negative impact a place or environment is subjected to, such as interference from outside a specific space, climate change, environmental change, etc. Meanwhile, sensitivity refers to the factors that render humans vulnerable to negative impacts, the degree to which humans react to stimuli, and the conditions that may leave humans at risk of such adverse effects. On the other hand, adaptability is defined as the capability to adjust to or resist potential harm, withstand outside interference, and continue functioning [22,42,44]. This study defines ERS exposure as "factors that cause direct and indirect negative impacts on urban ecosystems", sensitivity as "factors that respond easily (or quickly, or significantly) to exposure", and adaptability as "factors that resist exposure and reduce or cope with potential impacts". The concepts of exposure, sensitivity, and adaptability of urban ERS indices are outlined in Table 2.

Table 2. Three dimensions for urban ERS vulnerability diagnoses.

	CSC	FMC	HSRC
Exposure	Areas exposed to risk of carbon storage capacity loss	Areas exposed to flooding risk	Areas exposed to thermal environment deterioration
Sensitivity	Areas prone to carbon storage capacity loss	Areas prone to flood reaction	Areas prone to reaction from thermal environment deterioration
Adaptability	Areas where degraded carbon storage capacity can be restored or managed	Areas where flooding can be mitigated or managed	Areas where the deteriorated thermal environment can be restored or managed

Based on previous studies, subcomponents that elaborate upon each dimension of exposure, sensitivity, and adaptability per urban ERS index were further developed. The subcomponents were selected, and duplication of the meanings and multi-collinearity were avoided while considering data obtainability. The value ranges of the subcomponents were determined considering absolute minimum and maximum values of subcomponents, references or threshold values, and data distribution of cities similar to the study area. The selected subcomponents and their ranges are presented in Tables 3–5.

Table 3. Subcomponents for the CSC vulnerability diagnosis.

Dimensions	Subcomponents	Units	References	Scale (Min–Max)
Exposure	Vegetation cover ratio (inversed value)	%	[45,46]	0–100%
	Impervious surface area ratio	%	[47,48]	0–100%
	Ongoing and proposed development projects *	-	[47,48]	0–1
Sensitivity	Soil erosion potential	Class	[49,50]	0–3 grade
	Vegetation age class (aged vegetation)	Class	[46,51]	1–9 grade
	Long-term unexecuted facility area ** ratio	%	[52,53]	0–100%
Adaptability	Green space area ratio	%	[46,48]	0–100%
	Number of street trees	Number	[54,55]	0–132
	Budget ratio of greening and ecosystem restoration projects	%	[46,56,57]	1.07–5.74%

* Subcomponents for CSC exposure seem to be neutral in value. However, human-induced urban developments in the study area often deteriorate ecological value and the functionality of the areas substantially. Therefore, the exposure aspect for CSC in this study implies adverse impacts on urban ERS. ** Long-term unexecuted facility area: Areas designated by local governments for use as parks and green spaces but that have not been implemented for a long time. Most of them are left as vacant areas filled with trees and grassland. The areas are likely to be changed to built-up areas, and if developed, they can cause carbon storage reduction due to the large amount of soil loss.

Table 4. Subcomponents for the FMC vulnerability diagnosis.

Dimensions	Subcomponents	Units	References	Scale (min-max)
Exposure	Number of days of heavy rainfall	Days	[58,59]	0–8 days
	Impervious surface area ratio	%	[58,59]	0–100%
	Annual flooding frequency	Times	[60,61]	0–8 times
Sensitivity	Lowland area	m	[62–64]	11.30–148.16 m
	Proximity to water body	m	[58,65]	0–100 m
	Population density	Persons/ha	[59,66]	0–2997 persons/ha
Adaptability	Flood treatment facility capacity	m^3/min	[58,59]	0–3000 m^3/min
	Green space area ratio	%	[63,65]	0–100%
	River improvement ratio	%	[60,67]	0–100%

Table 5. Subcomponents for the HSRC vulnerability diagnosis.

Dimensions	Subcomponents	Units	References	Scale (Min–Max)
Exposure	Number of days of extreme heat	Days	[68,69]	0–20 days
	Number of days of tropical nights	Days	[69,70]	0–18 days
	Impervious surface area ratio	%	[71–73]	0–100%
Sensitivity	Vulnerable age population ratio (age > 65 or age < 5)	%	[69,74]	0–100%
	Low-income population ratio	%	[69,74]	0–10.13%
	Health-related vulnerable population ratio (cardiovascular, respiratory and cerebrovascular)	%	[75,76]	0–15.78%
Adaptability	Accessibility to emergency medical and rescue facilities	m	[77,78]	0–2080 m
	Proximity to green spaces and water bodies	m	[79–81]	0–447.21 m
	Accessibility to cooling centers	m	[76,82]	0–300 m

2.3. Diagnosis of Urban ERS

To link the study results with urban planning, all analytic data and results were prepared with the resolution of a 100 m × 100 m grid (polygon), which is the same size as the reference grid resolution for land suitability assessment used for urban management planning in South Korea. The analytic methods to prepare subcomponents have been presented in Appendix A. Because the selected subcomponents have different measurement units and characteristics, the process of normalizing the subcomponents for the urban ERS vulnerability assessment is required. The normalization of subcomponents is calculated by using the Min-Max method (Equation (2)), and all of the subcomponents are converted to the same range (0–1).

$$I = \frac{X - min(X)}{max(X) - min(X)} \tag{2}$$

X: Raw score of the indicator, *min (X)*: Minimum value of the indicator, *max (X)*: Maximum value of the indicator.

Based on an expert survey conducted in 2021, the relative importance (weight) of the ERS vulnerability assessment index was determined via the use of an analytic hierarchy process (AHP) analysis. The questionnaire involved respondents consisting of 60 professionals who are engaged in urban planning and environment management, including 30 academics and researchers and 30 government officials in the city of Suwon. Detailed information on the expertise and responsibilities of the expert group and survey results are given in the Appendix B.

Having used the developed diagnosis method, subcomponents, and relative importance (weight), the vulnerability of urban ERSs was evaluated. Three subcomponents per dimension were used, which were normalized in the range of 0–1 so that exposure, sensitivity, and adaptability had a distribution ranging between 0–1. Thus, the vulnerability assessment results would have a range of -1 (if exposure and sensitivity are both 0 and adaptability is 1) to 2 (if exposure and sensitivity are both 1 and adaptability is 0).

2.4. Prioritization of Vulnerable Areas of Urban ERS and Suggestions for Spatial Managements

Hot spot analysis (Getis-Ord Gi*) [83] was applied to confirm where the areas of high vulnerability are spatially clustered (Equation (3)). Hot spot analysis could determine locations of statistically significant hot spots and cold spots. Thus, the urban ERS vulnerability results by Equation (1) were inputted as x_j in Equation (3) to identify spatial clustering of high or low urban ERS vulnerability. A Z score (Gi*) and p-value for each grid were returned by hot spot analysis. According to the Z score, when Z > 2.58, which corresponds to the 99% confidence level, it is regarded as a significant high value spatial clustering. In this study, areas with a Z score > 2.58 were determined as significant aggregations of high urban ERS vulnerability (priority areas).

$$G_i^* = \frac{\sum_{j=1}^{n} w_{i,j}x_j - \overline{X}\sum_{j=1}^{n} w_{i,j}}{S\sqrt{\frac{\left[n\sum_{j=1}^{n} w^2_{i,j} - \left(\sum_{j=1}^{n} w_{i,j}\right)^2\right]}{n-1}}} \quad (3)$$

$$\overline{X} = \frac{\sum_{j=1}^{n} x_j}{n},\ S = \sqrt{\frac{\sum_{j=1}^{n} x_j^2}{n} - (\overline{X})^2}$$

where xj is the vulnerability value of grid j, and wij is the spatial weight between grid elements i and j (adjacent is 1, non-adjacent is 0), and n is the total number of grids.

From each urban ERS, eight combinations were derived from the upper and lower value maps of exposure, sensitivity, and adaptability results (Table 6). The resulting maps of eight combinations were then overlaid with the priority area maps.

Table 6. Combination of the vulnerability assessment results.

Type	Exposure	Sensitivity	Adaptability	Type	Exposure	Sensitivity	Adaptability
A	▲	▲	▲	E	▼	▲	▲
B	▲	▲	▼	F	▼	▲	▼
C	▲	▼	▲	G	▼	▼	▲
D	▲	▼	▼	H	▼	▼	▼

▲: Higher than mean, ▼: Lower than mean.

Overlaid results can provide a basic clue as to what kinds of dimensions should be mainly considered in improving priority areas. However, to establish concrete and practical measures to reduce urban ERS vulnerability in priority areas, subcomponents that increase ERS vulnerability should be further investigated. If the mean value of any subcomponents in priority areas became higher than the value corresponding to a Z score of 2.58 in the study area, the subcomponents were classified as major factors to cause high ERS vulnerability. Thus, normalized values corresponding to a Z score of 2.58 in the study area were confirmed for each subcomponent. The mean value of each subcomponent for each priority area was calculated, and the mean values were compared with normalized values corresponding to a Z score of 2.58 in the study area. For example, if the normalized value of the impervious surface ratio corresponding to a Z score of 2.58 in the study area was 0.63, and the mean value of impervious surface ratios in priority areas was 0.83, the impervious surface ratio turned out to be a major factor to increase urban ERS vulnerability. The spatial management strategies including land use management and GI installation in priority areas were then developed based on characteristics of the identified subcomponents.

3. Results

3.1. The Results of Urban ERS Diagnosis

The vulnerabilities of CSC, FMC, and HSRC were diagnosed with a resolution of 100 m × 100 m grid. CSC vulnerability showed a range of −0.63 to 1.05, with the mean value being 0.26. In addition, the vulnerability score distribution for FMC ranged from −0.56 to 0.87, with the mean value being 0.15. Finally, the vulnerability of HSRC ranged from −0.59 to 1.18, with the mean value being 0.15. Among the three kinds of dimensions, exposure was found to be a major dimension in increasing vulnerability of all urban ERS. On the other hand, sensitivity was relatively low compared with other dimensions. Such results mean that measures to reduce exposure should be prepared with priority to reduce urban ERS vulnerability in the study area (Table 7).

Table 7. Descriptive statistics of vulnerability diagnosis of urban ERS.

Indices	Vulnerability	Mean	Median	Min	Max.	S.D.
CSC	Exposure	0.47	0.49	0.04	1.00	0.18
	Sensitivity	0.07	0.00	0.00	0.60	0.09
	Adaptability	0.28	0.19	0.08	0.69	0.20
	Vulnerability	0.26	0.36	−0.63	1.05	0.32
FMC	Exposure	0.34	0.34	0.03	0.81	0.15
	Sensitivity	0.09	0.01	0.00	0.80	0.15
	Adaptability	0.28	0.26	0.00	0.70	0.18
	Vulnerability	0.15	0.19	−0.56	0.87	0.31
HSRC	Exposure	0.38	0.38	0.06	0.73	0.17
	Sensitivity	0.16	0.00	0.00	0.81	0.22
	Adaptability	0.39	0.43	0.00	0.88	0.20
	Vulnerability	0.15	0.15	−0.59	1.18	0.42

The study area is a basin type in which mountainous areas surround the city, with urban development occurring mainly in the center. This topographic condition had an influence on the vulnerability diagnosis results. The areas with relatively high exposure and the most negative impacts on urban ERS vulnerability were mainly distributed in the central part of the study area. The results of the sensitivity assessment of three urban ERS indices show spatially different distribution patterns. It was found that the sensitivity of CSC is high in areas with great soil erosion potential. The highly sensitive areas of FMC are distributed in lowland areas near the streams. In addition, in the case of HSRC, sensitivity is high in residential areas at the center of the densely populated area. As a result, the areas that have relatively high urban ERS vulnerability are distributed in the central part of the study area for all the three urban ERS indices (Figures 3–5).

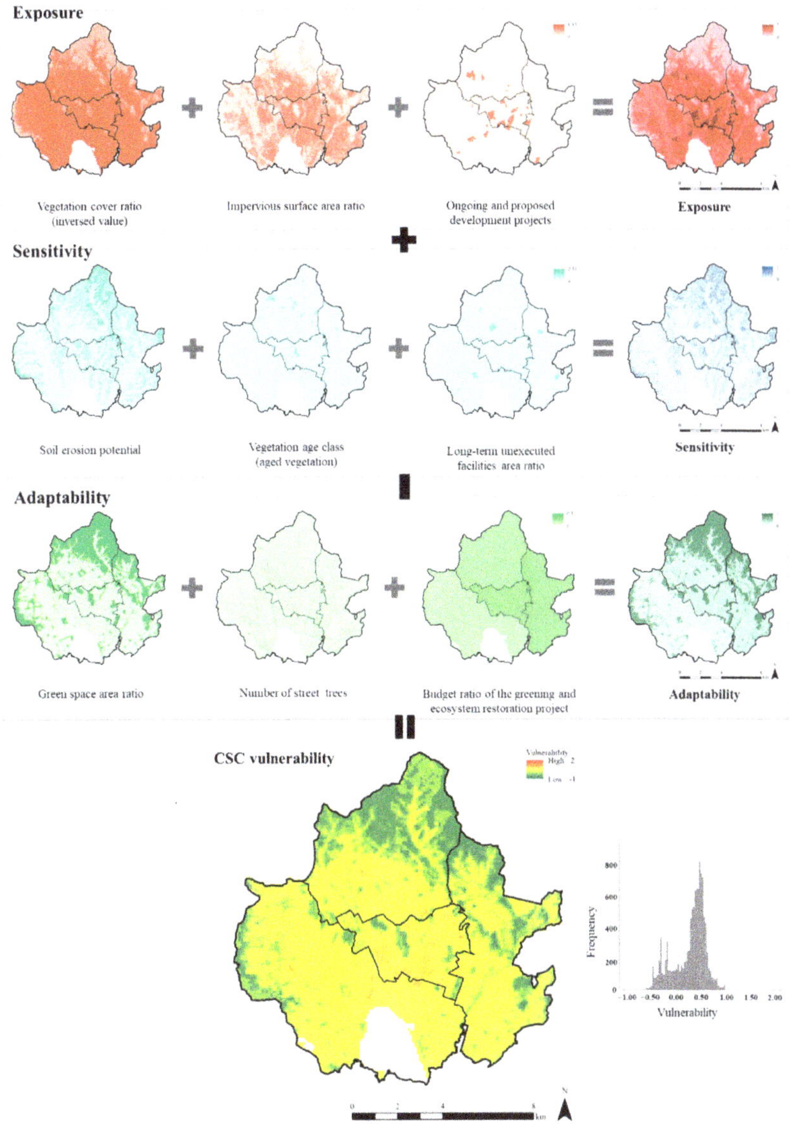

Figure 3. CSC vulnerability diagnosis.

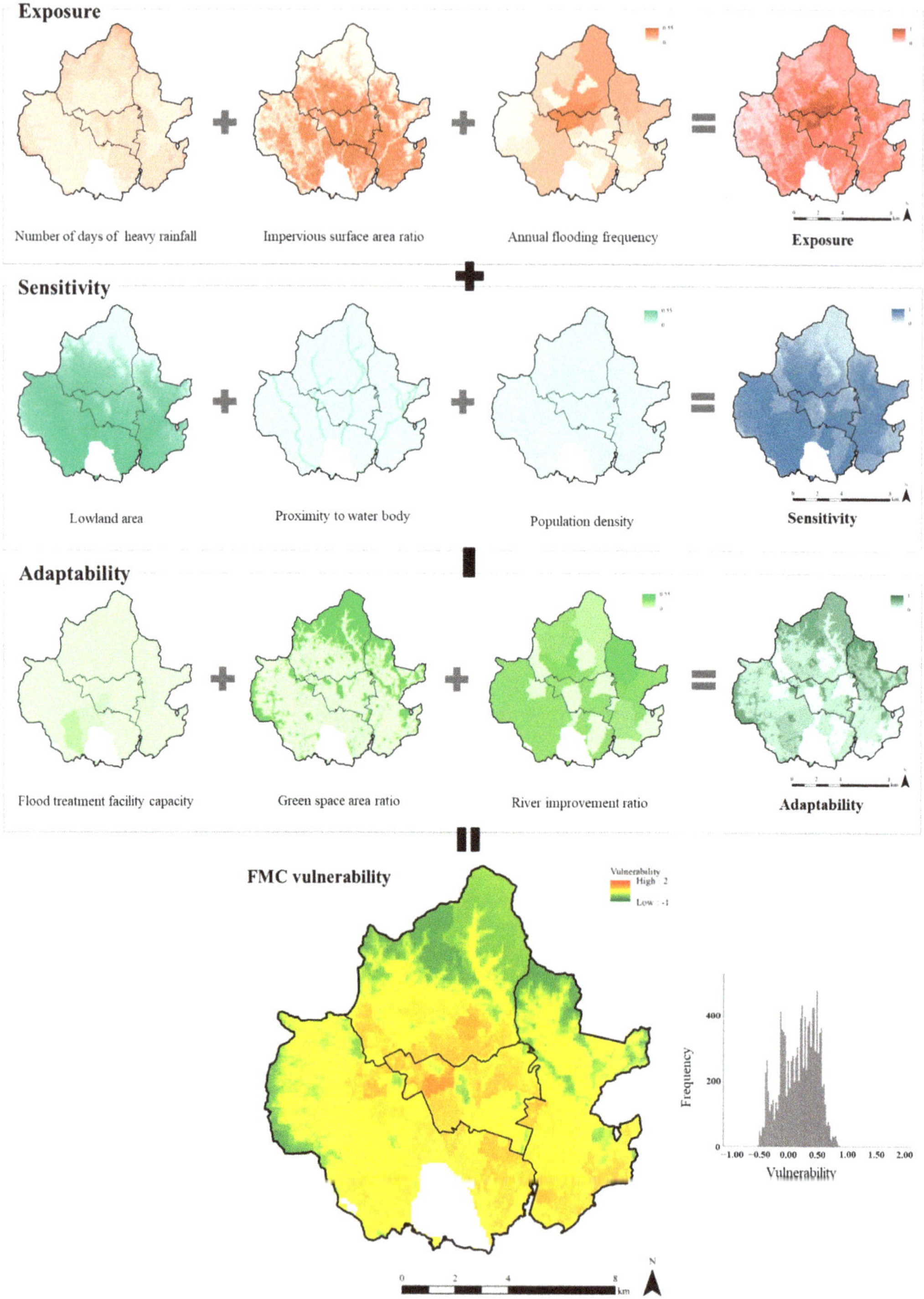

Figure 4. FMC vulnerability diagnosis.

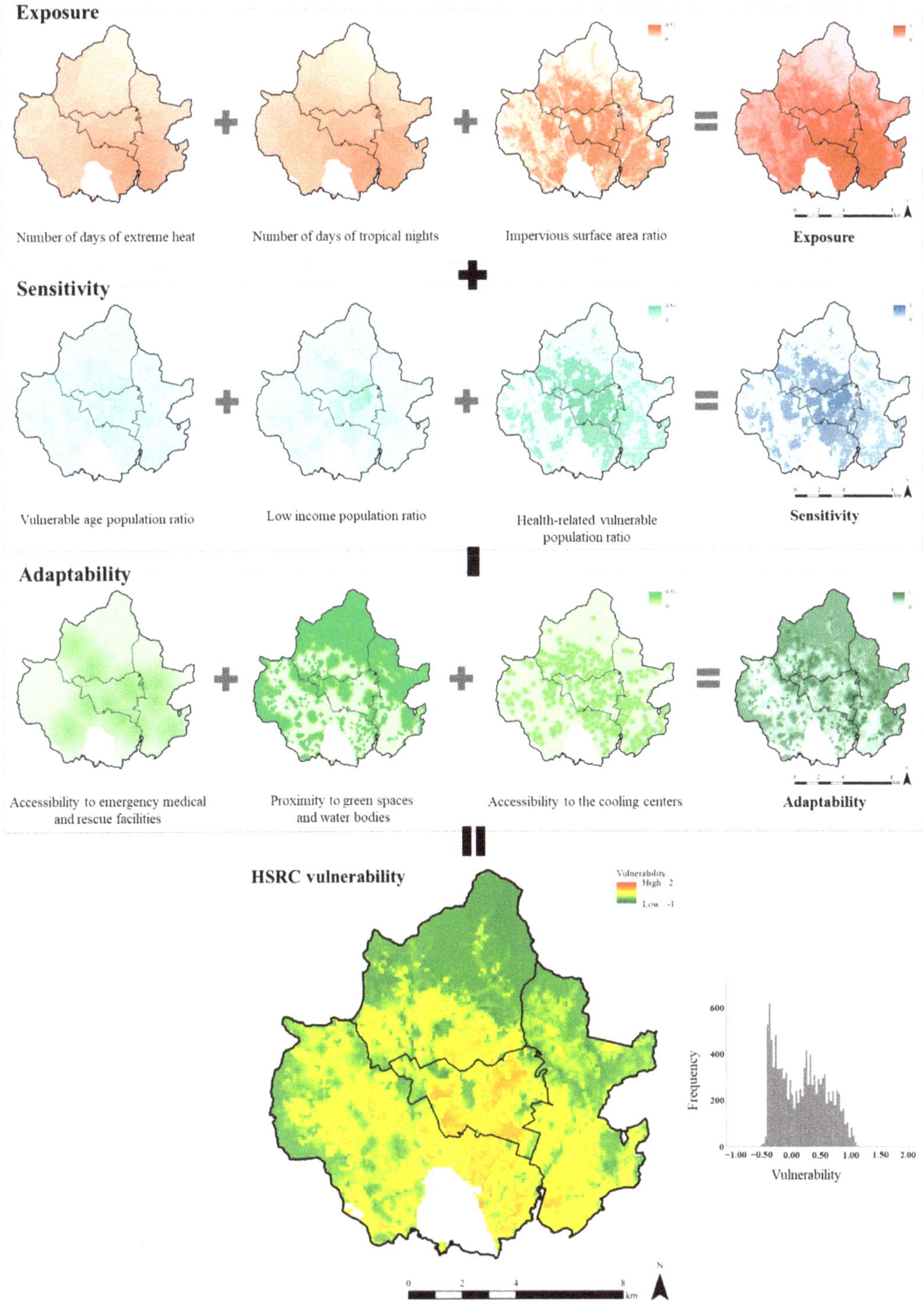

Figure 5. HSRC vulnerability diagnosis.

3.2. The Results for Priority Areas of High Urban ERS Vulnerability

As mentioned, urban spaces where improvement is most urgently needed among the vulnerable areas were identified through the hot spot analysis. The total priority areas in the diagnosis results of CSC, FMC, and HSRC comprised 1213 ha (10.02%), 2309 ha (19.32%), and 2321 ha (19.14%) of the study area, respectively (Table 8). It was confirmed that hot spot areas were mainly distributed in highly urbanized central parts of the study area (Figure 6a). The combination types of exposure, sensitivity, and adaptability were analyzed to investigate the spatial and societal problems to determine why priority areas have high vulnerability (Figure 6b). Type B of CSC, FMC, and HSRC, which is expected to be the most vulnerable, was found to be distributed across 12.98% (1488 ha), 11.91% (1365 ha) and 34% (2228 ha), respectively, of the study area.

Table 8. The spatial combinations and vulnerability in the priority areas.

Indices	The Priority Area (ha)	Spatial Combinations Type	Area (%)	Vulnerability (Mean)	Dimensions (Mean) Exposure	Sensitivity	Adaptability
CSC	1213 ha	Type B	5.29 ha (43.61%)	0.68	0.67	0.15	0.00
		Type D	6.84 ha (56.39%)	0.60	0.73	0.14	0.13
FMC	2309 ha	Type A	2.19 ha (9. 36%)	0.52	0.47	0.40	0.35
		Type B	7.94 ha (33.95%)	0.55	0.44	0.29	0.18
		Type D	8.76 ha (37.45%)	0.53	0.59	0.02	0.08
		Type F	4.5 ha (19.24%)	0.52	0.25	0.38	0.12
HSRC	2321 ha	Type A	2.45 ha (10.57%)	0.55	0.57	0.51	0.53
		Type B	16.9 ha (72.94%)	0.80	0.56	0.46	0.22
		Type D	3.43 ha (14.81%)	0.47	0.59	0.00	0.12
		Type F	0.39 ha (1.68%)	0.65	0.36	0.51	0.22

Figure 7 shows the overlaid results of priority areas with a spatial combination map. Based on the three urban ERS indices, Type B and Type D were identified as having the greatest distributions in the priority areas (Table 8). It was found that 100% of the priority areas of CSC belonged to Type B and Type D. In addition, more than 70% of the priority areas were confirmed to belong to Type B and type D in the case of HSRC and FMC. In addition, as presented in the satellite image in Figure 7, it was found that the priority areas were mainly filled with many buildings and roads. Such urbanized areas have a very high impervious area ratio, while there are few natural elements that can improve the urban ERS. Therefore, the amount of carbon emissions is very high due to energy consumption, and the CSC is relatively insufficient (Figure 7a). In addition, such areas increase the volume of surface runoff by reducing the amount of water infiltration into the ground. Because streams in those areas run faster and higher during heavy rainfall, the probability of flood risk is drastically increased (Figure 7b). Such urbanized areas degrade the thermal environment by retaining longwave thermal radiation and reducing evaporation. Furthermore, massive anthropogenic heat is added directly, resulting from human activities including modes of transportation and the use of air conditioners, and therefore, the heat stress of citizens is increased (Figure 7c).

3.3. Suggestions for Spatial Management Measures for the Priority Areas

As mentioned, the major subcomponents in the priority areas that increase ERS vulnerability were investigated (Table 9). In all ERS vulnerability indices except Type F of FMC, the impervious surface area ratio was proved to be the major factor increasing urban ERS vulnerability.

Because the priority areas have already been highly urbanized, further urbanization is likely to continue in the future. In this regard, existing green spaces and small parks should be preferentially conserved to prevent additional ERS degradation. In the long term, urban design should be highly considered to reduce building coverage ratios and increase green space ratios when redevelopments are carried out. In particular, urban development in potentially high soil erosion areas (Type B in CSC) could cause heavy carbon storage loss. The developments in lowland areas (Type A and Type F in FMC) could also bring about additional flood damage during heavy rainfall. The local government should prohibit drastic land use changes in those areas. If urban developments are inevitable, alternatives to minimize the vulnerability of urban ERS should be prepared.

The priority areas are filled with many buildings and roads, and as a result, introducing new large urban forests and parks to decrease the impervious surface ratio is not feasible. Therefore, applicable areas for additional small GI, including street trees, green roofs, porous paving material, etc., should be analyzed and applied in the priority areas. It has already been determined that introducing GI has multiple effects that enhance urban ERS. Additional carbon storage and infiltration of rainfall is possible if planting is ensured in priority areas. At the pedestrian level, trees have the advantage of reducing the amount of sensible heat due to the shadow effect. If the local government of a city provides financial incentives including subsidies and tax credits for residents to plant on their private land, such multiple effects will be further increased.

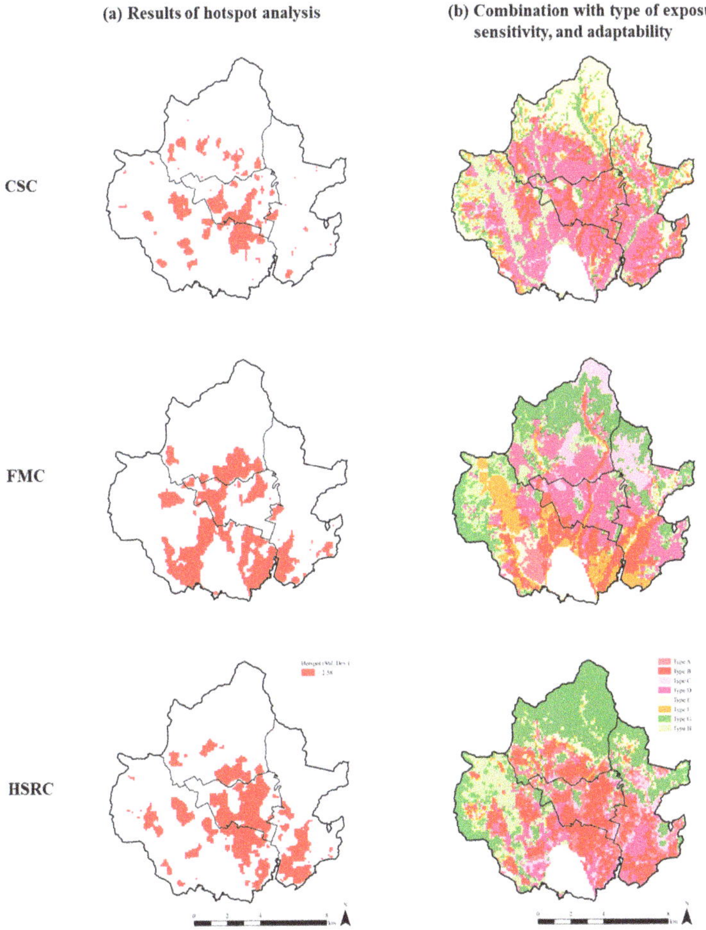

Figure 6. Results of hotspot analysis and combination with type of exposure, sensitivity, and adaptability.

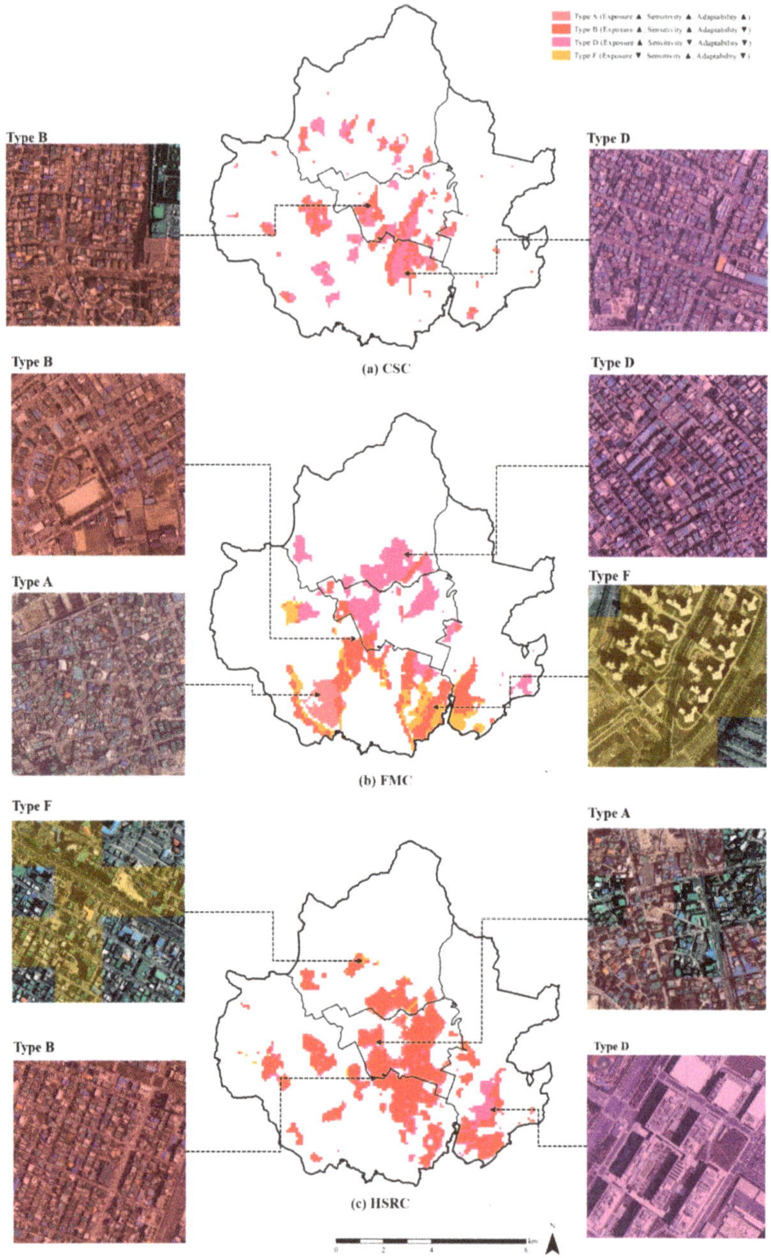

Figure 7. Overlaid results of priority areas with spatial combination maps (a: CSC; b: FMC; c: HSRC). Note: The uncolored areas in the satellite imagery (Type A of HSRC, Type B of CSC, Type F of FMC and HSRC) mean that those areas do not belong to a presented spatial type.

Table 9. Mean values of subcomponents in the priority areas.

Indices	Dimensions	Subcomponents	Type A	Type B	Type D	Type F
CSC	Exposure	Vegetation cover ratio (inversed value)	-	**0.99**	**0.99**	-
		Impervious surface area ratio		**0.87**	**0.86**	
		Ongoing and proposed development projects		0.14	0.31	
	Sensitivity	Soil erosion potential		**0.35**	0.00	
		Vegetation age class (aged vegetation)		0.00	0.00	
		Long-term unexecuted facilities area ratio		0.00	0.00	
	Adaptability	Green space area ratio		**0.00**	**0.01**	
		Number of street trees		**0.05**	**0.04**	
		Budget ratio of greening and ecosystem restoration projects		0.45	0.42	
FMC	Exposure	Number of days of heavy rainfall	0.14	0.18	0.19	0.17
		Impervious surface area ratio	**0.81**	**0.78**	**0.89**	0.37
		Annual flooding frequency	0.38	0.29	**0.58**	0.19
	Sensitivity	Lowland area	**0.68**	0.46	0.01	**0.58**
		Proximity to water body	0.02	0.07	0.01	**0.16**
		Population density	0.03	**0.48**	0.07	0.04
	Adaptability	Flood treatment facility capacity	0.30	**0.04**	**0.01**	**0.01**
		Green space area ratio	**0.08**	**0.01**	**0.02**	**0.03**
		River improvement ratio	0.71	0.51	**0.21**	**0.31**
HSRC	Exposure	Number of days of extreme heat	**0.35**	**0.35**	**0.42**	0.21
		Number of days of tropical nights	**0.61**	**0.60**	**0.70**	**0.37**
		Impervious surface area ratio	**0.89**	**0.86**	**0.77**	**0.60**
	Sensitivity	Vulnerable age population ratio	**0.19**	**0.17**	0.00	**0.21**
		Low-income population ratio	0.32	0.26	0.00	0.33
		Health-related vulnerable population ratio	**0.88**	**0.82**	0.00	**0.85**
	Adaptability	Accessibility to emergency medical and rescue facilities	0.53	0.49	0.39	0.37
		Proximity to green spaces and water bodies	0.60	**0.03**	**0.01**	**0.05**
		Accessibility to cooling centers	0.44	0.39	**0.15**	0.41

Bold: Main factors that cause high ERS vulnerability (mean value > Z score 2.58 in study area).

Finally, social welfare services provided by local governments are also needed to improve HSRC. It has been confirmed that there are many health-related groups (suffering from cardiovascular, respiratory, and cerebrovascular problems) and age-related groups (age > 65 or age < 5) vulnerable to extreme heat events that live in Type A Type B, and Type F areas. As a result, it is necessary to enhance monitoring and disaster information services for the chronically ill in those areas. Therefore, it is necessary to apply not only traditional measures including green infrastructure, but also to introduce welfare policies to improve HSRC. Moreover, local governments should consider providing visiting-care services because it may be difficult for the chronically ill who are elderly and living alone to obtain disaster information.

4. Discussion and Conclusions

Improvement of ERS in urban areas could be achieved through an organic combination with urban planning. However, the complexity of analysis of urban ERS has become a barrier for practical application to urban planning [84]. In particular, in the urban planning process, improvements to maximize the effects of ERS should be established, as there may be limited budgets and time constraints [85]. This study developed a relatively clear framework to diagnose urban ERS vulnerability and prioritize vulnerable areas. The results from the urban ERS diagnosis and for priority areas provide key information where improvement should be applied as a priority. The fact that the priority areas are mainly distributed in highly urbanized areas seems to align with observations from previous studies. However, as presented in Figure 7, although the priority areas of each dimension were partially similar, the spatial distribution of each dimension was different. Furthermore, as presented in Table 9, the main subcomponents increasing urban ERS vulnerability in each dimension and their intensity were different. Such results imply that affordable and distinct measures for each dimension and spatial type should be applied to reduce urban ERS vulnerability.

The overlaid results for the priority areas and combination map enabled the identification of the current issues of the priority areas and the formulation of customized improvement measures. Urban planners will be able to control the urban development type and intensity in the priority areas so that urban ERS will not be seriously damaged in the process of establishing land use planning. Additionally, as presented in the results, concrete and practical measures including GI applications to reduce urban ERS vulnerability in priority areas could be prepared by investigating major subcomponents that increase ERS vulnerability. This study linked the results of urban ERS diagnosis and urban planning by providing a stepwise process to reduce urban ERS vulnerability through the diagnosis of urban ERS vulnerability in the preparation of measures.

Most vulnerability assessment studies have analyzed vulnerability on a large scale (1–5 km^2 or greater) due to problems in obtaining data, and subsequently, only identify the approximate locations and causes of vulnerability. While such an analysis can serve as a reference point for macro-spatial planning, it lacks the specificity necessary for micro-spatial planning [86,87]. Spatially distinguished information to explain ERS is necessary in urban planning to enhance urban ERS [88,89]. This study used GIS spatial analysis methods such as overlay, proximity analysis, etc., to assess the micro-spatial data and construct most of the analysis subcomponents with a 100 m $\times$ 100 m resolution. The results of these specific analyses were also compiled to diagnose the vulnerability of the ERS in urban areas, which provided a basis for the establishment of specific improvement plans.

However, this study has the following limitations. Due to an absence of analytic data availability, this study did not consider air purification, noise reduction, and waste treatment, which are also important aspects of ERS. With the advent of low-cost and high-efficiency sensors, many sensors measuring urban ERS have been put into trial operation in cities in recent years. If such measurements can be gathered and analyzed, it may be possible to determine the vulnerability of other ERS in the future. The results of the vulnerability diagnosis still present only the relatively vulnerable areas and do not provide an absolute answer to the question of whether or not the areas in question are severely problematic. The usefulness of this study will be further enhanced if the results can be verified through the acquisition of relevant data and long-term monitoring. In addition, as priority areas were mainly distributed in highly urbanized areas under the developed analysis framework, another limitation of this study is that measures for suburban areas that require systematic management of urban ERS were not presented. If an analysis framework that can clearly identify areas sensitive to urban ERS changes (such as suburban or rural areas) is developed through further studies, more systematic and effective urban ERS management will be possible.

As the climate change crisis grows, interest in ERS has been steadily increasing in the urban planning field. Climate regulation and natural disaster regulation are closely related to maintaining a safe urban environment. Based on an assessment of both natural and social factors, this study diagnosed the vulnerability of ERS. By conducting a detailed study using GIS spatial analysis, it was possible to ensure the scientific nature and concreteness of the diagnosis results. Furthermore, the spatial and societal problems in priority areas were also investigated to suggest spatial management measures. The developed framework to diagnose and prioritize urban ERS could be utilized in urban planning to improve the health of urban ecosystems.

Author Contributions: This article is the result of the joint work by all the authors. K.O. supervised and coordinated work on the paper. Conceptualization, K.O.; methodology, K.O.; validation, K.O. and D.L.; formal analysis, D.L. and J.S.; data curation, D.L. and J.S.; writing—original draft preparation, D.L.; writing—review and editing, K.O. and D.L.; visualization, D.L. and J.S. All authors have read and agreed to the published version of the manuscript.

Funding: This research was funded by the Korea Ministry of Environment (MOE) grant number 2019002760002.

Data Availability Statement: Not applicable.

Acknowledgments: This work was conducted with the support of the Korea Environment Industry and Technology Institute (KEITI) through its Urban Ecological Health Promotion Technology Development Project, and funded by the Korea Ministry of Environment (MOE) (2019002760002).

Conflicts of Interest: The authors declare no conflict of interest.

Appendix A Analytic Methods for Raw Data Preparation

Table A1. CSC.

Dimensions	Subcomponents	Analytic methods
Exposure	Vegetation cover ratio (inversed value)	GIS overlay analysis application to distribute the vegetation cover by the grid in the biotope map of the study area and results were inversed
	Impervious surface area ratio	GIS overlay analysis application to distribute the ratio of impervious surface areas in the biotope map by the grid
	Ongoing and proposed development projects	GIS spatial query application to identify ongoing and proposed development projects
Sensitivity	Soil erosion potential	GIS overlay application to distribute the erosion data in soil maps by the grid
	Vegetation age class (aged vegetation)	GIS overlay application to input tree ages in biotope map by the grid
	Long-term unexecuted facility area ratio	GIS overlay application to calculate the area occupied by the long-term unexecuted facilities in the grid
Adaptability	Green space area ratio	GIS overlay application to calculate the area occupied by the green spaces in the grid
	Number of street trees	GIS spatial query application to calculate the number of street trees in the grid
	Budget ratio of greening and ecosystem restoration projects	Allocation of ratio of budget for each region corresponding to park and green space development, ecosystem restoration, and tree planting projects by the grid

Table A2. FMC.

Dimensions	Subcomponents	Analytic Methods
Exposure	Number of days of heavy rainfall	Extraction of the number of days of heavy rainfall from measurement data, with calculation carried out via interpolation in the GIS
	Impervious surface area ratio	GIS overlay analysis application to distribute the ratio of impervious surface areas in the biotope map by the grid
	Annual flooding frequency	GIS spatial query application to input the inundation frequency by the grid
Sensitivity	Lowland area	Extraction of the areas below the planned flood level from DEM data after calculating the average flood level based on the planned flood level caused by rivers
	Proximity to water body	Multiple ring buffer tool application in GIS to calculate the distance from water bodies
	Population density	Application of the source material
Adaptability	Flood treatment facility capacity	Distribution of regional statistical data into the grid
	Green space area ratio	GIS overlay application to calculate the area occupied by the green spaces in the grid
	River improvement ratio	Reallocation of regional statistical data into the grid

Table A3. HSRC.

Dimensions	Subcomponents	Analytic Methods
Exposure	Number of days of extreme heat	A temperature of 33 °C or above during summer days is selected from the measurement data, and calculation is carried out via interpolation in the GIS
	Number of days of tropical nights	Extraction of days with the lowest temperature of 25 °C or above on summer days from the measurement data, and calculation via interpolation in the GIS
	Impervious surface area ratio	GIS overlay analysis application to distribute the ratio of impervious surface areas in the biotope map by the grid
Sensitivity	Vulnerable age population ratio (age > 65 or age < 5)	Application of source material
	Low-income population ratio	Reallocation of regional statistical data by the grid
	Health-related vulnerable population ratio (cardiovascular, respiratory and cerebrovascular)	Equal distribution of regional statistical data by the grid
Adaptability	Accessibility to emergency medical and rescue facilities	Calculation of the distance from emergency medical and rescue facilities using the multiple ring buffer tool application in the GIS
	Proximity to green spaces and water bodies	Calculation of the distance from green spaces and water body boundaries using the multiple ring buffer tool application in the GIS
	Accessibility to cooling centers	Calculation of the distance from cooling centers using the multiple ring buffer tool application in the GIS

Appendix B

Table A4. Survey Respondents' Fields of Expertise.

Classification		Total		Expert		Local Government Officers	
		Frequency (Persons)	Ratio (%)	Frequency (Persons)	Ratio (%)	Frequency (Persons)	Ratio (%)
Fields of expertise	Environment	12	20.0	4	13.3	8	26.7
	Architecture	5	8.3	2	6.7	3	10.0
	Landscape/ecosystem	19	31.7	10	33.3	9	30.0
	Urban	15	25.0	11	36.7	4	13.3
	Administration	5	8.3	1	3.3	4	13.3
	Civil engineering	3	5.0	2	6.7	1	3.3
	Transportation	1	1.7	0	0	1	36.
Total		60	100	30	100	30	100

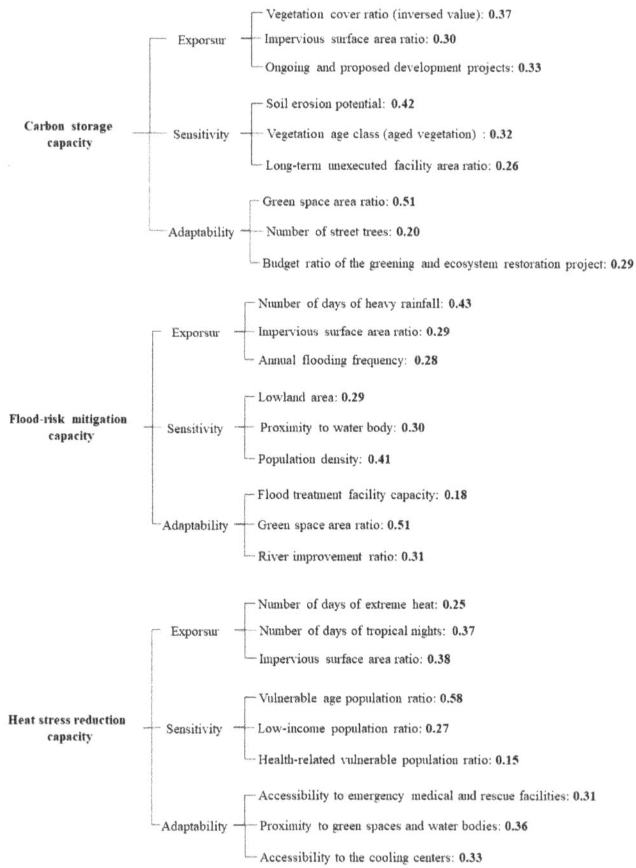

Figure A1. The Relative Importance (Weight) of Subcomponents to Diagnose Urban ERS. Note: All survey results met the consistency ratio (CR) * < 0.1. * The CR is the most representative index to identify whether respondents answered the survey with consistency. Generally, when the CR is less than 0.1, the researcher can judge if the survey results meet the consistency standard. In this study, the CR values of each dimension were identified, and results that met the standard were applied to calculate the relative importance (weight).

References

1. Costanza, R.; d'Arge, R.; De Groot, R.; Farber, S.; Grasso, M.; Hannon, B.; Limburg, K.; Naeem, S.; O'neill, R.V.; Paruelo, J. The value of the world's ecosystem services and natural capital. *Nature* **1997**, *387*, 253–260. [CrossRef]
2. De Groot, R.S. *Functions of Nature: Evaluation of Nature in Environmental Planning, Management and Decision Making*; Wolters-Noordhoff BV: Groningen, The Netherlands, 1992.
3. Millennium Ecosystem Assessment (MA). *Ecosystem and Human Well-Being: A Framework for Assessment*; World Resources Institute: Washington, DC, USA, 2003.
4. Baró, F.; Gómez-Baggethun, E.; Haase, D. Ecosystem service bundles along the urban-rural gradient: Insights for landscape planning and management. *Ecosyst. Serv.* **2017**, *24*, 147–159. [CrossRef]
5. Langemeyer, J.; Wedgwood, D.; McPhearson, T.; Baró, F.; Madsen, A.L.; Barton, D.N. Creating urban green infrastructure where it is needed–A spatial ecosystem service-based decision analysis of green roofs in Barcelona. *Sci. Total Environ.* **2020**, *707*, 135487. [CrossRef] [PubMed]
6. García-Pardo, K.A.; Moreno-Rangel, D.; Domínguez-Amarillo, S.; García-Chávez, J.R. Remote sensing for the assessment of ecosystem services provided by urban vegetation: A review of the methods applied. *Urban For. Urban Green.* **2022**, *74*, 127636. [CrossRef]
7. Ahern, J. Urban landscape sustainability and resilience: The promise and challenges of integrating ecology with urban planning and design. *Landsc. Ecol.* **2013**, *28*, 1203–1212. [CrossRef]
8. Dushkova, D.; Haase, D. Not simply green: Nature-based solutions as a concept and practical approach for sustainability studies and planning agendas in cities. *Land* **2020**, *9*, 19. [CrossRef]
9. Wang, S.; Zhuang, Y.; Cao, Y.; Yang, K. Ecosystem Service Assessment and Sensitivity Analysis of a Typical Mine–Agriculture–Urban Compound Area in North Shanxi, China. *Land* **2022**, *11*, 1378. [CrossRef]
10. Gómez-Baggethun, E.; Barton, D.N. Classifying and valuing ecosystem services for urban planning. *Ecol. Econ.* **2013**, *86*, 235–245. [CrossRef]
11. Sukhdev, P.; Wittmer, H.; Schröter-Schlaack, C.; Nesshöver, C.; Bishop, J.; Brink, P.t.; Gundimeda, H.; Kumar, P.; Simmons, B. *The Economics of Ecosystems and Biodiversity: Mainstreaming the Economics of Nature: A Synthesis of the Approach, Conclusions and Recommendations of TEEB*; UNEP: Ginebra, Suiza, 2010.
12. WHO. *Ecosystems and Human Well-Being: Health Synthesis: A Report of the Millennium Ecosystem Assessment*; World Health Organization: Geneva, Switzerland, 2005.
13. Arnold, J.; Kleemann, J.; Fürst, C. A differentiated spatial assessment of urban ecosystem services based on land use data in Halle, Germany. *Land* **2018**, *7*, 101. [CrossRef]
14. Cortinovis, C.; Geneletti, D. A framework to explore the effects of urban planning decisions on regulating ecosystem services in cities. *Ecosyst. Serv.* **2019**, *38*, 100946. [CrossRef]
15. Von Haaren, C.; Albert, C.; Barkmann, J.; de Groot, R.S.; Spangenberg, J.H.; Schröter-Schlaack, C.; Hansjürgens, B. From explanation to application: Introducing a practice-oriented ecosystem services evaluation (PRESET) model adapted to the context of landscape planning and management. *Landsc. Ecol.* **2014**, *29*, 1335–1346. [CrossRef]
16. Lee, D.; Oh, K.; Jung, S. Classifying Urban Climate Zones (UCZs) Based on Spatial Statistical Analyses. *Sustainability* **2019**, *11*, 1915. [CrossRef]
17. Weber, S.; Sadoff, N.; Zell, E.; de Sherbinin, A. Policy-relevant indicators for mapping the vulnerability of urban populations to extreme heat events: A case study of Philadelphia. *Appl. Geogr.* **2015**, *63*, 231–243. [CrossRef]
18. Beroya-Eitner, M.A. Ecological vulnerability indicators. *Ecol. Indic.* **2016**, *60*, 329–334. [CrossRef]
19. Cutter, S.L. Vulnerability to environmental hazards. *Prog. Hum. Geogr.* **1996**, *20*, 529–539. [CrossRef]
20. Van Straalen, N. Biodiversity of ecotoxicological responses in animals. *Neth. J. Zool.* **1994**, *44*, 112–129. [CrossRef]
21. Hong, W.; Jiang, R.; Yang, C.; Zhang, F.; Su, M.; Liao, Q. Establishing an ecological vulnerability assessment indicator system for spatial recognition and management of ecologically vulnerable areas in highly urbanized regions: A case study of Shenzhen, China. *Ecol. Indic.* **2016**, *69*, 540–547. [CrossRef]
22. Qiu, B.; Li, H.; Zhou, M.; Zhang, L. Vulnerability of ecosystem services provisioning to urbanization: A case of China. *Ecol. Indic.* **2015**, *57*, 505–513. [CrossRef]
23. Liu, W.; Zhan, J.; Zhao, F.; Yan, H.; Zhang, F.; Wei, X. Impacts of urbanization-induced land-use changes on ecosystem services: A case study of the Pearl River Delta Metropolitan Region, China. *Ecol. Indic.* **2019**, *98*, 228–238. [CrossRef]
24. Salas, J.; Yepes, V. Urban vulnerability assessment: Advances from the strategic planning outlook. *J. Clean. Prod.* **2018**, *179*, 544–558. [CrossRef]
25. Steenberg, J.W.N.; Duinker, P.N.; Nitoslawski, S.A. Ecosystem-based management revisited: Updating the concepts for urban forests. *Landsc. Urban Plan.* **2019**, *186*, 24–35. [CrossRef]
26. Maragno, D.; Dalla Fontana, M.; Musco, F. Mapping Heat Stress Vulnerability and Risk Assessment at the Neighborhood Scale to Drive Urban Adaptation Planning. *Sustainability* **2020**, *12*, 1056. [CrossRef]
27. Chan, K.M.A.; Shaw, M.R.; Cameron, D.R.; Underwood, E.C.; Daily, G.C. Conservation planning for ecosystem services. *PLoS Biol.* **2006**, *4*, e379. [CrossRef]
28. Egoh, B.; Reyers, B.; Rouget, M.; Bode, M.; Richardson, D.M. Spatial congruence between biodiversity and ecosystem services in South Africa. *Biol. Conserv.* **2009**, *142*, 553–562. [CrossRef]

29. Elderbrock, E.; Enright, C.; Lynch, K.A.; Rempel, A.R. A guide to public green space planning for urban ecosystem services. *Land* **2020**, *9*, 391. [CrossRef]
30. Semeraro, T.; Scarano, A.; Buccolieri, R.; Santino, A.; Aarrevaara, E. Planning of urban green spaces: An ecological perspective on human benefits. *Land* **2021**, *10*, 105. [CrossRef]
31. Chen, S.; Haase, D.; Xue, B.; Wellmann, T.; Qureshi, S. Integrating quantity and quality to assess urban green space improvement in the compact city. *Land* **2021**, *10*, 1367. [CrossRef]
32. Kopecká, M.; Szatmári, D.; Rosina, K. Analysis of urban green spaces based on Sentinel-2A: Case studies from Slovakia. *Land* **2017**, *6*, 25. [CrossRef]
33. Prybutok, S.; Newman, G.; Atoba, K.; Sansom, G.; Tao, Z. Combining co $ ting nature and suitability modeling to identify high flood risk areas in need of nature-based services. *Land* **2021**, *10*, 853. [CrossRef]
34. Towsif Khan, S.; Chapa, F.; Hack, J. Highly resolved rainfall-runoff simulation of retrofitted green stormwater infrastructure at the micro-watershed scale. *Land* **2020**, *9*, 339. [CrossRef]
35. UN-Habitat. *A New Strategy of Sustainable Neighbourhood Planning: Five Principles*; United Nations Human Settlements Programme: Nairobi, Kenya, 2014.
36. Van Kamp, I.; Leidelmeijer, K.; Marsman, G.; De Hollander, A. Urban environmental quality and human well-being: Towards a conceptual framework and demarcation of concepts; a literature study. *Landsc. Urban Plan.* **2003**, *65*, 5–18. [CrossRef]
37. Li, L.; Uyttenhove, P.; Van Eetvelde, V. Planning green infrastructure to mitigate urban surface water flooding risk–A methodology to identify priority areas applied in the city of Ghent. *Landsc. Urban Plan.* **2020**, *194*, 103703. [CrossRef]
38. Korea Ministry of Environment. *Urban Ecological Health Promotion Technology Development Project*; Korea Ministry of Environment (KME): Sejong, Korea, 2018.
39. Barnett, J.; Lambert, S.; Fry, I. The hazards of indicators: Insights from the environmental vulnerability index. *Ann. Assoc. Am. Geogr.* **2008**, *98*, 102–119. [CrossRef]
40. Turner, B.L.; Kasperson, R.E.; Matson, P.A.; McCarthy, J.J.; Corell, R.W.; Christensen, L.; Eckley, N.; Kasperson, J.X.; Luers, A.; Martello, M.L. A framework for vulnerability analysis in sustainability science. *Proc. Natl. Acad. Sci. USA* **2003**, *100*, 8074–8079. [CrossRef]
41. Shen, J.; Lu, H.; Zhang, Y.; Song, X.; He, L. Vulnerability assessment of urban ecosystems driven by water resources, human health and atmospheric environment. *J. Hydrol.* **2016**, *536*, 457–470. [CrossRef]
42. Jamshidi, O.; Asadi, A.; Kalantari, K.; Azadi, H.; Scheffran, J. Vulnerability to climate change of smallholder farmers in the Hamadan province, Iran. *Clim. Risk Manag.* **2019**, *23*, 146–159. [CrossRef]
43. IPCC. *2006 IPCC Guidelines for National Greenhouse Gas Inventories, Prepared by the National Greenhouse Gas Inventories Programme*; IGES: Kamiyamaguchi, Japan, 2006.
44. Brooks, N.; Adger, W.N.; Kelly, P.M. The determinants of vulnerability and adaptive capacity at the national level and the implications for adaptation. *Glob. Environ. Change* **2005**, *15*, 151–163. [CrossRef]
45. Asner, G.P.; Archer, S.; Hughes, R.F.; Ansley, R.J.; Wessman, C.A. Net changes in regional woody vegetation cover and carbon storage in Texas drylands, 1937–1999. *Glob. Change Biol.* **2003**, *9*, 316–335. [CrossRef]
46. Wu, J.; Chen, B.; Mao, J.; Feng, Z. Spatiotemporal evolution of carbon sequestration vulnerability and its relationship with urbanization in China's coastal zone. *Sci. Total Environ.* **2018**, *645*, 692–701. [CrossRef]
47. Ren, Y.; Yan, J.; Wei, X.; Wang, Y.; Yang, Y.; Hua, L.; Xiong, Y.; Niu, X.; Song, X. Effects of rapid urban sprawl on urban forest carbon stocks: Integrating remotely sensed, GIS and forest inventory data. *J. Environ. Manag.* **2012**, *113*, 447–455. [CrossRef]
48. Steenberg, J.W.; Millward, A.A.; Nowak, D.J.; Robinson, P.J.; Ellis, A. Forecasting urban forest ecosystem structure, function, and vulnerability. *Environ. Manag.* **2017**, *59*, 373–392. [CrossRef] [PubMed]
49. Jin, K.; Cornelis, W.; Gabriels, D.; Baert, M.; Wu, H.; Schiettecatte, W.; Cai, D.; De Neve, S.; Jin, J.; Hartmann, R. Residue cover and rainfall intensity effects on runoff soil organic carbon losses. *Catena* **2009**, *78*, 81–86. [CrossRef]
50. Li, Z.; Liu, C.; Dong, Y.; Chang, X.; Nie, X.; Liu, L.; Xiao, H.; Lu, Y.; Zeng, G. Response of soil organic carbon and nitrogen stocks to soil erosion and land use types in the Loess hilly–gully region of China. *Soil Tillage Res.* **2017**, *166*, 1–9. [CrossRef]
51. Escobedo, F.; Varela, S.; Zhao, M.; Wagner, J.E.; Zipperer, W. Analyzing the efficacy of subtropical urban forests in offsetting carbon emissions from cities. *Environ. Sci. Policy* **2010**, *13*, 362–372. [CrossRef]
52. Choi, J.; Lee, S. Change of Carbon Fixation and Economic Assessment according to the Implementation of the Sunset Provision. *Ecol. Resilient Infrastruct.* **2020**, *7*, 126–133.
53. Sung, W.; Choi, J.; Yu, J.; Kim, D.; Son, S. Impact Assessment of Vegetation Carbon Absorption and Economic Valuation Under Long-term Non-executed Urban Park Development. *J. Korea Acad. -Ind. Coop. Soc.* **2020**, *21*, 361–371.
54. Stoffberg, G.H.; van Rooyen, M.W.; van der Linde, M.J.; Groeneveld, H.T. Carbon sequestration estimates of indigenous street trees in the City of Tshwane, South Africa. *Urban For. Urban Green.* **2010**, *9*, 9–14. [CrossRef]
55. Tang, Y.; Chen, A.; Zhao, S. Carbon storage and sequestration of urban street trees in Beijing, China. *Front. Ecol. Evol.* **2016**, *4*, 53.
56. Lu, F.; Hu, H.; Sun, W.; Zhu, J.; Liu, G.; Zhou, W.; Zhang, Q.; Shi, P.; Liu, X.; Wu, X. Effects of national ecological restoration projects on carbon sequestration in China from 2001 to 2010. *Proc. Natl. Acad. Sci. USA* **2018**, *115*, 4039–4044. [CrossRef]
57. Park, H.; Oh, K.; Lee, S. Analysing effects of CO_2 absorption capability through enhancing urban green infrastructure in Seoul. *J. Korean Urban Manag. Assoc.* **2014**, *27*, 1–23.

58. Balica, S.; Douben, N.; Wright, N.G. Flood vulnerability indices at varying spatial scales. *Water Sci. Technol.* **2009**, *60*, 2571 2580. [CrossRef] [PubMed]
59. Munyai, R.B.; Musyoki, A.; Nethengwe, N.S. An assessment of flood vulnerability and adaptation: A case study of Hamutsha-Muungamunwe village, Makhado municipality. *Jamba* **2019**, *11*, 692. [CrossRef]
60. Ferdous, M.R.; Wesselink, A.; Brandimarte, L.; Di Baldassarre, G.; Rahman, M.M. The levee effect along the Jamuna River in Bangladesh. *Water Int.* **2019**, *44*, 496–519. [CrossRef]
61. Singh, P.; Sinha, V.S.P.; Vijhani, A.; Pahuja, N. Vulnerability assessment of urban road network from urban flood. *Int. J. Disaster Risk Reduct.* **2018**, *28*, 237–250. [CrossRef]
62. Kang, J.; Oh, K. Establishing Flood Vulnerability Assessment Indices for Climate Change Adaptation and its Application: The Case of theSeoul Metropolitan Area. *Korean Urban Manag. Assoc.* **2014**, *27*, 43–67.
63. Salata, S.; Ronchi, S.; Giaimo, C.; Arcidiacono, A.; Pantaloni, G.G. Performance-Based Planning to Reduce Flooding Vulnerability Insights from the Case of Turin (North-West Italy). *Sustainability* **2021**, *13*, 5697. [CrossRef]
64. Wang, H.-W.; Kuo, P.-H.; Shiau, J.-T. Assessment of climate change impacts on flooding vulnerability for lowland management in southwestern Taiwan. *Nat. Hazards* **2013**, *68*, 1001–1019. [CrossRef]
65. Tehrany, M.S.; Lee, M.; Pradhan, B.; Jebur, M.N.; Lee, S. Flood susceptibility mapping using integrated bivariate and multivariate statistical models. *Environ. Earth Sci.* **2014**, *72*, 4001–4015. [CrossRef]
66. Yang, W.; Xu, K.; Lian, J.; Bin, L.; Ma, C. Multiple flood vulnerability assessment approach based on fuzzy comprehensive evaluation method and coordinated development degree model. *J. Environ. Manag.* **2018**, *213*, 440–450. [CrossRef]
67. Veról, A.P.; Battemarco, B.P.; Merlo, M.L.; Machado, A.C.M.; Haddad, A.N.; Miguez, M.G. The urban river restoration index (URRIX)—A supportive tool to assess fluvial environment improvement in urban flood control projects. *J. Clean. Prod.* **2019**, *239*, 118058. [CrossRef]
68. Dong, W.; Liu, Z.; Zhang, L.; Tang, Q.; Liao, H.; Li, X.E. Assessing heat health risk for sustainability in Beijing's urban heat island. *Sustainability* **2014**, *6*, 7334–7357. [CrossRef]
69. Vescovi, L.; Rebetez, M.; Rong, F. Assessing public health risk due to extremely high temperature events: Climate and social parameters. *Clim. Res.* **2005**, *30*, 71–78. [CrossRef]
70. Fallmann, J.; Wagner, S.; Emeis, S. High resolution climate projections to assess the future vulnerability of European urban areas to climatological extreme events. *Theor. Appl. Climatol.* **2017**, *127*, 667–683. [CrossRef]
71. Mathew, A.; Khandelwal, S.; Kaul, N. Spatial and temporal variations of urban heat island effect and the effect of percentage impervious surface area and elevation on land surface temperature: Study of Chandigarh city, India. *Sustain. Cities Soc.* **2016**, *26*, 264–277. [CrossRef]
72. Oh, K.; Hong, J. The relationship between urban spatial elements and the urban heat island effect. *Urban Des. Inst. Korea* **2005**, *6*, 47–63.
73. Stewart, I.D.; Oke, T.R. Local climate zones for urban temperature studies. *Bull. Am. Meteorol. Soc.* **2012**, *93*, 1879–1900. [CrossRef]
74. Chow, W.T.; Chuang, W.-C.; Gober, P. Vulnerability to extreme heat in metropolitan Phoenix: Spatial, temporal, and demographic dimensions. *Prof. Geogr.* **2012**, *64*, 286–302. [CrossRef]
75. Reid, C.E.; O'neill, M.S.; Gronlund, C.J.; Brines, S.J.; Brown, D.G.; Diez-Roux, A.V.; Schwartz, J. Mapping community determinants of heat vulnerability. *Environ. Health Perspect.* **2009**, *117*, 1730–1736. [CrossRef]
76. Rinner, C.; Patychuk, D.; Bassil, K.; Nasr, S.; Gower, S.; Campbell, M. The role of maps in neighborhood-level heat vulnerability assessment for the city of Toronto. *Cartogr. Geogr. Inf. Sci.* **2010**, *37*, 31–44. [CrossRef]
77. Bae, M.; Kim, B.; Ban, Y. Analyzing the Spatial Distribution Characteristics of Urban Emergency Services Facilities—Focusing on Cheongju City. *Korea Environ. Policy* **2016**, *24*, 25–49. [CrossRef]
78. Zhu, Q.; Liu, T.; Lin, H.; Xiao, J.; Luo, Y.; Zeng, W.; Zeng, S.; Wei, Y.; Chu, C.; Baum, S. The spatial distribution of health vulnerability to heat waves in Guangdong Province, China. *Glob. Health Action* **2014**, *7*, 25051. [CrossRef]
79. Lee, D.; Oh, K.; Seo, J. An Analysis of Urban Cooling Island (UCI) Effects by Water Spaces Applying UCI Indices. *International J. Environ. Sci. Dev.* **2016**, *7*, 810. [CrossRef]
80. Sun, R.; Chen, L. How can urban water bodies be designed for climate adaptation? *Landsc. Urban Plan.* **2012**, *105*, 27–33. [CrossRef]
81. Suh, J.; Oh, K. Heat Mitigation Effects of Urban Space based on the Characteristics of Parks and their Surrounding Environment. *J. Korean Soc. Environ. Restor. Technol.* **2020**, *23*, 1–14.
82. Cho, H.; Lee, S. A Study on the Inducement Distance of Senior-Friendly Park and Evaluation of Green Service Area—Focused on the Pedestrian Aspect. *J. Korean Inst. Landsc. Archit.* **2019**, *47*, 1–9. [CrossRef]
83. Ord, J.K.; Getis, A. Local spatial autocorrelation statistics: Distributional issues and an application. *Geogr. Anal.* **1995**, *27*, 286–306. [CrossRef]
84. Sutherland, I.J.; Villamagna, A.M.; Dallaire, C.O.; Bennett, E.M.; Chin, A.T.; Yeung, A.C.; Lamothe, K.A.; Tomscha, S.A.; Cormier, R. Undervalued and under pressure: A plea for greater attention toward regulating ecosystem services. *Ecol. Indic.* **2018**, *94*, 23–32. [CrossRef]
85. Croeser, T.; Garrard, G.; Sharma, R.; Ossola, A.; Bekessy, S. Choosing the right nature-based solutions to meet diverse urban challenges. *Urban For. Urban Green.* **2021**, *65*, 127337. [CrossRef]

86. Kumar, P.; Geneletti, D.; Nagendra, H. Spatial assessment of climate change vulnerability at city scale: A study in Bangalore, India. *Land Use Policy* **2016**, *58*, 514–532. [CrossRef]
87. Yun, S.; Choi, B.; Jeon, E. A Study on Vulnerability Assessment to Climate Change in Siheung-si. *J. Clim. Change Res.* **2013**, *4*, 1–10.
88. Cowling, R.M.; Egoh, B.; Knight, A.T.; O'Farrell, P.J.; Reyers, B.; Rouget, M.; Roux, D.J.; Welz, A.; Wilhelm-Rechman, A. An operational model for mainstreaming ecosystem services for implementation. *Proc. Natl. Acad. Sci. USA* **2008**, *105*, 9483–9488. [CrossRef]
89. Maes, J.; Egoh, B.; Willemen, L.; Liquete, C.; Vihervaara, P.; Schägner, J.P.; Grizzetti, B.; Drakou, E.G.; La Notte, A.; Zulian, G. Mapping ecosystem services for policy support and decision making in the European Union. *Ecosyst. Serv.* **2012**, *1*, 31–39. [CrossRef]

land

MDPI

Article

Rediscovering the Scaling Law of Urban Land from a Multi-Scale Perspective—A Case Study of Wuhan

Qingsong He, Lingping Huang and Jing Li *

School of Public Administration, Huazhong University of Science and Technology, Wuhan 430074, China; heqingsong@hust.edu.cn (Q.H.); m202074959@hust.edu.cn (L.H.)
* Correspondence: lijing.99@hust.edu.cn; Tel.: +86-13387613430

Abstract: The law of urban scaling implies that there is a universally applicable nonlinear scaling relationship between population size and urban indicators, which is a method of quantitative analysis that can reflect the growth law and internal logic of the urban system. However, most present research is conducted at the municipal scale, and studies of scaling law in the inner-city system are scarce, especially from the perspective of compact urban form development. The goal of this paper is to discover the scaling law within urban systems from a multi-scale perspective. Through the empirical analysis of Wuhan, this paper examines the internal scale law of the urban system from the municipal and district scales. Moreover, we use the landscape expansion index to perform spatial autocorrelation analysis. In this way, we assess the relationship between the compactness of urban morphological development and the urban scaling law. The results indicate that the temporal scaling law on the city scale has a more significant linear law than the single-year scaling law. The analysis also shows the scaling law relationship within the inner-city system. Nevertheless, there is a deviation between the temporal scaling law and the cross-section scaling law. Namely, the time series development of a district does not follow the section scaling law of the urban system. Furthermore, the urban scaling law shows a negative correlation with the compactness of the urban form development. It is crucial to understand the current economic development and resource endowment of an urban system in the urbanization process, as it significantly contributes to urban development and regional coordinated planning.

Keywords: urban internal system; scaling law; multiscale; urban science

Citation: He, Q.; Huang, L.; Li, J. Rediscovering the Scaling Law of Urban Land from a Multi-Scale Perspective—A Case Study of Wuhan. *Land* **2022**, *11*, 914. https://doi.org/10.3390/land11060914

Academic Editors: Alessio Russo and Giuseppe T. Cirella

Received: 18 May 2022
Accepted: 13 June 2022
Published: 15 June 2022

1. Introduction

Since its reform, China's economic development has led to a rapid increase in the rate of urbanization, increasing from 17.92% in 1978 to 63.89% in 2021 [1]. At present, the total number of cities has reached 687, while the urban population has grown from 170 million in 1978 to 902 million in 2021. By 2035, the urbanization rate is expected to reach 75% to 80%, with nearly 400 million new urban residents [2]. Nevertheless, extensive and rapid urbanization has caused several problems for China's urban development, namely the intensification of human–land conflicts, the disharmony between social and economic growth and the carrying capacity of resources and the environment, and regional imbalances [3]. Even though urbanization cannot be avoided, it requires finding a balance of urban development in urban scientific research. At present, China is vigorously promoting a new type of urbanization, giving a leading role to central cities and urban circles that promote regional coordinated development and improve the quality of urbanization. As a typical complex system, the city has interrelated and relatively independent subunits, with similar social and interactive networks [4]. The urban population is the basic and most critical component within the urban system. It is also a key determinant of urban production activities, with urban indicators being reliable indicators of urban development [5,6]. A significant number of empirical studies have noted the scaling law relationship between population and urban indicators. The law refers to the quantitative analysis method that reflects the growth law and the internal logic of the urban system [1,5,7,8]. It has already attracted much attention and application in the field of new urban science.

Bettencourt et al. [8] were the first to discuss the concept of the scaling law in urban systems. Since then, a number of studies have focused on the importance of urban population size in urban systems, focusing mainly on the concept of urban scaling law [7,9,10], causes and mechanism of formation [11–15], verification [16–21], application [22–25], and questioning [26–30]. The abovementioned research was conducted at the municipal scale. Thus, academic circles generally agree that the urban scaling law appears to show a nonlinear scaling relationship between the urban population scale and different urban indicators during the process of urban expansion.

The process of urbanization in China has unique characteristics, with uneven regional resource endowment and distribution that hinders the integrated development of cities. Understanding the relationship between urban indicators and the size of an urban population is crucial for grasping a city's size, its state of urban economic development, and to promoting the development of inter-regional macroeconomic co-urbanization. In addition, the urban scaling law reflects the results of many indicators such as nature, economy, and policy. It also illustrates the process of urban evolution, reflects the future trend of the city's development, and deepens people's understanding of China's urbanization from the spatial scale. In addition, research on the urban scaling law in China is still relatively new. At present, it is mostly focused on the scaling relationship between the built-up urban area and the population scale, with most studies focusing on the municipal scale. Therefore, little is known about the internal scaling law on the micro scale, with an even more obvious lack of multi-scale research and comparative analysis. The interior of the city can be divided into smaller units, such as districts and patches.

As far as this study is aware, there are so far only studies of the internal scaling law of Shanghai [7] at the district scale. Furthermore, this particular research analyzed the temporal and cross-sectional scaling law from a single district scale perspective. Thus, with the exception of municipal and district scales, existing research is yet to address the patch scale. The patch is the most basic component unit of a landscape pattern. It is a spatial entity that differs significantly in nature or appearance from its surroundings, based on scale effects. By observing the evolution of the patch, we may be able to analyze the change in the compactness of the urban form. Various studies have examined the correlation between urban form compactness and spatial vitality [31], transportation and commuting [32–34], and air quality [35,36]. The abovementioned research leads us to perform a quantitative study of the relationship between the landscape sprawl index and the scalar law of different urban elements at the patch scale [37,38]. There is still a lack of empirical research on whether urban internal units adhere to the scaling law. Based on the fitting results of the scaling exponent, we are able to examine the current economic development and resource endowment of cities, and thus provide a scientific basis for regionally coordinated development.

Due to the research gaps discussed above, the research objectives of this study are as follows: (1) Firstly, we examine the difference between the time series scaling law of municipal cities and the theoretically expected single time point scaling law; (2) secondly, the paper analyzes whether the scaling law also exists in the urban system; (3) we analyze the relationship and characteristics of the temporal scaling law and the cross-sectional scaling law in each region; and (4) lastly, we aim to examine the correlation between the compactness of urban form development and the urban scale law. The interconnectedness between the objectives is provided in Figure 1.

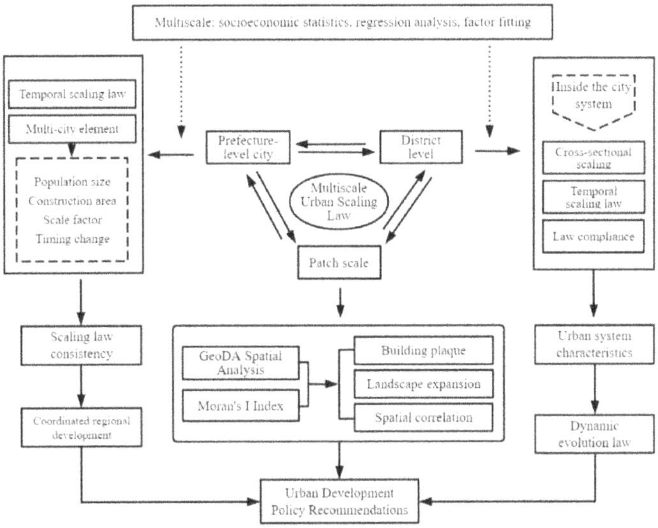

Figure 1. Flowchart of the study.

The process flow diagram expresses the basic idea of this research. Socio-economic statistical data of the city of Wuhan and each municipal district from 2005 to 2020 were collected, and the regression analysis method for calculating the scaling law exponent was determined. At the municipal scale, the temporal scaling law of the main urban indicators of Wuhan is being studied. The municipal city time series scalar law accordance with the theoretically expected single time point scalar law is checked. Therefore, the existing conclusions of the urban scaling law can be used to guide regional coordinated development. On the district scale, we first test whether the same scaling law exists within the inner-city system, and then calculate the cross-sectional scaling law and the temporal scaling law for 13 municipal districts. Finally, we focus on whether the temporal scaling law follows the cross-sectional scaling law in order to investigate the characteristics and evolution of the inner-city system. On the patch scale, based on GeoDA software, we conduct a bivariate global spatial autocorrelation analysis of the temporal scaling exponent and the landscape expansion index of each municipal district. Moran's I, the autocorrelation index, is then used to verify the relationship between the urban scaling law and the compactness of urban form development. Finally, based on the results of the multi-scale urban scaling law, we put forward policy suggestions for urban development.

Compared to existing research, this paper has a number of contributions. Firstly, it contributes to the understanding of spatial scale innovation. Spatial scale is a measure of the spatial size used to conduct research and is one of the key research issues in urban geography. The existence of spatial correlations within complex urban systems means that the results of detailed analyses of research objects will vary across spatial scales. At present, the urban scaling law does not include the patch scale. Based on remote sensing images and big data, the landscape expansion index is introduced in order to investigate the relationship between the compactness of urban form development and the urban scaling law. Secondly, the results of this study will provide a better theoretical understanding of the urban scaling law. Namely, this paper selects various urban indicators from the city, district, and patch scale in order to support the existing findings of the urban scaling law. At the same time, we are taking an innovative perspective to deepen the dynamic analysis of the scaling law within the inner-city system. The study also provides a theoretical basis for understanding the relationship between the urban scaling law and the temporal development of a city.

Wuhan is strategically important for the development of central China. Based on empirical analysis of Wuhan, this paper studies the urban scale law from the multi-scale perspective of the city, district, and patch. Based on the scaling law relationship between the urban population scale and the urban indicators, this study aims to understand the dynamics of the urban population development scale and economic development trend, guide regional coordinated development, allocate urban resources and new construction, as well as quantify the development and evolution law of the urban system.

2. Method and Data

2.1. Urban Scaling Law

The urban scaling law was introduced into the urban system from biological research. Foreign scholars used data from different cities around the world, such as European cities, Indian cities, Mexico, and former Spanish settlements. The abovementioned research confirmed the universality of the urban scaling law in different regions and periods from the municipal and settlement scale. Bettencourt et al. [8] stated that urbanization has increased the pace of urban life with increasing population. By comparing biological and urban systems and using data on relevant urban infrastructure, personal needs, and socio-economic indicators, the urban scale law is verified. In other words, the urban indicators and the urban population scale are expressed as a power law function of the scaling exponent β. Based on the theoretical and empirical evidence, scholars divide the relationship between the scaling exponent and 1 into urban indicators and further divide it into super-linear, sub-linear, and linear, thus establishing a systematic framework for urban scaling law. The power law function formed between urban indicators and the urban population size is presented in Formula (1) as follows:

$$Y(t) = Y_0 N(t)^{\beta} \tag{1}$$

In the formula above, $N(t)$ represents the urban population size at time t, while $Y(t)$ stands for the urban indicators corresponding to the urban population size at time t. Next, Y_0 is a standardized constant, while the index β is the urban scaling law. The size of β and 1 is used to measure the relationship between the urban population size and urban indicators. It is divided into the following three types [8]: (1) Linear relationship ($\beta \approx 1$), in which urban indicators (e.g., number of jobs, household water) are related to individual needs. In general, the increase in indicators is basically consistent with the population increase; (2) sub-linear relation ($\beta < 1$), in which the indicators (e.g., total road length, number of gas stations) are related to urban infrastructure. The increase in indicators is less than the increase in population size, which reflects economies of scale. The relatively large population of cities allows residents to share infrastructure. For example, city A has 3 times the population of city B, but city A does not need 3 times the total length of roads of city B; and (3) super-linear relation ($\beta > 1$), in which the growth rate of indicators (e.g., GDP, local general public budget revenue, and knowledge output) interacting with urban social and economic development is greater than that of the urban population. Social interaction grows super-linearly with population growth, reflecting the characteristics of the urban economic agglomeration effect. In this study, we simultaneously considered the logarithms on both sides of Equation (1), and applied the least square method to obtain the "scaling law" regression equation:

$$logY(t) = \beta logN(t) + logY_0 \tag{2}$$

After taking the logarithms of the city and population size indicators, a linear fitting was performed to obtain the scale exponent β. The least square method for linear fitting is simple and easy to operate. It is the most commonly used fitting method in the study of the urban scaling law.

2.2. Landscape Expansion Index

The landscape expansion index (LEI) was proposed by Liu Xiaoping [39] and is used to quantitatively describe the different types of dynamic expansion and spatial pattern distribution, thus reflecting information on the spatial layout and dynamic changes of a landscape. The calculation process requires the following steps: Firstly, information on the expansion of urban construction land in different time periods is extracted from multi-temporal remote sensing data. Then, a buffer zone is established using the GIS spatial analysis function. Lastly, the urban landscape expansion index is calculated [40,41]. The degree of dispersion of the new patch can be obtained from the intersection area of the new patch buffer and the old patch. The higher the index value, the more compact the location relationship between the newly expanded patch and the original city. This provides a more intuitive spatial meaning and can well explain the spatial and temporal pattern evolution of urban land expansion. The expression formula is as follows:

$$LEI = \frac{A_0}{A_0 + A_V} \times 100 \tag{3}$$

In Formula (3), LEI represents the landscape expansion index of the newly added patch, while A_0 represents the intersection area of the newly added patch buffer and the old patch. Next, A_V indicates the intersection area of the newly added patch buffer and other patches, except for the old patch. The LEI can range from 0 to 100, and this study sets the LEI thresholds for three expansion types [39]: (1) when $0 < LEI < 2$, it is an enclave expansion; (2) when $2 \leq LEI \leq 50$, it is an edge expansion; (3) and when $50 < LEI \leq 100$, it is a filling expansion. Marginal sprawl and enclave sprawl represent the urban "diffusion" process that usually results in a more discrete urban form, while infill sprawl leads to a more compact urban form by reducing intra-city voids.

Spatial autocorrelation analysis originated from biometrics, and it is now one of the basic methods for analyzing the statistical distribution of spatial data [42]. GeoDA is an open-source software designed to analyze spatial autocorrelation. Using bivariate Moran's I analysis, it explores the spatial correlation and dependence of two elements. The LEI examined in this paper has spatial characteristics on the patch scale. Compared to Pearson's test, bivariate Moran's I has a higher efficiency and applicability rate. Thus, this study used GeoDA software to analyze both the temporal scaling exponent of each district and the LEI of each district from 2005 to 2020. Furthermore, the study conducted a bivariate global spatial autocorrelation analysis between them. Moran's I verified the relationship between the urban scaling law and the compactness of urban form development.

2.3. Study Area and Data Source

The research period of this study is from 2005 to 2020. The research area is Wuhan, a Chinese megalopolis with a population of over 13 million. The city is an important transportation hub, industrial, scientific, and educational base, as well as a crucial point of economic development in central China. The city has 13 municipal districts under its jurisdiction, including the Jiang'an District, Jianghan District, Qiaokou District, Hanyang District, Wuchang District, Qingshan District, Hongshan District, Caidian District, Jiangxia District, Huangpi District, Xinzhou District, Dongxihu District, and HanNan District. In addition, there are several functional areas of Wuhan. To unify the analysis of research units and urban indicators, the statistical data of urban indicators of the Wuhan Economic and Technological Development Zone is merged into HanNan District. Furthermore, data for the East Lake High-tech Development Zone is merged into the Hongshan District, and data of the Wuhan New Chemical Industry Zone is merged into the Qingshan District.

This paper defines the urban population as the residential population in the urban area of the city. The statistical scope of the remaining urban indicators data is defined as the municipal area within the statistical yearbook. Taking into account statistical limitations and factors such as representativeness of indicators and data availability, 18 urban indicators, including urban construction land and residential land, were selected at the municipal scale. At the regional scale, eight urban indicators such as GDP and local general public budget revenue were selected. Data regarding urban indicators are from the "*China Urban Construction Statistical Yearbook*", the "*Hubei Statistical Yearbook*", the "*Wuhan Statistical Yearbook*", the "*Wuhan Yearbook*", as well as from the statistical yearbooks and bulletins of social and national economic development of various districts.

3. Results

3.1. Temporal Scaling Law of Urban Indicators in Wuhan

At the municipal scale, studying the temporal scaling relationship between major urban indicators and urban population size can help measure the inherent time series growth law of China's urban system. Taking the regional GDP, the local general public budget revenue, and the urban population size as examples, a linear fitting regression is performed on the double logarithm of urban indicators and population size (Figure 2). The ß value of the GDP scale exponent is 2.953 and the goodness-of-fit R^2 is 0.866, while the p value is less than 0.01. This indicates that if the population is twice its original size, GDP will increase by 7.744 ($2^{2.953}$) times. Furthermore, the local general public budget revenue scale exponent ß is 3.461, while the goodness-of-fit R^2 is 0.793 and the p value is less than 0.01. In other words, if the population is twice the original size, the local general public budget revenue will increase 11.012 ($2^{3.461}$) times. In the double logarithmic coordinate system, the two urban indicators show a significant linear correlation with population size, and a solid fitting effect. With respect to social economy, there is a super-linear scale relationship between GDP and population size, reflecting the strong agglomeration effect of social economy in Wuhan during the period of 2005–2020. In China's rapid urbanization process, such a relationship also suggests a significant advantage for larger cities when it comes to economy. Moreover, in terms of public finance, the temporal scaling exponent of local general public budget revenue is close to 3.5. This result indicates that the exponent in question is in a super-linear relationship with the population size, which even exceeds the GDP scale exponent. As the public finance scale exponent has the maximum value of all major urban indicators studied, it suggests the presence of the phenomenon of urban fiscal polarization, which is peculiar to urbanization in China. The growth rate of fiscal revenues in larger cities is increasing from year to year and is significantly higher than the increase in population.

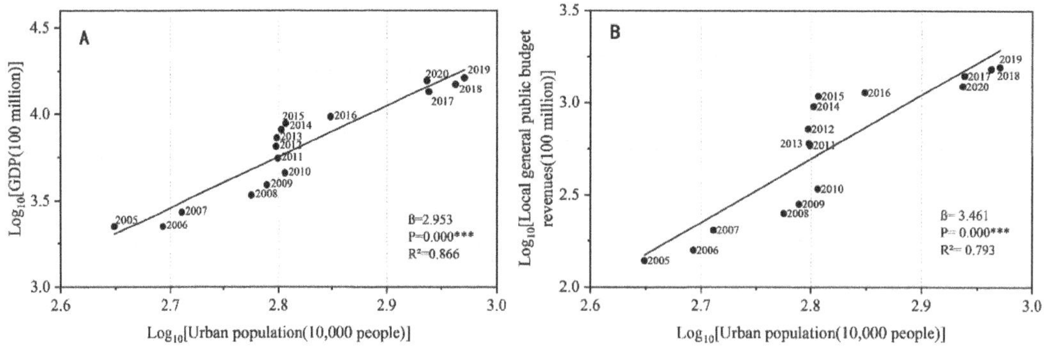

Figure 2. Temporal scaling laws of GDP and local general public budget revenues. (**A**) Temporal scaling law between GDP and urban population in Wuhan from 2005 to 2020; (**B**) Temporal scaling law between local general public budget revenues and urban population in Wuhan from 2005 to 2020; *** $p < 0.001$.

For the temporal scaling exponent of each indicator and the urban population size (Figure 3) within the same linear fitting processing, the average goodness-of-fit R^2 is measured to be 0.689, while the p values are less than 0.05. These results show that there is a significant scaling relationship between urban indicators and population size, and it can be used to reveal simple laws in the evolution of complex urbanization. Except the negative β index of industrial wastewater discharge, the β index of other indicators is positive and shows a positive correlation with population size.

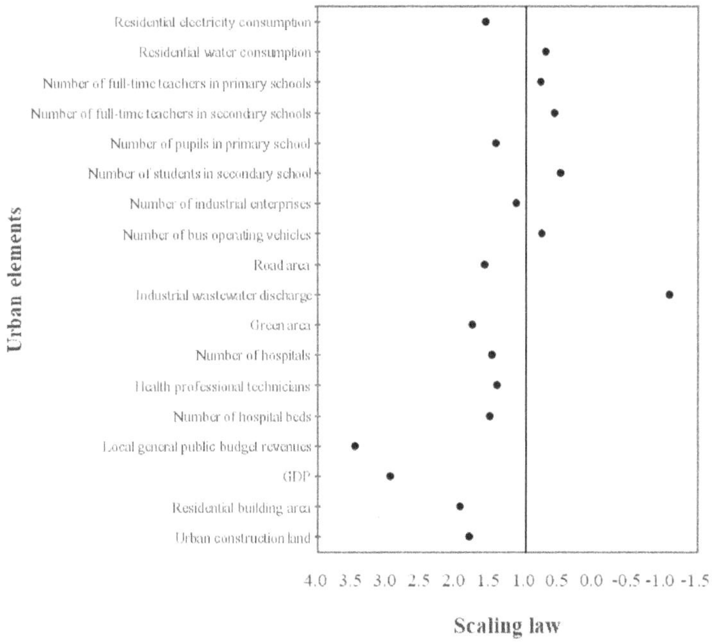

Figure 3. Temporal scaling laws of urban indicators in Wuhan.

In terms of land consumption, the β scale factor is between 1.8 and 1.9 for the urban construction land area and the residential construction land area, which shows a significant super-linear relationship with population size. This reveals the high demand for construction land and residential land in China. As the population size of an area increases, so will the required residential building area. Green spaces, roads, and population size form a super-linear relationship. This suggests that during urban expansion there is an adequate supply of land for roads, which makes for good transportation accessibility in large cities. In addition, against the background of high-density urban construction, a substantial increase in urban green area shows the remarkable importance of ecological greening in large cities.

In terms of educational indicators, the increase in the number of teachers in schools is smaller than the population growth. Moreover, the temporal scaling exponents of the number of full-time middle school teachers and full-time primary school teachers are both on the right side of the $\beta = 1.0$ indicator line. Namely, secondary schools have the lowest number of full-time teachers. This fact highlights the current issue of teacher shortages in larger cities, which in turn causes competition for educational resources. This leads to vicious competition for urban education resources, and leads to allocation and utilization of other resources, such as school district housing supply, zoning of schools for school-age children, etc. There is a significant difference in the temporal scaling law between primary and secondary school students. The temporal scaling exponent of the number of students in middle school is the same as the theoretical expectation, i.e., β is less than 0.5. On the contrary, the number of students enrolled in elementary school has a super-linear scalar relationship with the urban population size. These findings relate to the mandatory nine-year study period defined by the state, the impact of the two-child policy, and the Wuhan economic agglomeration, which drives foreign resident population. Thus, the number of primary school students has increased, increasing the need for educational resources in larger cities as well.

With respect to medical facilities, the temporal scaling exponent β of the number of hospital beds, practicing physicians, and hospitals is between 1.4 and 1.5. These findings indicate a super-linear relationship between these factors and population size, which is contrary to theoretical expectations ($\beta < 1$). These results also reflect the economic development of large Chinese cities, with a massive increase in medical resources and a concentration of high-quality medical resources in large cities.

In relation to infrastructure, the number of buses within a city shows a sublinear relationship with the population size, indicating that the development of bus transportation in large cities is adequate and at the level of demand. However, as subway travel in larger cities is more convenient in practice, the bus increase rate is actually lower than the population growth rate.

In terms of industrial enterprises, there is a negative super-linear relationship between industrial wastewater discharge and the population size, and a positive super-linear relationship between the number of industrial enterprises and the population size. This reflects the "structural adjustment and pollution reduction" in the industries of big cities with a declining trend in industrial wastewater discharge.

In relation to personal needs, the residents' water consumption shows a sub-linear relationship, while electricity consumption shows a super-linear relationship. In the development of a city, with the increase of population, the consumption of water resources for domestic use also increases greatly. However, it is maintained in relative balance with the population increase. On the contrary, the residents' electricity consumption shows a significant linear increase. These findings indicate that there is significant consumer demand for electricity in larger cities, with supply and demand for electricity resources being greater than demand for water resources.

3.2. *Scaling Law of Urban Indicators for Wuhan Districts*

3.2.1. The Cross-Sectional Scaling Law for Each District

In order to explore the scaling law of the internal units of an urban system at the district level, this paper selects eight representative urban indicators.

The scaling law of major urban indicators was calculated for 13 districts in Wuhan from 2005 to 2020 (Figure 4). The results are classified as follows: (1) $\beta \approx 1$: the cross-sectional scalar exponents in the number of hospital beds, health technicians, and retail sales of social consumer goods changed more moderately. Before 2014, the exponent increase rate was higher than the rate of population growth. This showed a fluctuating trend of small amplitude and a weakening of the scale economies. However, after 2014, overall scale indicators exhibited a downward trend. Thus, the scale economy of urban indicators is strengthened, with the possibility of causing tensions among related resource indicators; (2) $0.5 < \beta < 1$: the cross-sectional scaling exponent of GDP fluctuated around 0.7, which was not in line with theoretical expectations. Based on the cross-sectional scale, this measurement indicates that the economic output of each district had a strong agglomeration effect. Nevertheless, the trend of socio-economic development within each district differs, while the internal distribution of economic resources remains unbalanced; (3) $0.4 < \beta < 1.2$: the cross-sectional scaling exponent of investments in real estate development varied greatly. It reached its minimum of 0.485 in 2010 and its maximum of 1.144 in 2017. In general, this exponent shows a downward trend from 2005 to 2010, and an upward trend after 2010, as more capital accumulated in the real estate economy. Furthermore, this exponent points to the fact that government real estate policies in larger cities are variable and regulatory, as the real estate market differs between districts due to their geographical location, political status, social economy, and other factors; and (4) $\beta \approx 0.5$: the cross-sectional scaling exponents of local general public budget revenue, total output value of the construction industry, and added value of the construction industry were stable with the value around 0.5 from 2005 to 2015. Moreover, these exponents showed a significant upward trend after 2016. In addition, the scale exponents of total output value and added value of the construction industry were both higher than 1 after 2016. This indicates a growth trend identical to that of investment in real estate development and reflects the trend of rapid development within the real estate market. It is worth noting that all the scale exponents of urban indicators showed a small decline in the period 2019–2020 due to the impact of the COVID-19 pandemic.

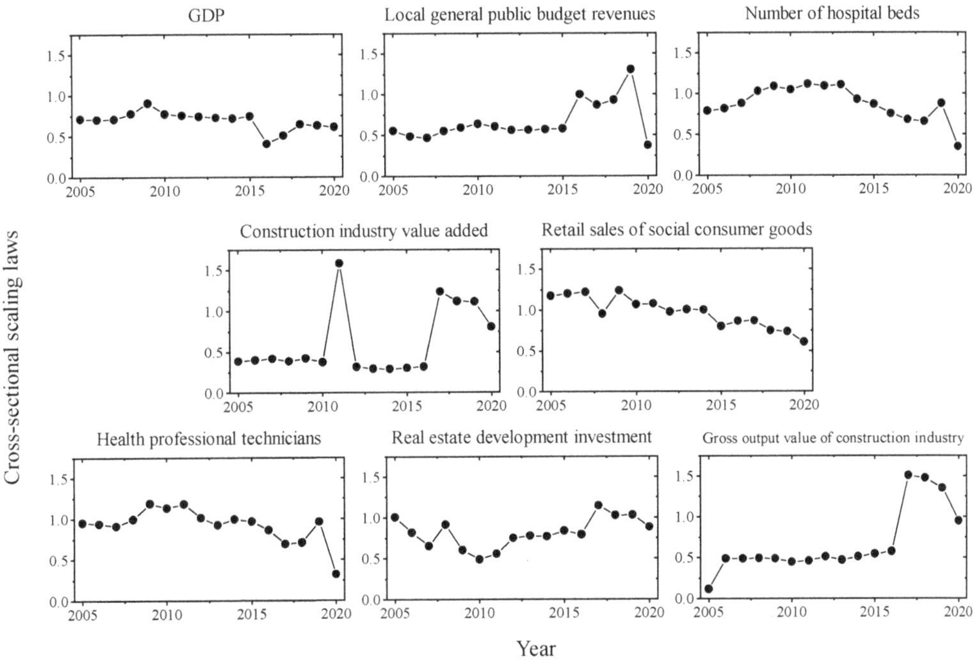

Figure 4. Cross-sectional scaling law of each region.

3.2.2. Temporal Scaling Law of Each District

In order to calculate the temporal scaling law of the main urban indicators for each district from 2005 to 2020, a temporal scale analysis was carried out in 13 Wuhan districts (Figure 5). With the exception of the Caidian District and the Jiangxia District, the temporal scaling exponent for health technology was less than 1, while the other urban indicators' temporal scaling exponent was found to be greater than 1. Comparing the temporal scaling exponent of zoning with the sectional scaling exponent of zoning indicates that the span of the Y-axis values and the difference between maximum and minimum values are large. It also points to the fact that the linear relationship between different urban indicators and population size differs significantly for each municipal district. In the example of GDP, the temporal scaling exponent of the Xinzhou District was 21.17, while the temporal scaling exponent of the Qiaokou District was 1.98. The GDP of each district has a super-linear relationship with the population size. In contrast, the cross-sectional scaling exponent of Wuhan's GDP from 2005 to 2020 was less than 1, indicating a sublinear relationship. This shows that there are fundamental differences between temporal scaling laws and cross-sectional scaling laws within the urban system, and that the linear relationship between urban indicators and population differs across different spatial and temporal scales. Due to the impact of the epidemic in 2020, the cross-sectional scaling exponents have shown a downward trend. However, the temporal scaling exponents of each district were not affected. In general, these indicators and population size continued to show a super-linear relationship, reflecting the development trend and law of each district over time.

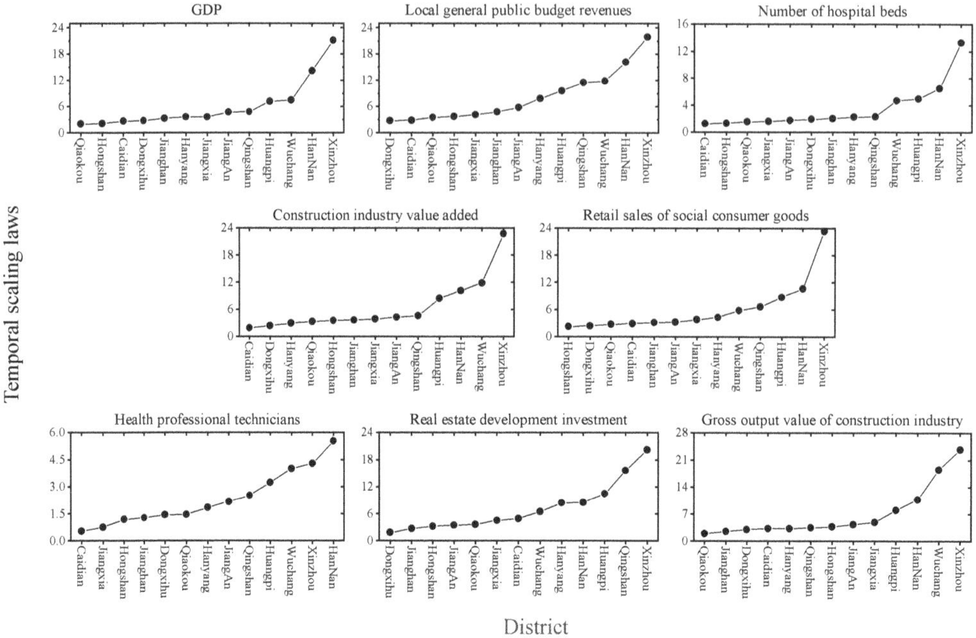

Figure 5. Temporal scaling law of each region.

The temporal scaling exponents of other urban indicators were further analyzed. Therefore, indicators such as medical facilities, hospital beds, and health technical personnel showed little difference. For example, in the Xinzhou District, the hospital beds indicator was at a maximum of 13.28, while in the HanNan District, the health technical personnel exponent was at a maximum of 5.23. Compared to other economic indicators, all indicators of medical facilities and population size show a relatively small growth rate. The temporal scaling exponents of other economic indicators fluctuated between 1.0 and 24.0. However, the values of most municipal districts ranged between 1.0 and 6.0, thus showing the strong effect of economic agglomeration for larger cities and the trend of prosperous development over time. Nevertheless, the analysis of temporal scaling exponents showed that the economic scaling exponent remained stable at around 0.5 before 2015, showing an upward trend only after 2016. These findings further support the assumption that the development law of a single municipal district differs from the cross-sectional development. Furthermore, this indicates that the sequential development of a single district does not follow the cross-sectional scaling law of the urban system.

3.3. Correlation Analysis between Landscape Expansion Index and Scaling Law

At the patch scale, based on an in-depth understanding of the scaling law relationship between the urban population size and urban indicators, we first calculated the temporal scaling exponents and the *LEI* of each district from 2005 to 2020, and then conducted a spatial autocorrelation analysis. In this way, the relationship between the compactness of urban form development and the urban scaling law is checked, and the development of law within the urban system can be obtained.

First, data on land use between 2005 and 2020 helped to identify new patches of construction land in each district. To understand urban development in each Wuhan district, the landscape expansion index for each district (*GLEI*) was calculated [43] (Figure 6). The lowest *GLEI* value was in the HanNan District, measured at 3.56, while the highest *GLEI* value was in the Jianghan District, measured at 47.25. These findings indicate that the Jianghan District has the most compact urban form development, while the HanNan District is the least compact. Secondly, GeoDa tested the spatial autocorrelation between the temporal scaling exponents of urban indicators and *GLEI*. The Moran's I index of urban indicators was also calculated [44,45] (Table 1). The Moran's I index of the eight major urban indicators was negative, and the p values of the other indicators were less than 0.1, except for the health technician factor. This indicates that the temporal scaling exponent is negatively correlated with *GLEI*. Taking as an example the number of hospital beds in the HanNan District, which has the least compact urban form, this indicator had a temporal scaling exponent of 6.47. In other words, it was neither compact nor did it have economy of scale. On the other hand, in the Jianghan District, which had the most compact urban form, the temporal scaling exponent of hospital beds was 1.99, indicating a compact economy of scale. The degree of compactness of the urban form had a significantly negative impact (19.67%) on the scaling exponent of hospital beds. Namely, when *GLEI* is higher, the urban form development is more compact. Furthermore, the scaling exponent of urban indicators is smaller, and the scale economy of social development is more significant. Finally, a higher *GLEI* and a smaller scaling exponent cause more efficient distribution and circulation of various resources within the system.

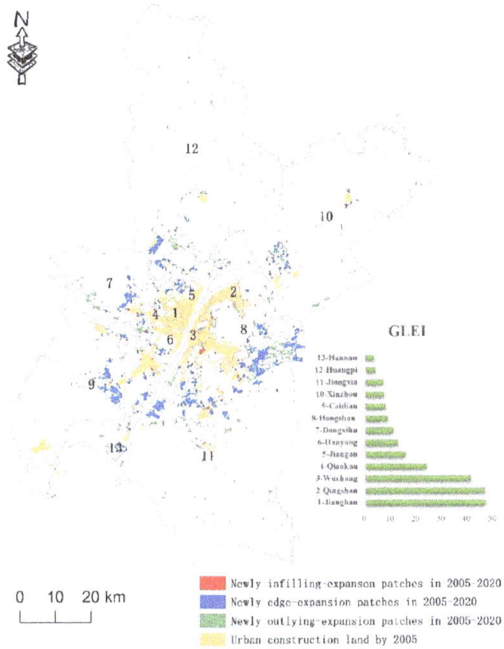

Figure 6. Distribution of new types of expansion patches in each district of Wuhan and calculated GLEI.

Table 1. Results of the spatial autocorrelation test.

	GDP	Local General Public Budget Revenues	Number of Hospital Beds	Health Professional Technicians	Real Estate Development Investment	Gross Output Value of Construction Industry	Construction Industry Value Added	Retail Sales of Social Consumer Goods
Moran's I	−0.2207	−0.1697	−0.1967	−0.1213	−0.1687	−0.161	−0.1638	−0.223
p	0.05	0.10	0.05	0.12	0.08	0.08	0.10	0.07

4. Comparative Analysis and Policy Implications of Scaling Law

When the city-level temporal scale law is compared with previous research, it appears to be in line with the sub-linear, linear, and super-linear law of Bettencourt et al. However, due to the temporal span, the urban indicators are shown to be more sensitive in correlation with population size.

In terms of social economy, the gross regional product, local general public budget revenue, and population size should form a super-linear relationship, with the β exponent above 1.3. The temporal scaling exponents of Wuhan's two indicators are 2.953 and 3.461, while the β value is significantly higher than 1.3. This indicates a strong agglomeration effect for larger cities undergoing urbanization, rapid development, and a sharp increase in the size of urban population. Simultaneously, it has a super-linear relationship with the high exponent, on the temporal scaling law.

When the results of the zoning temporal scaling law and cross-sectional scaling law are compared with the internal scaling law of Shanghai [7], it seems that the scaling law is present in the urban system, while the cross-sectional and temporal scaling exponents do not appear to follow the same law. Due to differences between the sampled cities and some statistical factors, the scale exponents of some urban indicators still differ. In terms of the cross-sectional scaling law, the cross-sectional scaling exponent of Shanghai's GDP from 2005 to 2017 is greater than 2, while in Wuhan it is around 0.7. This may be due to the fact that in 2017, the permanent population of Shanghai exceeded 24 million, while in Wuhan it was only 8.68 million. Therefore, the difference in the urban population size results in the difference in the scale exponents. As the urbanization process is closely related to the Chinese leadership, the urbanization rate and the major differences between provinces and cities also depend on the government. Therefore, cities with a larger population have a more significant linear relationship with the growth rate of urban indicators when the scale exponent value is higher. With respect to the temporal scaling law, the temporal scaling exponents of the GDP of each district in Shanghai are between −4 and 15. On the other hand, in Wuhan they are between 1.9 and 22, making the change of the exponents of each district greater in Wuhan than in Shanghai. These results are due to the different resource endowments and population scale of each district within the urban system. Each district's urban indicators are unevenly distributed, resulting in significant temporal differences in scaling exponents between districts.

Rapid urbanization in China differs from that in developed countries, making the law of the urban scale in China quite unique. Namely, the agglomeration of the urban population encourages economies of scale and the agglomeration of other indicators. The crucial difference in the scale exponent results also reflects the imbalance of resources both between and within Chinese cities. Based on the scale exponent results, this paper proposes a number of policies for regional coordinated development. First, Shanghai is a mega-city, while Wuhan is a super-city. Nevertheless, both are among the most prominent Chinese cities in terms of urban population and economic development. However, in terms of the cross-sectional scaling exponent of GDP, Shanghai is much larger than Wuhan, indicating significant differences in China's urban development. To resolve this discrepancy, policies should promote urban–rural integration and a more equitable and efficient flow of resources. Moreover, they should play a radiant role for big cities, while working to increase the population size and urban indicators of small and medium-sized cities. Secondly, even though larger cities have enough tax revenues to ensure infrastructural development, their scale exponent is still far smaller than the socio-economic indicator. Therefore, to ensure the scale exponent and efficient infrastructural development, it is necessary to focus on different infrastructure indicators. Furthermore, as China's urban construction occupies a large amount of land, the scale exponent of land use is large. Therefore, it is important to pay attention to the construction of infrastructure in order to ensure economies of scale, as well as the efficient operation of facilities. Thirdly, the urban scale law has a significant negative correlation with the compactness of urban form development. Namely, from 2005 to 2020, $GLEI$ values in Wuhan districts ranged from 0 to 50, indicating marginal expansion.

It is important to pay attention to the outward expansion of the city and strengthen the leading position of urban planning in urban construction. Then, planning can be used as a blueprint to promote compact urban development, making economies of scale of social development increasingly visible.

5. Discussion and Conclusions

By examining the Chinese city of Wuhan, this study explores the scale law of urban population size and urban indicators from the city, district, and patch levels. Research shows that China's urban development is in line with the characteristics of the scalar law. However, rapid urbanization in China has led to imbalances in resource endowments and urban development of cities with different population sizes. The numerical difference in the scale exponent between cities is large and sensitive. On the other hand, the increase in socio-economic benefits from population clustering in Chinese cities is more significant, and the efficiency of the use of urban infrastructure indicators is becoming apparent. The results have several implications. First, there is an evident scale law relationship within the inner-city system. At the municipal district level, the city is divided into district-level administrative units. This is done to improve the reliability of statistical data and to clearly define internal urban indicators and population size. In the process of urban expansion, there is a correlation between the population size and the indicators of the metropolitan system and the inner-city system. Secondly, the temporal scale law and the cross-section scale law did not adhere to theoretical predictions. The generally accepted scaling law relationship is expressed as a power law function of the scaling exponent β. Furthermore, the index β of infrastructure elements is sub-linear with values around 0.8, the index β of individual needs is linear and measured to be 1, and the index β of socio-economic relatedness is super-linear and measured at around 1.2. Nevertheless, the temporal and cross-sectional scaling laws differ significantly from the established scaling law relationships in time and space. These discrepancies suggest that the scaling laws within the urban system theoretically follow the meaning of the exponent β. However, the interpretation of its β should be based on the development process within a single urban system. Thirdly, the temporal scale law at the city level shows a more significant linear law than the single year scale exponent. In long-term urban development, the law of temporal scale is more sensitive to urban population growth and indicator density growth. Fourthly, there are fundamental differences between the temporal scale law and the cross-sectional scale law in the urban system. Furthermore, the temporal development of the city does not follow the cross-sectional scale law of the urban system. Namely, the temporal scale law of the district reflects the individual development of the inner city, while the cross-sectional scale law indicates the overall development trajectory of the city in a particular time node. The cross-sectional exponent is more stable than the temporal exponent. Lastly, there is a significant negative correlation between the urban scale law and the development of urban form compactness. In other words, as urban morphological development becomes more compact, the smaller the urban scaling exponent is, the more prominent the scale economy of social development becomes.

This paper studies the urban internal scaling law on multiple levels. However, the sample size does not sufficiently represent the characteristics of the entire Chinese urban system. Examination of the cross-sectional and temporal scaling law of each district in Wuhan is limited by the current statistical caliber and the quality of statistical data. Therefore, the number of indicators in cities at the district level is smaller than the number of indicators on the city level, hindering the complete unification of the research indicators. A comparative study at the district and municipal scales needs to be further explored.

Author Contributions: Conceptualization, Q.H.; methodology, Q.H. and J.L.; software, L.H.; validation, Q.H. and L.H.; formal analysis, L.H.; resources, Q.H.; data curation, Q.H.; writing—original draft preparation, L.H.; writing—review and editing, J.L. and Q.H.; supervision, J.L. All authors have read and agreed to the published version of the manuscript.

Funding: This research was funded by National Natural Science Foundation of China (42001334), Key Laboratory of New Technology for Construction of Cities in Mountain Area, Ministry of Education, Chongqing University (ID: LNTCCMA-20210101) and Independent Innovation Fund for Young Teachers of Huazhong University of Science and Technology (ID. 2021WKYXQN031).

Informed Consent Statement: Not applicable.

Data Availability Statement: The data presented in this study are available on request from the corresponding author.

Conflicts of Interest: The authors declare no conflict of interest.

References

1. Jiao, L.M.; Lei, W.Q.; Xu, G.; Xu, Z.B.; Zhou, Z.Z. Scaling Law and Spatial-Temporal Characteristics of Scaling Factors in Chinese Cities. *Acta Geogr. Sin.* **2020**, *75*, 2744–2758.
2. Wang, K.; Lin, C.H.; Wu, C.Y. Trends and Planning Options after 60% Urbanization rate in China. *Urban Plan.* **2020**, *44*, 9–17.
3. Liu, R.Z. Urbanization in China: Problems, Reflections and Transformation. *J. Zhengzhou Univ. (Philos. Soc. Sci. Ed.)* **2013**, *46*, 68–72.
4. Berry, B.J.L. Cities as systems within systems of cities. *Pap. Reg. Sci.* **1964**, *13*, 146–163. [CrossRef]
5. Bettencourt, L.M.; Lobo, J.; Strumsky, D.; West, G.B. Urban scaling and its deviations: Revealing the structure of wealth, innovation and crime across cities. *PLoS ONE* **2010**, *5*, e13541. [CrossRef] [PubMed]
6. Schläpfer, M.; Bettencourt, L.M.; Grauwin, S.; Raschke, M.; Claxton, R.; Smoreda, Z.; West, G.B.; Ratti, C. The scaling of human interactions with city size. *J. R. Soc. Interface* **2014**, *11*, 20130789. [CrossRef]
7. Xu, G.; Xu, Z.; Gu, Y.; Lei, W.; Pan, Y.; Liu, J.; Jiao, L. Scaling laws in intra-urban systems and over time at the district level in Shanghai, China. *Phys. A Stat. Mech. Appl.* **2020**, *560*, 125162. [CrossRef]
8. Bettencourt, L.M.; Lobo, J.; Helbing, D.; Kühnert, C.; West, G.B. Growth, innovation, scaling, and the pace of life in cities. *Proc. Natl. Acad. Sci. USA* **2007**, *104*, 7301–7306. [CrossRef]
9. Fu, J.C.; Li, G.; Zhao, H.; Zhang, J.Y. Allometric relationship between urban population and built-up area in China: A case study of 652 cities. *China Land Sci.* **2015**, *29*, 46–53.
10. Lei, D.; Hao, W.; Hongrui, Z. The definition of city boundary and scaling law. *Acta Geogr. Sin.* **2017**, *72*, 213–223.
11. West, G. *Scale: The Universal Laws of Growth, Innovation, Sustainability, and the Pace of Life in Organisms, Cities, Economies, and Companies; Environment and Planning B: Urban Analytics and City Science*; Penguin Press: New York, NY, USA, 2016.
12. Yakubo, K.; Saijo, Y.; Korošak, D. Superlinear and sublinear urban scaling in geographical networks modeling cities. *Phys. Rev. E* **2014**, *90*, 22803. [CrossRef] [PubMed]
13. Zhang, J.; Li, X.; Wang, X.; Wang, W.X.; Wu, L. Scaling behaviours in the growth of networked systems and their geometric origins. *Sci. Rep.* **2015**, *5*, 9767. [CrossRef] [PubMed]
14. Ribeiro, F.L.; Meirelles, J.; Ferreira, F.F.; Neto, C.R. A model of urban scaling laws based on distance dependent interactions. *R. Soc. Open Sci.* **2017**, *4*, 160926. [CrossRef] [PubMed]
15. Li, R.; Dong, L.; Zhang, J.; Wang, X.; Wang, W.X.; Di, Z.; Stanley, H.E. Simple spatial scaling rules behind complex cities. *Nat. Commun.* **2017**, *8*, 1841. [CrossRef]
16. Bettencourt, L.M.; Lobo, J. Urban scaling in Europe. *J. R. Soc. Interface* **2016**, *13*, 20160005. [CrossRef]
17. Sahasranaman, A.; Bettencourt, L.S.M.A. Urban geography and scaling of contemporary Indian cities. *J. R. Soc. Interface* **2019**, *16*, 20180758. [CrossRef]
18. Cesaretti, R.; Lobo, J.; Bettencourt, L.M.; Ortman, S.G.; Smith, M.E. Population-Area Relationship for Medieval European Cities. *PLoS ONE* **2016**, *11*, e0162678. [CrossRef]
19. Ortman, S.G.; Cabaniss, A.H.; Sturm, J.O.; Bettencourt, L.M. Settlement scaling and increasing returns in an ancient society. *Sci. Adv.* **2015**, *1*, e1400066. [CrossRef]
20. Lobo, J.; Bettencourt, L.M.; Smith, M.E.; Ortman, S. Settlement scaling theory: Bridging the study of ancient and contemporary urban systems. *Urban Stud.* **2020**, *57*, 731–747. [CrossRef]
21. Zünd, D.; Bettencourt, L.M. Growth and development in prefecture-level cities in China. *PLoS ONE* **2019**, *14*, e0221017. [CrossRef]
22. Alves, L.G.; Ribeiro, H.V.; Lenzi, E.K.; Mendes, R.S. Distance to the Scaling Law: A Useful Approach for Unveiling Relationships between Crime and Urban Metrics. *PLoS ONE* **2013**, *8*, e69580. [CrossRef]
23. Ribeiro, H.V.; Hanley, Q.S.; Lewis, D. Unveiling relationships between crime and property in England and Wales via density scale-adjusted metrics and network tools. *PLoS ONE* **2018**, *13*, e0192931. [CrossRef] [PubMed]
24. Lobo, J.; Bettencourt, L.M.; Strumsky, D.; West, G.B. Urban Scaling and the Production Function for Cities. *PLoS ONE* **2013**, *8*, e58407. [CrossRef] [PubMed]
25. Jiao, L.; Xu, Z.; Xu, G.; Zhao, R.; Liu, J.; Wang, W. Assessment of urban land use efficiency in China: A perspective of scaling law. *Habitat Int.* **2020**, *99*, 102172. [CrossRef]
26. Leitao, J.C.; Miotto, J.M.; Gerlach, M.; Altmann, E.G. Is this scaling nonlinear? *R. Soc. Open Sci.* **2016**, *3*, 150649. [CrossRef]

27. Arcaute, E.; Hatna, E.; Ferguson, P.; Youn, H.; Johansson, A.; Batty, M. Constructing cities, deconstructing scaling laws. *J. R. Soc. Interface* **2015**, *12*, 20140745. [CrossRef]
28. Emanuele, S.; Vishal, S. Rich and Poor Cities in Europe. An Urban Scaling Approach to Mapping the European Economic Transition. *PLoS ONE* **2016**, *11*, e159465.
29. Gudipudi, R.; Rybski, D.; Lüdeke, M.K.; Zhou, B.; Liu, Z.; Kropp, J.P. The efficient, the intensive, and the productive: Insights from urban Kaya scaling. *Appl. Energy* **2019**, *236*, 155–162. [CrossRef]
30. Muller, N.Z.; Jha, A. Does environmental policy affect scaling laws between population and pollution? Evidence from American metropolitan areas. *PLoS ONE* **2017**, *12*, e0181407. [CrossRef]
31. Fang, R.; He, Q.S.; Xie, P.; Cairen, Z.M. Spatial dynamics of enclave patches in Urban agglomerations: A case study of Three Major Urban agglomerations in China. *Geogr. Inf. World* **2021**, *28*, 40–45.
32. Aulia, P.L.; Taki, H.M.; Wartaman, A.S. Impact of transportation on urban compactness index in South Tangerang City, Indonesia. *IOP Conf. Series. Earth Environ. Sci.* **2021**, *737*, 12052. [CrossRef]
33. Su, Q. Urban Spatial Expansion, Urban Compactness, and Average Travel Demand in the US Urbanized Areas. *Int. J. Reg. Dev.* **2020**, *7*, 1. [CrossRef]
34. Zia, M.M.; Jeonghyun, R. Effects of Urban Compactness and Residential Density on Trip Generation: Focusing on Work Trips in Seoul, Korea. *J. Korean Soc. Transp.* **2017**, *35*, 1–10.
35. Xu, C.; Haase, D.; Su, M.; Yang, Z. The impact of urban compactness on energy-related greenhouse gas emissions across EU member states: Population density vs. physical compactness. *Appl. Energy* **2019**, *254*, 113671. [CrossRef]
36. Yuan, M.; Huang, Y.; Shen, H.; Li, T. Effects of urban form on haze pollution in China: Spatial regression analysis based on PM2.5 remote sensing data. *Appl. Geogr.* **2018**, *98*, 215–223. [CrossRef]
37. Wei, W.; Liping, X.; Min, Z.; Minghao, O.; Haiyue, F. Response of landscape index grain-size effects in different patch types: A case study of Wuxi city. *Acta Ecol. Sin.* **2016**, *36*, 2740–2749.
38. Fan, C.; Myint, S. A comparison of spatial autocorrelation indices and landscape metrics in measuring urban landscape fragmentation. *Landsc. Urban Plan.* **2014**, *121*, 117–128. [CrossRef]
39. Liu, X.; Li, X.; Chen, Y.; Qin, Y.; Li, S.; Chen, M. Landscape Expansion Index and Its Applications to Quantitative Analysis of Urban Expansion. *Acta Geogr. Sin.* **2009**, *64*, 1430–1438.
40. Qingsong, H.E.; Mingjun, W.; Zhuoma, C. Analysis of Urban Compactness Based on Multi-dimensional Landscape Expansion Index. *Geomat. World* **2021**, *28*, 58–65.
41. He, Q.; Song, Y.; Liu, Y. Diffusion or coalescence Urban growth pattern and change in 363 Chinese cities from 1995 to 2015. *Sustain. Cities Soc.* **2017**, *35*, 729–739. [CrossRef]
42. Moran, P.A.P. Notes on Continuous Stochastic Phenomena Author. *Biometrika* **1950**, *37*, 17–23. [CrossRef] [PubMed]
43. Jiao, L.; Mao, L.; Liu, Y. Multi-order Landscape Expansion Index: Characterizing urban expansion dynamics. *Landsc. Urban Plan.* **2015**, *137*, 30–39. [CrossRef]
44. Zhang, S.L.; Zhang, K. Comparative study on Moran index and G coefficient of global spatial autocorrelation. *J. Sun Yat-Sen Univ. (Nat. Sci. Ed.)* **2007**, *4*, 93–97.
45. Xiong, Y.; Bingham, D.; Braun, W.J.; Hu, X.J. Moran's I statistic-based nonparametric test with spatio-temporal observations. *J. Nonparametric Stat.* **2019**, *31*, 244–267. [CrossRef]

land

Article

Relational Values of Cultural Ecosystem Services in an Urban Conservation Area: The Case of Table Mountain National Park, South Africa

Gregg C. Brill [1], Pippin M. L. Anderson [2] and Patrick O'Farrell [3,*]

[1] Department of Geography, University of Victoria, Victoria, BC V8W 2Y2, Canada; greggbrill@gmail.com
[2] Department of Environmental and Geographical Science, University of Cape Town, Rondebosch 7701, South Africa; pippin.anderson@uct.ac.za
[3] Fitzpatrick Institute of African Ornithology, DST/NRF Centre of Excellence, Department of Biological Sciences, University of Cape Town, Rondebosch 7701, South Africa
* Correspondence: patrick.ofarrell@uct.ac.za

Abstract: This paper assesses how residents of a developing city in the Global South, recognize and value the multiple diverse cultural ecosystem services associated with freshwater ecosystems, as provided by different landscape features originating in an urban protected area. This objective was achieved by establishing who benefits from freshwater ecosystem services, uncovering the spatial and temporal relationships these beneficiaries have with landscape features, and determining the relational nature of ecosystem service values, benefits and trade-offs as experienced by the different users. Recreation, aesthetic and existence services were valued highest by respondents. People who live closer to the park use, and benefit from, the park's freshwater ecosystems more frequently than those living further away. Park visitors want ease of access in terms of distance to specific freshwater ecosystems, and then once there, they want a diversity of activity options, such as recreation opportunities, as well as places to reflect and meditate. This study of cultural ecosystem services improves our understanding of social-ecological systems in urban areas by exploring the relationships between park and people which can guide management to ensure equitable and sustainable ecosystem service provision to all city residents.

Keywords: cultural ecosystem services; freshwater features; urban protected area; Table Mountain National Park; Cape Town; Global South

Citation: Brill, G.C.; Anderson, P.M.L.; O'Farrell, P. Relational Values of Cultural Ecosystem Services in an Urban Conservation Area: The Case of Table Mountain National Park, South Africa. *Land* **2022**, *11*, 603. https://doi.org/10.3390/land11050603

Academic Editors: Alessio Russo and Giuseppe T. Cirella

Received: 29 March 2022
Accepted: 19 April 2022
Published: 20 April 2022

1. Introduction

As urban environments have expanded and continue to grow [1], so too has the recognition of human dependence on nature for the provision of vital goods and services [2,3]. National parks within cities are critical components of urban ecosystems, holding valuable green (e.g., grasslands and forests) and blue (e.g., rivers and wetlands) infrastructure, which includes freshwater systems and their functioning. Freshwater ecosystems, in particular, provide a variety of cultural ecosystem services (ESs), ranging from spiritual enlightenment to recreation [4,5]. These urban parks are key elements of the social ecological landscape where biophysical, social, economic and cultural factors are inextricably intertwined [6].

Cities and urban parks represent a new class of ecosystem shaped by the dynamic interactions between ecological and social systems [7] and as such need to be considered as social-ecological systems (SES), which are composed of organized assemblages of humans and nonhuman life forms in a spatially determined geophysical setting [8]. The functioning of SES is complex, heterogeneous and often structured by dynamic processes [9] with environmental variables influencing natural systems across space and time. Understanding the complexity of ecological systems is crucial if we are to comprehend human interactions that result in critical changes in the goods and services an ecosystem can provide.

Cities, such as Cape Town, which have seen rapid expansion in recent decades [10], should be considered as SES as there are multiple demands on urban land use to meet development and conservation needs [11,12]. When projecting the expansion of current cities, one must become more proactive in allowing for reconciliation between human development and environmental sustainability [12–15]. Understanding the links between urban protected areas and the cities in close proximity to them, as well as the relationships between ESs and people, particularly within African contexts, requires an SES approach [16–18].

Urban parks play a vital role in generating ESs that are important for human well-being, a role which is becoming increasingly well recognized [2,19–25]. Nature-based recreation provides an opportunity for both physical and mental stimulation and relaxation [26–28], and numerous studies have shown that access to water in urban natural spaces correlates with higher levels of physical activity [23,28–30]. Additionally, the need to experience nature and escape from the city constitutes an important reason for people to visit urban parks [31]. Relaxation, quietude, nature appreciation, stress relief, combatting fatigue, social interactions, and aesthetic enjoyment are stated as important reasons for using blue infrastructure in urban natural spaces and protected areas [3,22,23]. However, understanding the nuances, nature and workings of cultural ESs, especially in an urban context in the Global South, has yet to be adequately achieved [32,33]. Rapid urban growth in sub-Saharan Africa makes work in this area particularly pertinent [19,34].

While cultural ESs have been included in all seminal typologies relating to ESs [3,35,36], they have received less attention in the mounting body of empirical ecosystem service (ES) research and assessments [20,23,32,37–42]. Ecosystem service research, and more specifically a spatial understanding of the working of services, has grown substantially in the past decade [43–47]. Several dedicated tools and applications (e.g., SolVES) have been created to develop a better spatial understanding [47,48] of ES delivery as well as being used to establish and enhance key ES frameworks (e.g., the Common International Classification of Ecosystem Services (CICES), Intergovernmental Science-Policy Platform on Biodiversity and Ecosystem Services (IPBES), etc.) [49]. Several techniques have also been developed for the localization of cultural ESs valued by urban residents through participatory mapping [40,48,50–54]. The identification of locations of high ES delivery has been helpful in understanding the spatial determinants of ES provision, and its associated value to society [42,47,55]. Further, mapping techniques are playing an important role in informing landscape management, and offer researchers an opportunity to capture cultural benefits and trade-offs at a landscape level [46,56].

Cultural ESs are heterogeneous in space and evolve through time [57]. This heterogeneity over different scales suggests that the properties of ESs that people regard as useful or valuable may change over time or across different points in a landscape, even if the ecological system itself remains in a relatively constant state [16]. Being spatially explicit is important to consider the spatial heterogeneity of ES flows and of the social-ecological values and trade-offs that can be assigned to ESs [13,50,58–60].

These socio-ecological relationships have an immense impact on the relational valuation of ES [61], defined as the preferences, principles, and virtues of human-nature relationships [62–64]. Current conservation and land-use planning tools, including ES frameworks, tend to measure instrumental values (i.e., the value of an entity as merely a means to an end [64–70], whereas we should seek to value and protect relationships amongst components of ecosystems, including humans. Dominant discourses and approaches emphasize the dichotomy between instrumental (anthropocentric) versus intrinsic (non-anthropocentric) dimensions of nature [71]. Instead, a relational value perspective tends to emphasize the value of the interactions between people and nature and those among individuals in society. Thus, relational values fill a gap left by inadequacies and ambiguities in the common application of the instrumental valuation paradigm [72]. Recently, the IPBES (2015) included the category of relational values in its conceptual framework, in addition to instrumental values. This high-level acknowledgment of the way in which values are articulated and assigned reveals possible areas for action [72], as well as pointing to how values are shared and negotiated among groups of people, enabling them to act together as a community and creating a sense of belonging [61,62]. This recognition of how important relational values are to communities has a significant bearing on both ES policy and practice.

The value of protected areas is often judged in terms of their importance to society [73]. However, this importance, and the complex ways in which people relate to and interact with protected areas, cannot be captured by instrumental and intrinsic value framings alone. Rather, our understanding of the role of urban protected areas in society needs to take account of people's relational values concerning nature [73–75].

The overarching aim of this study was to develop an understanding of the cultural ESs, namely the underlying landscape factors and features determining their relational value and importance as perceived by users of a national park in a metropolitan setting, and the related benefits associated with use. To this end, we focused on freshwater ecosystems and features within the Table Mountain National Park (TMNP). We were guided by two research questions. Firstly, we investigated which cultural services and disservices emerge from different users in relation to freshwater features in the landscape, and secondly what the spatial factors are that may influence how users of cultural ESs value freshwater features in the TMNP. Our samples provided overviews of the diversity of perspectives within built and blue infrastructure and nuanced insight into the complex phenomena underlying water-related cultural ESs according to the state of water features in the TMNP, the section in which these features are located, and the accessibility of these features.

2. Materials and Methods

2.1. Study Site

Instituted in 1998, the TMNP located in the city of Cape Town is South Africa's only large urban national park and is the most visited of all South Africa's national parks [76], receiving over 4 million visitors per year of which 2.8 million are residents of Cape Town, visiting for outdoor recreational purposes such as hiking, biking, swimming, picnicking, bird watching, wildlife viewing, etc. [77,78] (See Table A1). Non-residents, comprising national and international visitors, make up the remaining 1.2 million visitors to the TMNP annually. The park covers an area of 221 km^2 and is almost entirely surrounded by the metropolitan area of Cape Town (Figure 1). It largely functions as an open-access system, as most of the park is unfenced [79], with 20 formal access points of which six are payment points (Table A2 in Appendix D). The park has an established network of over 500 km of trails for hiking, running and walking, as well as overnight accommodation facilities [77].

The park is comprised of three distinct sections: Northern (Table Mountain), Central (Silvermine) and Southern (Cape of Good Hope) sections. The Northern and Central sections are mountainous and are separated by developed urban areas on intervening terrain. The Southern section comprises most of the land on the southern peninsula of Cape Town.

TMNP has obtained World Heritage Site status in line with its global importance as a hotspot of biodiversity for higher plants and invertebrates. The park contains several natural and human-made freshwater landscape features, including rivers and streams, wetlands, pools, dams, reservoirs and waterfalls, 39 of which are included in this study. One of the main reasons why the park area has been protected is because of its early recognition as an important mountain catchment area for freshwater provision [80] for the residents of Cape Town, and it still performs this function today.

The City of Cape Town is located on the south-western tip of southern Africa (see Figure 1), occupies an area of roughly 2461 km^2 and has a population of 3.7 million people [81]. Population density in the City is measured at between 1530 [81] and 3950 people per square kilometer [82]. The broader metropolitan area of Cape Town is spatially distinct, from a development perspective [82], with areas of extreme wealth and poverty [83]. On the western and eastern flanks of the park, very affluent households reside, whereas further east of the park, poverty prevails in the poorer communities [84].

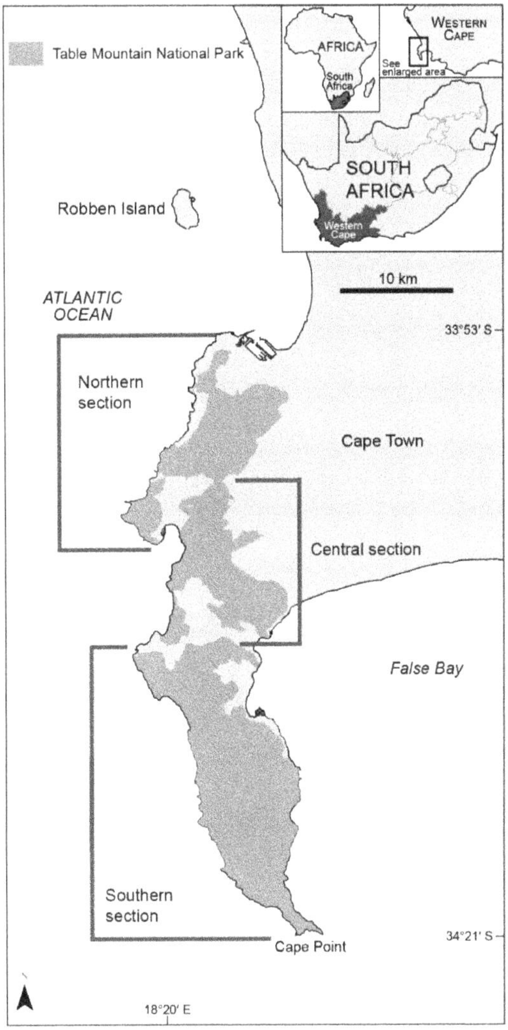

Figure 1. The Table Mountain National Park and its discontinuous sections along the Cape Peninsula, located in the Western Cape province of South Africa.

2.2. Data Gathering and Analyses

In partnership with the TMNP tourism unit, we identified 45 organized groups that use the park on a regular basis, all of which were contacted to take part in this study. User groups consisted of hiking clubs, walking groups, trail running groups, religious orders, research organizations, etc. Contact details for each of the user groups were provided by South African National Parks (SANParks), the state agency responsible for the management of governance of national parks. Emails were sent out to the administrators of each group inviting them and their group to take part in an online survey. Organized (formal) user groups who were contacted formed the majority of the sample population of this study, although the area is also a popular recreational destination for individuals and informal user groups (e.g., dog walkers). To elicit broader participation, posters that requested participation in the study were developed and displayed at shopping malls, libraries, and community centers throughout Cape Town, and at a number of the entrance points to the park. Additionally, newspaper articles written by the study researchers were published in two free Cape Town community papers containing contact details for further participation. This study assessed the perceptions of ESs from formal and informal user groups, as these groups may appreciate cultural services in the TMNP differently [85,86].

Structured questionnaires were used [23,87] (Appendix A) to research the role and importance of an urban park. The online survey was constructed using KwikSurveys (www.kwiksurveys.com; 19 September 2017) and was active for eight months. The survey captured respondent details and asked 17 open-ended questions relating to the demographics of park users, historical and current access patterns and user behavior, present and past water-related activities, residents' proximity to the park, willingness to travel to the park and its freshwater features, as well as questions based on water-related impacts and management. Open-ended questions were designed to enrich understanding rather than draw definitive conclusions from quantitative summaries [38]. Water-related activities in this study relate to all undertakings recognized by survey respondents to be either in or near freshwater. A snowball methodology was adopted in which survey participants were encouraged to send a link to the survey to other users of the park or suggest other individuals who could be contacted to participate in the study. Measures were taken to ensure that respondents could not participate in the survey and mapping exercise more than once, which included registering for the survey using an email address. Further, we checked respondent details in spreadsheets for any duplication. A hard-copy version of the survey was also available and sent to members of the public who selected this option. A statement regarding compliance of relevant ethical standards and appropriate rules and guidelines regarding participation of human subjects was included in the terms and conditions of the online and hard-copy survey.

The online survey included a mapping application that pre-identified and numbered the park's freshwater landscape features [23,43,53]. The survey's use of a physical map focused discussion of cultural ES values on specific blue infrastructure, which helped make complex intangible concepts more concrete to respondents familiar with maps as expressions of place [38,40,55,59]. We focused on five cultural services, namely aesthetics and existence; cognitive development, learning and scientific discovery; cultural and historical; recreation; and spiritual and religious. A sixth category was added to include any disservice (negative) values assigned to particular water features. Survey respondents were asked to assign value to these freshwater features according to the six cultural ES categories. Values were derived from the number of responses a water feature receives across the six ES categories. Respondents were asked to select between five and ten freshwater features per ES category. This selection allowed for the more nuanced capturing of difference between features and showed a greater range of preferences for ES categories across the various water bodies. An option for non-applicability was also offered. The number of respondents who selected an ES category per water feature was totaled to achieve an overall cultural ES score. Negative values were subtracted from the overall score of a freshwater feature. The mapping application was written in Ruby On Rails 3 and source code was developed using Jetbrains RubyMine v6. The questionnaire and mapping application are included in Appendix A (see Figures A1 and A2).

We examined the location of park entry points, residential proximity to the TMNP, and the distances users travel to access the park and its water features, as critical properties of these water features. The scoring flowchart adopted in this study is shown in Figure A3 (Appendix B). The highest number of points a water body could achieve was 15. Features that scored ten or less were ranked as not easily accessible, while those that scored 11 or higher were deemed easily accessible. These accessibility scores were used to determine how the level of access may influence ES values.

Cultural service scores were compared using the Kruskal-Wallis H test for state and section, and the Mann-Whitney U test for access. Both tests are rank-based nonparametric tests. An alpha threshold of 0.05 was used Visual inspection of the boxplots (histograms) for each of the six cultural services for each factor suggested some deviation from normality.

Equality of variances of groups were assessed using the Levene's test of homogeneity of variances. Pairwise post hoc comparisons were used to further explore differences between groups.

3. Results

3.1. Characteristics of Respondents

We received 265 complete entries (survey and mapping exercises). Incomplete entries were discounted. There was a relatively even distribution of male and female respondents, totaling 139 (52.5%) and 126 (47.5%), respectively. The average age of male respondents was 43.7 ± 16.2 years, while that of females was 42.6 ± 14.1 years. The youngest respondent was 18, while the oldest was 82 years old. The majority (208 respondents; 57.4%) of participants have visited the TMNP for less than 30 years. Over 50% reported visiting the park weekly (139 respondents; 52.5%) and to visiting the park for less than 30 years (152 respondents; 57.4%). The largest proportion of study participants (148 respondents; 57.8%) claimed to visit the park with family or friends, in a larger group, and alone. Activities undertaken in and around the park's water features are presented in Table A1 (Appendix C). Activities such as hiking and walking were differentiated in the survey, as there are dedicated user groups that classify themselves according to these two activities, while running and trail running were combined in a single group.

3.2. Analyses of Ecosystem Services Features

Respondents assigned preference values to each of the 39 freshwater features in the park according to the six categories of cultural ESs. On average, people chose to select seven water features per ES category. The ES values assigned to recreation ranked as the largest (1866 responses; 40.9%), and with those for aesthetics and existence second (1573 responses; 34.5%) (Table A3 in Appendix E). Responses for cognitive development, learning and scientific discovery (334 responses; 7.3%), cultural and historical (367 responses; 8.1%), and spiritual and religious (275 responses; 6.0%) categories were lower, with all three categories amounting to only 21.4% (976 responses) of the total number of responses. There were 143 (3.1%) responses relating to disservice values. Responses from trail runners, hikers, and walkers as well as members of some religious orders suggest that conflicts with dogs and their owners create significant tension in the usage of some water features in the park. Other tensions were reported between trail runners and mountain bikers who both use the same trails in the park.

3.2.1. Influence of the State of Freshwater Features

In a comparison of the overall cultural ES values by water feature category, the 17 rivers, streams and waterfalls, which were categorized as natural (flowing) systems, received the greatest number of survey responses, some 44.0% (1881 responses). These features were specifically valued for their aesthetics and existence, promotion of cognitive development, learning and scientific discovery, and provision of spiritual and religious services. The 15 dams and reservoirs recorded a slightly lower overall total (1872 responses; 43.8%). This result is in part bolstered by the high recreational, and cultural and historical ES values attributed to these features, although these features were noted as holding little cognitive development, learning and scientific discovery value. Human-made features recorded the highest number of negative-value responses (98 responses). The natural (stationary) systems (encompassing wetlands and pools) reported only six negative values from survey respondents, and showed the lowest values across all ES categories, except for cognitive development, learning and scientific discovery. Overall, natural (stationary) systems contributed only 12.1% (519 responses) of the total number of responses.

The cognitive development, learning and scientific discovery, cultural and historic, recreation and negative categories did not show equal variances when comparing the state of water body groups (Table A3). Tests for difference indicate that cognitive development, learning and scientific discovery, negative, cultural and historical values and negative categories showed significant differences when tested against state of the water body. Post-hoc testing indicates significant differences between human-made and natural water features in all three state of water feature categories. Built infrastructure showed a higher mean number of responses than blue infrastructure in recreation and negative value categories.

3.2.2. Influence of Section

The Northern section of the park, with 16 water features, produced the highest overall service tally, accounting for 54.8% (2340 responses) of the cultural ESs provided for freshwater. Notably, the Northern section is an important location for cultural and historical services, as well as holding high values for spiritual and religious ESs (Table A3). The five water features in the Central section yielded 1090 responses (25.5%), although the mean number of responses indicates that this section holds major ES importance across five of the six categories. Values for cultural and historical services were ranked second highest in this section. The 18 features in the Southern section reported only 842 responses (19.7%) and recorded the lowest mean values across all ES categories.

Levene's test showed that the cultural and historical, recreation and spiritual and religious categories did not satisfy the assumption of equality of variance. Tests for difference showed that all categories showed highly significant differences for all ES categories against section ($p < 0.01$), except for the cognitive categories. Post-hoc comparisons between sections are presented in Table A3, with the Southern section having consistently smaller mean ranks corresponding to lower scores.

3.3. *Accessibility of Park and Features*

3.3.1. Park Access Points

Of the 20 entry gates, 12 are located in the Northern section, five in the Central section, and three in the Southern section. The gates at Silvermine accounted for 27.4% (167 responses) of the overall access to the park. Only three other sites amassed more than 10% contribution, namely Cape Point (92 responses; 15.1%), Newlands Forest (75 responses; 12.3%), and Kirstenbosch (74 responses; 12.1%). Combined, these four areas accounted for 66.9% (408 responses) of the total access points to the TMNP reported by park users. Three out of four of these locations are pay-points. Sixteen respondents claimed to access the park solely outside of official entry points. Almost 81% (214 respondents) of survey takers admitted to entering the park outside of formal access sites.

3.3.2. Proximity to the Park

A total of 51 suburbs as linked to respondents' residential addresses were recorded, 29 (56.7%) of which abutted the park (Figure 2). Over 80% (41 suburbs; 80.4%) of suburbs were within 5 km of the park. Only one suburb reported in this study fell outside of a 20 km radius from the park boundary. Of the suburbs recorded by survey responses, the majority (47 suburbs; 92.1%) are categorized as either middle- or upper-income communities based on annual household income figures.

3.3.3. Distances Travelled to the Park and Freshwater Features

Respondents were asked how far they were prepared to travel to access the park, measured as the maximum distance per return trip in kilometers from their home to a park access point. The average maximum distance was measured as 19.3 ± 19.5 km. The shortest distance reported was 1 km and the largest distance recorded as 150 km (round trip).

The majority of study participants (171 respondents; 64.5%) travelled between one and nine kilometers on their return trips to visit freshwater features in the TMNP. There was a marked decline in the number of survey takers who walked for 20 km or more to water, with 93.2% (247 respondents) of park users finding their preferred freshwater features within this distance. Distances of more than 30 km were reported by just 4 (1.5%) park users. The average return trip to water was 6.8 ± 8.1 km with the lowest recorded as 100 m and the largest distance at 65 km. These figures relate directly to water-related excursions and not to other land-based activities in the park.

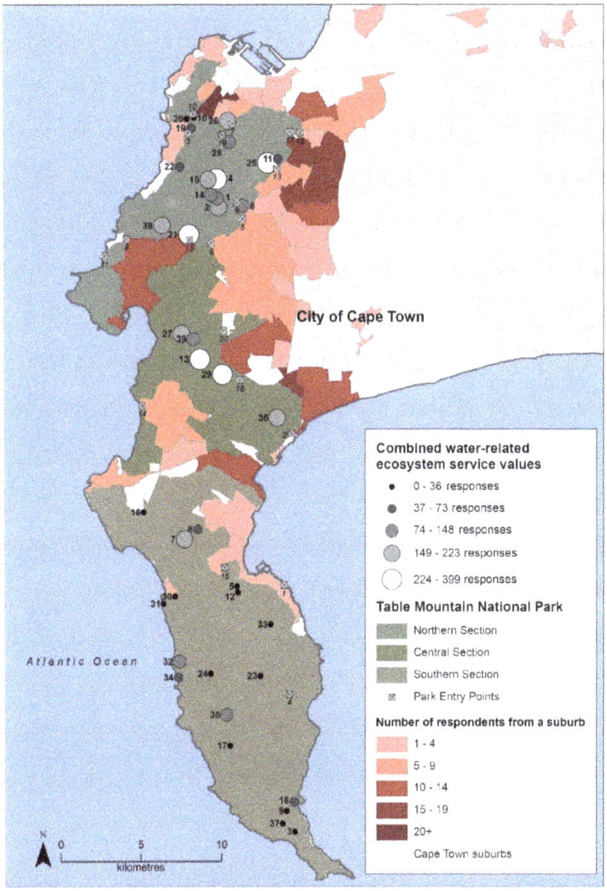

Figure 2. Combined total cultural ecosystem service responses by users of the Table Mountain National Park freshwater features. Water features are numbered from 1 to 39 and relate to Table A4 (Appendix F). The numbered crossed boxes refer to park entry points. Park entry points and numbered water features are listed in Tables A1 and A2, respectively. Suburbs are color-coded according to the number of survey participants residing there.

The Levene's test scored for level of accessibility indicated that the aesthetics and existence, cultural and historical, recreation and spiritual and religious categories did not show an equality of variance. Tests for difference indicated that all ES categories proved highly significant ($p < 0.01$) when scored by accessibility, except for the cognitive development category. Easily accessible water features scored highest across all of the cultural ES categories (Table A3). In each case, ease of access to water features is a significant driver in why certain freshwater features are valued higher than others in the TMNP.

4. Discussion

4.1. Water as a Cultural Ecosystem Service

This study demonstrates that freshwater ecosystems and ecosystem features in the TMNP are of major cultural ES value, and is aligned with similar studies [3–5,20,88–91]. The highest values measured related to recreation services. This finding is supported by several authors [2,26] who suggest that the recreational aspects of all urban ecosystems are the highest valued ESs in cities. High aesthetic and existence values attributed to freshwater features indicate that visitors to the park came to appreciate nature, admire the beauty of the landscape, and value the existence of a natural-state system within an urban setting. This is in line with other findings which state that nature appreciation and aesthetic enjoyment were central to urban residents using urban natural spaces and protected areas [3,22,23,56]. Although the values for cognitive development, learning, and scientific discovery, spiritual and religious, and cultural and historical services did not rank as highly as the values achieved for recreation, and aesthetics and existence categories, these are still critical considerations for holistic valuations of cultural ESs. A possible justification for why these categories did not rank as highly is that ESs are often benefit dependent [57], and it is difficult to express the socio-cultural importance or value of these services [92]. Uncertainty and lack of familiarity with the ESs under valuation or the terminology used raises concerns about the factors that influence an individual's expressed preferences. Assigned values should incorporate a person's perception of the service, their held values and associated preferences, and the context of the valuation [86,93–95]. Another possibility for why these three categories did not achieve the same scores as the recreation and aesthetics and existence categories could be due to the respondent sample. The TMNP management records only physical activity groups as these are often formalized. Individuals who use the park mainly for uses which fall in to the three lower-valued categories may not have been adequately captured. However, some cultural services such as spiritual and religious values are not intuitively associated with any particular landscape attribute [43,50,85] or feature, despite some individuals, groups or communities holding strong spiritual connections with water and water-related features [65].

Identified Trade-Offs and Tensions

Urban ecosystems contribute to urban well-being but can also negatively impact humans [2,13,58,60]. In this study negative values were assigned sparingly, most notably to built infrastructure. Additionally, survey responses indicate that conflicts with dog owners were reported by trail runners, hikers, walking groups and religious orders. Generally, respondents were obliging and accepting of other park users, yet conflicts were evident between a number of activities, most notably trail runners and mountain bikers. It is thus imperative to consider that landscape features offer multiple services to various groups and individuals, based on peoples' agendas, visions and uses of nature [96,97], and are often the natural sites of tensions and conflicts between users. Some authors [13] suggest that the quantification of trade-offs among ecosystem services and their interactions with human well-being are pressing areas for research, policy derivation and management.

4.2. Linking Cultural Services to the Landscape Features

4.2.1. State of Water Features

This study shows that natural (flowing) systems are critical blue infrastructural components in the landscape, providing a multitude of services, and are ranked particularly high in the aesthetics and existence and spiritual and religious categories. This is supported in other work [98] which state that natural rivers and waterscapes are sources of inspiration and deep spiritual value, and their intrinsic beauty enhances the quality of life and the landscape through which they flow. These natural systems are also important spaces for cognitive development, learning and scientific discovery, and also hold recreational importance [26]. Globally, a wide range of sports and activities are based in and around rivers and streams [99]. The negative values attributed to the natural (flowing) water features are low, indicating that respondents either favor natural systems or tensions are limited or reduced in these waterways.

Although some respondents assigned negative values to dams and reservoirs in the TMNP, this built infrastructure scored a combined cultural ES total almost equal to that of the natural (flowing) water features. They are rated highly as both aesthetic and recreational areas, making frequent participation in outdoor activities and other health and well-being benefits possible [27,28,51,86]. Many of these dams and reservoirs are easily accessible by well-used trails and paths, and are popular swimming sites, supporting the literature on the critical role of infrastructural requirements necessary to access ESs in urban environments [100,101]. Human-made interventions are critical elements of the landscape, providing multiple and key resources.

4.2.2. Proximity to Park

The location of parks relative to their potential users is an important factor in the assessment of accessibility [28,102,103], and is an important factor affecting peoples' park visitation [21,104]. In this study, people who live closer to a park use it more frequently, on average, than people who live farther away [28]. Communities bordering on the TMNP have the potential to benefit from close contact with a natural area, uninterrupted views and easy access to nature [78]. Properties not bordering on the park, but within walking distance or a short commute of it, enjoy the benefit of being able to access recreational and other nature-based opportunities with ease. Studies reveal that residents expressed that the maximum distance to a recreation area should not exceed one kilometer [105], be so close that they can be accessed within a five-minute walk [23], or be within five kilometers of a park when travelling by car [29,88,106]. This study confirms this view where the majority of users were prepared to travel up to 10 km (round-trip) to access the TMNP. Having to travel large distances has been indicated as one of the main reasons not to use urban green areas, often explored as the distance-decay model [103,107,108] and the blue infrastructure contained within them [109]. In contrast, other studies suggest that it is unlikely that the total travel distance is a significant barrier to park use [110]. Individuals with a greater level of attachment to the TMNP or a higher level of commitment to gaining benefits from the park might be more willing to travel greater distances to reach preferred areas in the park. This is evident in the willingness of survey participants to travel to the park's furthest flung Southern section to benefit from water-related services surrounding cognitive development, learning and scientific discovery.

4.2.3. Levels of Accessibility

With the greatest number of access points in the Northern and Central sections, it is to be expected that these entrances would be favored. This is primarily due to three factors, namely proximity to the majority of the users of the park, the variety of water features found in these sections, and the level of access to these water features. Although the Southern section has the greatest number of water features, this section has the fewest number of human-made and natural (flowing) systems, which this study shows to be favored. Formal access points to this section are limited to three, all of which are pay points.

Findings show that paying fees to access parks, or in the case of the TMNP certain areas in the park, does not cause a dramatic reduction in demand [111]. This is evident here where the most reported-as-used entrance points to the park charge entry fees. These sites do not offer additional amenities to non-pay entry points and are not managed differently. Pay-points may be perceived to reduce localized crime (e.g., car break-ins), as measures are in place to ease security concerns. Further justification for why respondents are prepared to pay to access certain areas in the park may relate to their attitude toward the environment, a site's attributes or attractions, and the cultural significance of an area [112]. In this regard, distance to these sites does not appear to be a limiting factor.

When examining the distance from entry points to freshwater features, it becomes clear that certain entry points are favored over others due to their location in the landscape and the water features near them. Most notable are the values for cognitive development, scientific learning and discovery which are not as spatially bounded as those for recreation and aesthetics and existence, which are located nearest to populous areas. The distance people will travel to reach these natural refuges for specific service benefits is evidence of the significant value of these landscapes and resources [113]. Studies suggest that these findings may be indicative because of other features of these environments, such as their remoteness or lack of human construction [114]. This appreciation could lead to rising demand for wilderness and remoteness in the future [115] and managers should engage more with park users across different socio-economic circumstances to understand this particular preference, which will become more critical with population growth in Africa, and harder to secure.

4.2.4. Bolstering Understanding of Relational Values

The design and implementation of ES-based processes and the mainstreaming of these results in decision-making is somewhat limited [73]. We do not yet have sufficient tools to assess and apply relational values in protected areas, nor a strong understanding of the contexts in which applying relational values in practice would be most appropriate [73,116]. ES studies, like this one, do however help capture the broad importance of a range of cultural ESs, alongside relational values [62,75]. In developing city settings, capturing people's perceptions of the importance of ESs is important yet challenging, because people may hold different ES values [75], or multiple values for the same landscape feature. This experience is seen around the world [116–119], including examples from developing countries and Global South contexts [75,113,120–122].

Adopting and developing the concept of relational values will be essential to ensure equitable provision of benefits, appropriate management of landscapes providing cultural ESs, and the recognition of ESs in supporting health and well-being, as well as other urban challenges [63]. Here, it is important to note that well-being and other cultural ES benefits received cannot simply be reduced to benefit accrual and that, instead, much derives from positive agency [123] and a direct connection with landscapes and landscape features. This change in framing fundamentally alters the way in which managers should identify and interpret the feedback between how people value nature, and people's interactions with, and perceptions of, protected areas [73], particularly in the Global South [124].

4.2.5. Management and Governance of Protected Areas and ES Provision

In several geographies and with many local and indigenous communities in the Global South, urban parks are essential repositories of ESs, notably provisioning services, providing urban residents with food, medicine, fuel, fibre and fodder [125,126]. These provisioning services may be less critical to survival or cultural identity in Global North contexts. Understanding ES provision, outside of just cultural services, is essential if parks are to be considered places of inclusion and empowerment, and managed as effective SES.

One of the greatest challenges to urban park managers is to balance the tension between providing for the diverse uses and values of landscapes and landscape features while simultaneously preserving the unique qualities of these places [96,97]. In some contexts, protected areas are managed for biodiversity goals, without considering the social influences and impacts on natural systems [127]. This is somewhat the case in TMNP, where the priority is still focused on landscape and biodiversity conservation, rather than managing the people that impact the landscape. However, because of the dynamic socio-ecological relationships between people and park, there is a need to integrate different disciplines and ways of managing and governing urban protected areas to better understand urban SES [128]. In developing countries, where projections indicate significant increases in urban populations, the need to establish planning processes based on contemporary SES approaches is vital to meet the challenges of sustainability, resilience and equity in the twenty-first century [128,129].

While park managers will still be required to protect landscapes, thus promoting ES generation and provision, city planners also need to consider the values and benefits of these protected areas, even if outside of their direct mandate, due to the nature of ES flows from park to urban residents and the influences that residents have on ESs and landscapes within an urban protected area. To support the interconnections between people and park, there is a need to enhance partnerships and communication around the protection of landscapes to ensure ES provision [127]. These partnerships should be across institutional levels [130] and should look to include public sector actors and those from academia, non-governmental organizations and civil society. Ecosystem services in an urban context provide key links for bridging planning, management and governance practices seeking transitions to more sustainable and resilient urban protected areas and cities [89,130]. Resilience goals for developing cities and surrounding protected areas, should explicitly incorporate the value of ESs in planning and governance [89].

Integration of ESs into decision-making and the inclusion of multiple stakeholders in the management and governance of urban systems is not without its challenges. Inadequate policies and practices [127], and fragmented, disjointed or unjust urban development and land management [130,131] affect urban conservation as well as residents' preference for ESs. Therefore, managing and governing a socially equitable urban national park is an institutional change process and should be based on the diversity in needs of urban residents who benefit from ESs generated in the park [131]. Understanding benefit and trade-off accrual should inform management and associated policy where social needs and desires must be met simultaneously to conservation agendas [7].

4.2.6. Study Limitations

In most studies involving human subjects and survey instruments, there are limitations. Other studies discuss issues with adopting the ES concept and frameworks into surveys and mapping exercises, as some respondents may not have fully conceptualized cultural ES benefits prior to their participation in surveys or may find it difficult to explain cultural ES concepts [32,38,87]. To address this, we provided a simplified definition of the ES concept to respondents in the text of the mapping exercise.

Although a definition of the ES was provided, further detail regarding the nuances of cultural ES categories may have been needed. In this study, spiritual and religious values may not have been adequately captured as respondents may not have articulated these ESs in the survey or mapping exercises. A further option could have been the understanding of the concept of spiritual and religious ESs, a possible shortcoming in the design of the questionnaire or the perceptions and understanding of those sampled [38,86,94].

A disproportionate lack of access to blue infrastructure has come to be seen as both a social and environmental injustice [30] and may reduce visitation among some individuals and communities, particularly those in economically disadvantaged groups [112,125]. Entrance fees to natural attractions have significant equity, economic, administrative and political implications [113] and could place constraints on some segments of society [112,132]. Although the park functions as an open-access system as most of the park is unfenced with few entry pay-points, travel costs may still influence the willingness or ability to visit the park. In this regard, some communities may have been unintentionally excluded from this study. Future studies of this nature should look to include groups that may have historically been ignored or prevented from participating due to personal circumstances. Particular attention should be applied when trying to source information from informal user groups, particularly those not known to park management or those without adequate or formal channels through which to be contacted.

4.2.7. Areas for Future Research

There are several areas for consideration when considering future research areas. Of immediate concern is the lack of scientifically robust information detailing the status and changes in ecosystems, the drivers of change, and the consequences of management responses [133–135]. Existing information may be fragmented and incomplete, incomparable from one place to another, highly technical or unsuitable for practitioners or policymakers [136–138]. Research is needed to develop innovative methodologies, through transdisciplinary efforts [13,139,140] that ensure more comprehensive analyses and results that further our understanding of the values of urban ES [16,33,141,142], how ES generation and provision may change over time, or the temporal changes in relational values of ESs. This research is essential to addressing gaps in the literature.

Within the spatial elements of further research, mapping and modelling approaches are needed to spatially explore ES generation, provision and perceptions. These may include socio-ecologically informed comparative modeling to promote sustainable urban policy transitions [143], or place-based SES research to analyze patterns in perceptions of ESs [90]. Further exploration into social (e.g., gender, education) and local ecological (e.g., land use and climate) characteristics will be needed to elucidate people's perceptions of which ESs are important.

As part of the interactions between people and nature, and how these may change over time, we should further explore different values and valuation approaches. The valuation of cultural ES is increasingly seen as a crucial element of inclusive research and robust decision-making [144] across multiple levels of governance and management. This is important in the context of areas demarcated as conservation areas. A key rationale for further research is the ongoing need to estimate or quantify the value of urban protected areas to assist city residents in benefitting from ESs and nature-based solutions (NBS), in order to be able to reveal the various social and economic benefits of managing and conserving these protected areas for equitable benefit accrual.

A growing area of research is identifying the nexus between ESs and NBS [91,145,146]. Addressing urban challenges with nature-based approaches can improve and protect ESs. Yet urban planning has not efficiently integrated such approaches to managing land use and landscapes [91]. Design and planning around NBS in urban settings, targeted at the needs of local systems, requires knowledge about the causal relationships and nexuses between NBS, ESs and urban challenges [145]. Within this space, we should investigate the roles that ESs and NBS can play in meeting urban challenges. We should also explore how these concepts can help transition to a green infrastructure city, help unpack stacked benefits to meet multiple objectives, and improve long-term societal, economic and environmental resilience. A hurdle here is the challenge of turning the concept of ESs into a practical tool in the formulation of day-to-day policies [147]. To address this challenge of ES integration, planning could better reconcile interests between nature conservation and urban planning, and ES supply and demand mapping may be a useful tool for such purposes [147].

5. Conclusions

It is crucial to understand how cultural services hold meaning and value to people in different contexts, as well as how these services connect to or interact across biophysical and SES. These connections, and the values they hold or yield, will be critical to enhancing our understanding of the ES concept. Considering growing population numbers in urban settings demands placed on cultural ESs will grow. This study provides insights into how visitors to urban conservation areas recognize, use and place importance on urban cultural ESs, which to date has not been adequately explored, particularly within an African developing country context. This study shows significant use and appreciation of the cultural ESs offered by the TMNP with respect to freshwater features. People use the park to undertake a variety of water-related physical activities, to enjoy the aesthetics of the environment, to reflect and learn, and for spiritual reasons, all of which are supported by findings in the global literature. This study highlights that both the built and blue infrastructure are of similar importance in producing cultural services, being assigned similar values across all the ES categories, meeting the needs of diverse user groups. This finding is significant and should be addressed by managers of the park to ensure sustainable and equitable use of freshwater features. There is further possibility of opening other freshwater features to ensure a greater spread of cultural services across the park. The results indicate that, while users were prepared to travel substantial distances both to the park and within the park to access cultural services, more easily accessed features and sections of the park had higher visitation rates. Paying to enter at access points was not found to be a deterrent. Numerous other socially and culturally informed factors can serve as barriers of access and, while they are not addressed in this study, the high value of cultural uses for communities adjacent to the park is evident, with less representation from further neighborhoods suggesting that other barriers are in place. It can be argued that built and natural freshwater infrastructure are critical cultural features of this landscape providing multiple and key services, in this urban context, with varying degrees of use and varying tensions between the different user groups noted. Management of the park should embrace the findings that cultural services are important and that these ESs can be enhanced by creating more opportunities across both built and blue infrastructure.

Author Contributions: Conceptualization, G.C.B., P.M.L.A. and P.O.; methodology, G.C.B., P.M.L.A. and P.O.; software, G.C.B.; validation, G.C.B., P.M.L.A. and P.O.; formal analysis, G.C.B.; investigation, G.C.B.; writing—original draft preparation, G.C.B., P.M.L.A. and P.O.; writing—review and editing, G.C.B., P.M.L.A. and P.O. All authors have read and agreed to the published version of the manuscript.

Funding: This research was funded by a doctoral fellowship grant from GreenMatter.

Institutional Review Board Statement: The study was conducted in accordance with the Declaration of Helsinki, and approved by the Ethics Committee of the Science Faculty at the University of Cape Town (approved November 2013) for studies involving humans.

Informed Consent Statement: Informed consent was obtained from all subjects involved in the study.

Data Availability Statement: Not applicable.

Conflicts of Interest: The authors declare no conflict of interest.

Appendix A. Table Mountain National Park Water-Related Cultural Ecosystem Service Questionnaire

Name	Age	Gender	M	F
Address				
Contact details				
How long have you been visiting TMNP?				
How often do you visit the park?				
How long have you been visiting the water in TMNP?				
How has your use of the parks water changed over time?				
Is your association with TMNP personal or professional?				
What types of activities have you previously undertaken in TMNP?				
What types of water-related activities have you previously undertaken in TMNP?				
What types of water-related activities do you take part in now?				
Do you use the parks water by yourself, with your family/friends or in a group?				
Do you consider the actions of other park users to have a potentially negative impact on the parks water?				
How far do you travel to access the park?				
How far do you travel in the park to access water bodies?				
Which park entry gates do you use?				
Do you access the park outside of entry gates?				
How do you feel about the way park officials manage water?				
Do you believe that current water management may impact on the future value of water in the park?				
Do you believe that negative water-related impacts will influence the water as it flows out of the parks boundaries?				
Can you suggest other people to contact to take part in this study? - Please provide contact details				
Other comments - Please use this space to elaborate on any of the questions or to bring information to the attention of the researchers				

Table Mountain National Park Water-Related Ecosystem Service Mapping Survey

This mapping survey is the last part of the survey on ecosystem services in Table Mountain National Park. This exercise consists of a set of questions about how users value water bodies in the park according to a particular cultural ecosystem service. Ecosystem services relate to the direct or indirect benefits (goods and services) that we receive from nature which contribute to our overall well-being. Categories include:

1. Recreation
2. Aesthetics/existence
3. Cultural/historical
4. Cognitive development/learning/scientific discovery
5. Spiritual/religious
6. Negative

All water bodies are numbered for easy reference as per below:

Water bodies in TMNP

Reservoirs/dams
1. Alexandra Reservoir
2. De Villiers Dam
3. Frans Dam
4. Hely Hutchinson Reservoir
5. Jackson Reservoir
6. Kirstenbosch Dam
7. Kleinplaas Dam
8. Lewis Gay Dam
9. Matroos Dam
10. Mocke Reservoir
11. Newlands Reservoir
12. Rawson Reservoir
13. Silvermine Dam
14. Victoria Reservoir
15. Woodhead Reservoir

Other

Vleis, pools and waterfalls
31. Duiwelsvlei
32. Groot rondevlei
33. Klawersvlei
34. Klein rondevlei
35. Nellies Pool
36. Sirkelsvlei
37. Skilpadvlei
38. Waterfall in Cecelia Forest
39. Waterfall on Prinskasteel River

Other

Rivers/streams
16. Bokramspruit River
17. Booiskraal River
18. Buffels River
19. Camps Bay Stream
20. Diepsloot
21. Disa Stream
22. Kasteelpoort River
23. Klaasjagers River
24. Krom River
25. Newlands Stream
26. Platteklip Stream
27. Prinskasteel River
28. Silver Stream
29. Silvermine River
30. Schusters River

Other

Listing of important water bodies

For each section, please list each water body used or valued in each cultural category. Please complete at least rankings 1–5. A park map with numbered water features is provided for geographical reference.

Recreational

Please select water bodies in the TMNP that have RECREATIONAL significance to you. The areas should allow for sporting, hobby and outdoor activities.

☐ Tick this box if you do not use water bodies in the park for this type of cultural ecosystem service.

	Water body name/number
1	
2	
3	
4	
5	
6	
7	
8	
9	
10	

Aesthetic or existence

Please select water bodies in the TMNP that have **AESTHETIC OR EXISTENCE** significance to you. These values may include view-points, areas of natural beauty, important areas for conservation etc.

☐ Tick this box if you do not use water bodies in the park for this type of cultural ecosystem service.

	Water body name/number
1	
2	
3	
4	
5	
6	
7	
8	
9	
10	

Cultural or historical

Please select water bodies in the TMNP that have CULTURAL OR HISTORICAL significance to you. Areas with significant cultural or historical value may be relevant to you or others living in Cape Town.

☐ Tick this box if you do not use water bodies in the park for this type of cultural ecosystem service.

	Water body name/number
1	
2	
3	
4	
5	
6	
7	
8	
9	
10	

Cognitive development, learning, scientific discovery

Please select water bodies in the TMNP that have COGNITIVE DEVELOPMENT, LEARNING OR SCIENTIFIC DISCOVERY significance to you. These water bodies must inspire you to understand more about the natural environment, conservation and/or species that live in and around a particular water body.

☐ Tick this box if you do not use water bodies in the park for this type of cultural ecosystem service.

	Water body name/number
1	
2	
3	
4	
5	
6	
7	
8	
9	
10	

Spiritual or religious

Please select water bodies in the TMNP that have SPIRITUAL OR RELIGIOUS significance to you. The water bodies ranked in this section should have a positive emotional and spiritual influence on you resulting in a feeling of awe and wonder.

☐ Tick this box if you do not use water bodies in the park for this type of cultural ecosystem service.

	Water body name/number
1	
2	
3	
4	
5	
6	
7	
8	
9	
10	

Negative values

Please select water bodies in the TMNP that have NEGATIVE significance to you. These areas may influence the values for other categories, may be unsightly or unsafe, or may just be negative spaces.

☐ Tick this box if you do not use water bodies in the park for this type of cultural ecosystem service.

	Water body name/number
1	
2	
3	
4	
5	
6	
7	
8	
9	
10	

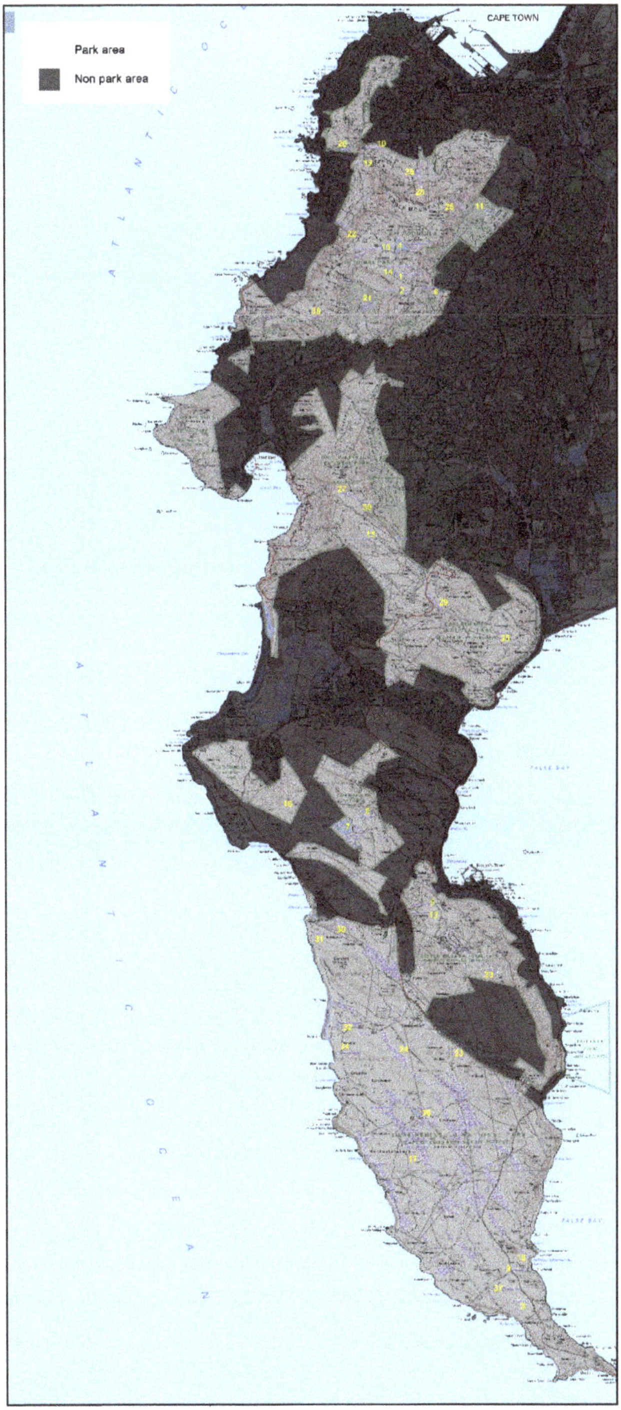

Figure A1. Map of the Table Mountain National Park with numbered water bodies relating to survey and mapping exercise on significance of water-related cultural ecosystem services.

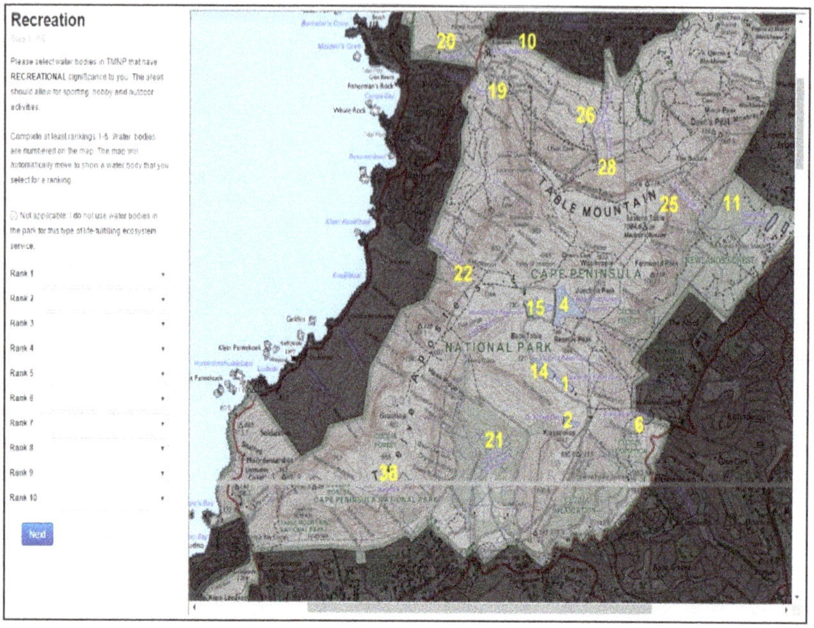

Figure A2. Screenshot of online mapping exercise detailing choice for water bodies in the Table Mountain National Park holding recreational value to survey respondents.

Appendix B. Scoring System for Accessibility of Water Bodies in the Table Mountain National Park

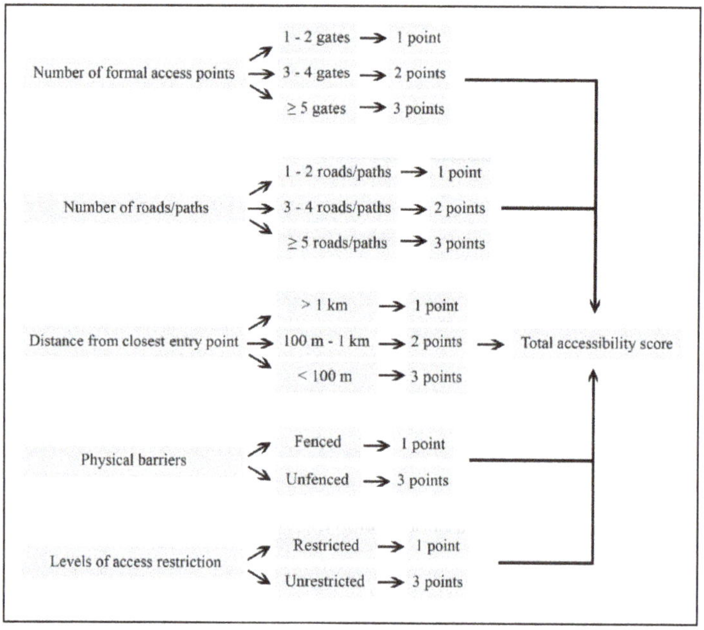

Figure A3. Flowchart highlighting how the scoring system for access points in the Table Mountain National Park was derived.

Appendix C. Water-Related Activities Undertaken in the Table Mountain National Park

Table A1. Number of responses for water-related activities in the Table Mountain National Park.

Activity	Number of Responses	Percentage Contribution
Hiking	211	79.6
Swimming	91	34.3
Running/trail running	80	30.2
Walking	77	0.0
Cycling/mountain biking	62	23.4
Dog walking	45	17.0
Picnics/braais	42	15.8
Climbing	36	13.6
Bird/wildlife watching	19	7.2
Camping/overnighting	14	5.3
Conservation management/volunteering	12	4.5
Nature appreciation	10	3.8
Photography/drawing	10	3.8
Alien clearing/vegetation clearing/rehabilitation	8	3.0
Horse riding	8	3.0
Caving	6	2.3
Research	6	2.3
Mapping/orienteering	4	1.5
Relaxation/meditation	4	1.5
Abseiling	3	1.1
Drinking water	3	1.1
Kloofing	3	1.1
Canoeing/kayaking	2	0.8
Emergency training	2	0.8
Fishing	2	0.8
Plant collecting	2	0.8
Educational activities	1	0.4
Guiding	1	0.4
Paragliding	1	0.4
Star gazing	1	0.4

Appendix D. Access Points to the Table Mountain National Park

Table A2. Access points to the Table Mountain National Park listed by section, type and class.

ID No.	Entry Gates Used	Section	Type	Class
1	Boulders	Southern	Pay point	Formal
2	Boyes Drive	Central	Non-pay point	Informal
3	Camps Bay	Northern	Non-pay point	Informal
4	Cape Point	Southern	Pay point	Formal
5	Cecilia Forest	Northern	Non-pay point	Formal
6	Constantia Nek	Central	Non-pay point	Formal
7	Deer Park	Northern	Non-pay point	Formal
8	Hout Bay	Northern	Non-pay point	Informal
9	Kirstenbosch	Northern	Pay point	Formal
10	Kloof Nek/Lions Head	Northern	Non-pay point	Formal
11	Llandudno	Northern	Non-pay point	Informal
12	Mostert's Mill	Northern	Non-pay point	Formal
13	Newlands Forest	Northern	Non-pay point	Formal
14	Noordhoek Beach	Central	Non-pay point	Informal
15	Orange Kloof	Northern	Pay point	Formal
16	Red Hill	Southern	Non-pay point	Formal
17	Rhodes Memorial	Northern	Non-pay point	Formal
18	Silvermine	Central	Pay point	Formal
19	Table Mountain Road	Northern	Non-pay point	Informal
20	Tokai Forest	Central	Pay point	Formal

Appendix E. Statistical Analyses

Table A3. Levene's test (degrees of freedom), mean rank, median, interquartile range, chi-squared value (degrees of freedom) and significance of cultural ecosystem service variables based on state of water features, section in which water feature are located, and access to water features in the Table Mountain National Park. An § indicates significant difference between human-made and natural (flowing) variables, an # indicates significant difference between access variables. An α indicates significant difference between Northern and Southern variables, an β indicates significant difference between Central and Southern variables. NS indicates not significant.

Cultural Ecosystem Service Category	State/Section/ Access	Variable	Levene's Test (df)	Mean Rank	Median	IQR	Chi-Square/U Value (df)	p-Value
Aesthetics and existence	State	Human-made Natural (flowing) Natural (stationary)	F(2,36) = 0.649; $p = 0.53$	18.57 22.29 17.50	29.00 31.00 25.00	63.00 68.50 51.00	$\chi^2(2) = 1.26$	NS
Cognitive development, learning and scientific discovery	State	Human-made §# Natural (flowing) § Natural (stationary) #	F(2,36) = 5.08; $p = 0.011$	10.40 23.71 31.57	2.00 7.00 12.00	4.00 10.00 13.00	$\chi^2(2) = 19.82$	<0.01
Cultural and historical	State	Human-made # Natural (flowing) Natural (stationary) #	F(2,36) = 14.65; $p < 0.001$	25.40 18.18 12.86	12.00 3.00 1.00	27.00 11.00 5.00	$\chi^2(2) = 6.63$	<0.05
Recreation	State	Human-made Natural (flowing) Natural (stationary)	F(2,36) = 3.72; $p = 0.034$	22.50 20.74 12.86	50.00 21.00 7.00	99.00 61.00 38.00	$\chi^2(2) = 3.54$	NS
Spiritual and religious	State	Human-made Natural (flowing) Natural (stationary)	F(2,36) = 0.95; $p = 0.397$	16.97 23.26 18.57	4.00 8.00 2.00	9.00 13.00 8.00	$\chi^2(2) = 2.59$	NS
Negative	State	Human-made # Natural (flowing) Natural (stationary) #	F(2,36) = 7.17; $p = 0.002$	13.47 22.82 27.14	6.00 1.00 0.00	10.00 5.00 2.00	$\chi^2(2) = 9.17$	0.01
Aesthetics and existence	Section	Northern α Central β Southern $\alpha\beta$	F(2,36) = 3.11; $p = 0.057$	24.94 32.00 12.28	36.00 72.00 9.00	16.30 54.00 16.30	$\chi^2(2) = 16.81$	<0.01
Cognitive development, learning and scientific discovery	Section	Northern Central Southern	F(2,36) = 0.20; $p = 0.822$	16.63 27.40 20.94	5.50 12.00 7.00	6.00 10.00 10.00	$\chi^2(2) = 3.67$	NS
Cultural and historical	Section	Northern α Central Southern α	F(2,36) = 5.64; $p = 0.007$	25.53 26.50 13.28	13.00 12.00 1.00	23.00 21.00 4.00	$\chi^2(2) = 11.80$	<0.01
Recreation	Section	Northern α Central β Southern $\alpha\beta$	F(2,36) = 5.32; $p = 0.009$	26.38 30.30 11.47	70.50 70.00 6.00	69.00 125.00 17.00	$\chi^2(2) = 19.17$	<0.01
Spiritual and religious	Section	Northern α Central β Southern $\alpha\beta$	F(2,36) = 6.56; $p = 0.004$	25.44 29.80 12.44	8.50 10.00 1.50	13.00 5.00 3.00	$\chi^2(2) = 15.40$	<0.01
Negative	Section	Northern α Central Southern α	F(2,36) = 3.03; $p = 0.061$	14.91 14.10 26.17	5.00 6.00 0.00	8.00 12.00 2.00	$\chi^2(2) = 10.31$	<0.01
Aesthetics and existence	Access	Easily accessible Not easily accessible	F(1,37) = 7.33; $p = 0.010$	29.22 13.59	71.50 13.00	68.80 23.00	U(1) = 36.50	<0.01
Cognitive development, learning and scientific discovery	Access	Easily accessible Not easily accessible	F(1,37) = 0.27; $p = 0.606$	21.59 18.89	7.00 7.00	9.00 10.00	U(1) = 158.50	NS
Cultural and historical	Access	Easily accessible Not easily accessible	F(1,37) = 11.69; $p = 0.002$	27.06 15.09	14.50 2.00	24.00 7.00	U(1) = 71.00	<0.01
Recreation	Access	Easily accessible Not easily accessible	F(1,37) = 5.79; $p = 0.021$	29.66 13.28	76.50 8.00	62.00 36.00	U(1) = 29.50	<0.01
Spiritual and religious	Access	Easily accessible Not easily accessible	F(1,37) = 5.23; $p = 0.028$	29.91 13.11	10.00 2.00	9.00 3.00	U(1) = 25.50	<0.01
Negative	Access	Easily accessible Not easily accessible	F(1,37) = 1.22; $p = 0.276$	14.25 24.00	5.00 0.00	7.00 5.00	U(1) = 92.00	<0.01

Appendix F. Water Features in the Table Mountain National Park

Table A4. Water features in Table Mountain National Park listed by section, state and level of accessibility. Natural (f) relates to flowing water features, such as rivers and streams; natural (s) refers to natural stationary water features such as pools, wetlands, etc.

ID No.	Water Body	Section	State	Accessibility (Score)
1	Alexandra Reservoir	Northern	Man-made	Not easily accessible (9)
2	De Villiers Dam	Northern	Man-made	Easily accessible (11)
3	Frans Dam	Southern	Man-made	Not easily accessible (9)
4	Hely Hutchinson Reservoir	Northern	Man-made	Easily accessible (12)
5	Jackson Reservoir	Southern	Man-made	Not easily accessible (9)
6	Kirstenbosch Dam	Northern	Man-made	Not easily accessible (10)
7	Kleinplaas Dam	Southern	Man-made	Easily accessible (11)
8	Lewis Gay Dam	Southern	Man-made	Not easily accessible (8)
9	Matroos Dam	Southern	Man-made	Not easily accessible (9)
10	Mocke Reservoir	Northern	Man-made	Not easily accessible (10)
11	Newlands Reservoir	Northern	Man-made	Not easily accessible (9)
12	Rawson Reservoir	Southern	Man-made	Not easily accessible (10)
13	Silvermine Dam	Central	Man-made	Easily accessible (13)
14	Victoria Reservoir	Northern	Man-made	Not easily accessible (9)
15	Woodhead Reservoir	Northern	Man-made	Easily accessible (11)
16	Bokramspruit River	Southern	Natural (f)	Not easily accessible (9)
17	Booiskraal River	Southern	Natural (f)	Not easily accessible (9)
18	Buffels River	Southern	Natural (f)	Not easily accessible (9)
19	Camps Bay Stream	Northern	Natural (f)	Easily accessible (12)
20	Diepsloot	Northern	Natural (f)	Easily accessible (13)
21	Disa Stream	Northern	Natural (f)	Easily accessible (13)
22	Kasteelpoort River	Northern	Natural (f)	Not easily accessible (10)
23	Klaasjagers River	Southern	Natural (f)	Not easily accessible (9)
24	Krom River	Southern	Natural (f)	Not easily accessible (9)
25	Newlands Stream	Northern	Natural (f)	Easily accessible (15)
26	Platteklip Stream	Northern	Natural (f)	Easily accessible (13)
27	Prinskasteel River	Central	Natural (f)	Easily accessible (11)
28	Silver Stream	Northern	Natural (f)	Easily accessible (13)
29	Silvermine River	Central	Natural (f)	Easily accessible (14)
30	Schusters River	Southern	Natural (f)	Not easily accessible (9)
31	Duiwelsvlei	Southern	Natural (s)	Not easily accessible (9)
32	Groot Rondevlei	Southern	Natural (s)	Not easily accessible (9)
33	Klawervlei	Southern	Natural (s)	Not easily accessible (9)
34	Klein Rondevlei	Southern	Natural (s)	Not easily accessible (9)
35	Nellies Pool	Central	Natural (s)	Not easily accessible (10)
36	Rawsons Reservoir	Southern	Man-made	Not easily accessible (10)
37	Skilpadvlei	Southern	Natural (s)	Not easily accessible (9)
38	Waterfall in Cecilia Forest	Northern	Natural (f)	Not easily accessible (10)
39	Waterfall on Prinskasteel River	Central	Natural (f)	Easily accessible (11)

References

1. United Nations, Department of Economic and Social Affairs, Population Division. *World Urbanization Prospects: The 2014 Revision*; (ST/ESA/SER.A/366); United Nations: New York, NY, USA, 2015.
2. Bolund, P.; Hunhammar, S. Ecosystem services in urban areas. *Ecol. Econ.* **1999**, *29*, 293–301. [CrossRef]
3. Daniel, T.C.; Muhar, A.; Arnberger, A.; Aznar, O.; Boyd, J.W.; Chan, K.M.A.; von der Dunk, A. Contributions of cultural services to the ecosystem services agenda. *Proc. Nat. Acad. Sci. USA* **2012**, *109*, 8812–8819. [CrossRef]
4. Jaung, W.; Carrasco, L.R. Using mobile phone data to examine weather impacts on recreational ecosystem services in an urban protected area. *Sci. Rep.* **2021**, *11*, 5544. [CrossRef]
5. Petroni, M.L.; Siqueira-Gay, J.; Gallardo, A.L.C.F. Understanding land use change impacts on ecosystem services within urban protected areas. *Landscape Urban Plan.* **2022**, *223*, 104404. [CrossRef]
6. Hjortsø, C.N.; Stræde, S.; Helles, F. Applying multi-criteria decision-making to protected areas and buffer zone management: A case study in the Royal Chitwan National Park, Nepal. *J. Forest Econ.* **2006**, *12*, 91–108. [CrossRef]

7. Du Plessis, C. Understanding cities as social-ecological systems. In Proceedings of the World Sustainable Building Conference, Melbourne, Australia, 21–25 September 2008; pp. 1–9.
8. Halliday, A.; Glaser, M. A Management Perspective on Social Ecological Systems: A generic system model and its application to a case study from Peru. *Hum. Ecol. Rev.* **2011**, *18*, 1–18.
9. Anand, M.; Gonzalez, A.; Guichard, F.; Kolasa, J.; Parrott, L. Ecological systems as complex systems: Challenges for an emerging science. *Diversity* **2010**, *2*, 395–410. [CrossRef]
10. Holmes, P.M.; Rebelo, A.G.; Dorse, C.; Wood, J. Can Cape Town's unique biodiversity be saved? Balancing conservation imperatives and development needs. *Ecol. Soc.* **2012**, *17*, 28. [CrossRef]
11. Anderson, P.; Elmqvist, T. Urban ecological and social-ecological research in the city of Cape Town: Insights Emerging from an Urban Ecology CityLab. *Ecol. Soc.* **2012**, *17*, 23. [CrossRef]
12. O'Farrell, P.; Anderson, P.; Culwick, C.; Currie, P.; Kavonic, J.; McClure, A.; Audouin, M. Towards resilient African cities: Shared challenges and opportunities towards the retention and maintenance of ecological infrastructure. *Glob. Sustain.* **2019**, *2*, E19. [CrossRef]
13. Carpenter, S.R.; Mooney, H.A.; Agard, J.; Capistrano, D.; Defries, R.S.; Díaz, S.; Dietz, T.; Duraiappah, A.K.; Oteng-Yeboah, A.; Pereira, H.M.; et al. Science for managing ecosystem services: Beyond the Millennium Ecosystem Assessment. *Proc. Natl. Acad. Sci. USA* **2009**, *106*, 1305–1312. [CrossRef] [PubMed]
14. Müller, N.; Ignatieva, M.; Nilon, C.H.; Werner, P.; Zipperer, W.C. Patterns and trends in urban biodiversity and landscape design. In *Urbanization, Biodiversity and Ecosystem Services: Challenges and Opportunities: A Global Assessment*; Elmqvist, T., Fragkias, M., Goodness, J., Güneralp, B., Marcotullio, P.J., McDonald, R.I., Parnell, S., Schewenius, M., Sendstad, M., Seto, K.C., et al., Eds.; Springer: Dordrecht, The Netherlands, 2013; pp. 123–174. [CrossRef]
15. Secretariat of the Convention of Biological Diversity. *Cities and Biodiversity Outlook—Actions and Policy*; Secretariat of the Convention of Biological Diversity: Montreal, QC, Canada, 2012.
16. Braat, L.C.; de Groot, R. The ecosystem services agenda: Bridging the worlds of natural science and economics, conservation and development, and public and private policy. *Ecosyst. Serv.* **2012**, *1*, 4–15. [CrossRef]
17. Duraiappah, A.K.; Asah, S.T.; Brondizio, E.S.; Kosoy, N.; O'Farrell, P.J.; Prieur-Richard, A.H.; Subramanian, S.M.; Takeuchi, K. Managing the mismatches to provide ecosystem services for human well-being: A conceptual framework for understanding the new commons. *Curr. Opin. Environ. Sustain.* **2014**, *7*, 94–100. [CrossRef]
18. Bux, Q.; Anderson, P.; O'Farrell, P.J. Understanding the local biodiversity and open space strategies in two South African cities. *Ecol. Soc.* **2021**, *26*, 4. [CrossRef]
19. Cilliers, S.; Cilliers, J.; Lubbe, R.; Siebert, S. Ecosystem services of urban green spaces in African countries-perspectives and challenges. *Urban Ecosyst.* **2013**, *16*, 681–702. [CrossRef]
20. de Lange, W.J.; Wise, R.M.; Forsyth, G.G.; Nahman, A. Integrating socio-economic and biophysical data to support water allocations within river basins: An example from the Inkomati Water Management Area in South Africa. *Environ. Model. Softw.* **2010**, *25*, 43–50. [CrossRef]
21. Gai, S.; Fu, J.; Rong, X.; Dai, L. Users' views on cultural ecosystem services of urban parks: An importance-performance analysis of a case in Beijing, China. *Anthropocene* **2022**, *37*, 100323. [CrossRef]
22. Jim, C.Y.; Chen, W.Y. Ecosystem services and valuation of urban forests in China. *Cities* **2009**, *26*, 187–194. [CrossRef]
23. Tyrväinen, L.; Mäkinen, K.; Schipperijn, J. Tools for mapping social values of urban woodlands and other green areas. *Landscape Urban Plan.* **2007**, *79*, 5–19. [CrossRef]
24. Gonzalez-Garcia, A.; Palomo, I.; Arboledas, M.; González, J.A.; Múgica, M.; Mata, R.; Montes, C. Protected Areas as a Double Edge Sword: An Analysis of Factors Driving Urbanization in Their Surroundings. *SSRN* **2021**, 51. Available online: https://papers.ssrn.com/sol3/papers.cfm?abstract_id=3962000 (accessed on 28 March 2022). [CrossRef]
25. Wessels, N.; Sitas, N.; O'Farrell, P.; Esler, K.J. Understanding community perceptions of a natural open space system for urban conservation and stewardship in a metropolitan city in Africa. *Environ. Cons.* **2021**, *48*, 244–254. [CrossRef]
26. Koh, Y.F.; Loc, H.H.; Park, E. Towards a "City in Nature": Evaluating the Cultural Ecosystem Services Approach Using Online Public Participation GIS to Support Urban Green Space Management. *Sustainability* **2022**, *14*, 1499. [CrossRef]
27. Nawrath, M.; Elsey, H.; Dallimer, M. Why cultural ecosystem services matter most: Exploring the pathways linking greenspaces and mental health in a low-income country. *Sci. Total Environ.* **2022**, *806*, 150551. [CrossRef]
28. Neuvonen, M.; Sievänen, T.; Tönnes, S.; Koskela, T. Access to green areas and the frequency of visits—A case study in Helsinki. *Urban For. Urban Green.* **2007**, *6*, 235–247. [CrossRef]
29. Cutts, B.B.; Darby, K.J.; Boone, C.G.; Brewis, A. City structure, obesity, and environmental justice: An integrated analysis of physical and social barriers to walkable streets and park access. *Soc. Sci. Med.* **2009**, *69*, 1314–1322. [CrossRef] [PubMed]
30. Dahmann, N.; Wolch, J.; Joassart-Marcelli, P.; Reynolds, K.; Jerrett, M. The active city? Disparities in provision of urban public recreation resources. *Health Place* **2010**, *16*, 431–445. [CrossRef]
31. Chiesura, A. The role of urban parks for the sustainable city. *Landsc. Urban Plan.* **2004**, *68*, 129–138. [CrossRef]
32. Chan, K.M.A.; Satterfield, T.; Goldstein, J. Rethinking ecosystem services to better address and navigate cultural values. *Ecol. Econ.* **2012**, *74*, 8–18. [CrossRef]
33. Dobbs, C.; Vasquez, A.; Olave, P.; Olave, M. Cultural Urban Ecosystem Services. In *Urban Ecology in the Global South*; Cities and Nature; Shackleton, C.M., Cilliers, S.S., Davoren, E., du Toit, M.J., Eds.; Springer: Dodrecht, The Netherlands, 2021. [CrossRef]

34. Anderson, P.; Okereke, C.; Rudd, A.; Parnell, S. Regional Assessment of Africa. In *Urbanization, Biodiversity and Ecosystem Services: Challenges and Opportunities*; Elmqvist, T., Fragkias, M., Goodness, J., Güneralp, B., Marcotullio, P., McDonald, R., Eds.; Springer: Dordrecht, The Netherlands, 2013; pp. 453–459. [CrossRef]
35. Costanza, R.; D'Arge, R.; de Groot, R.; Farberparallel, S.; Grasso, M.; Hannon, B.; van den Belt, M. The value of the world's ecosystem services and natural capital. *Nature* **1997**, *387*, 253–260. [CrossRef]
36. Millennium Ecosystem Assessment. *Ecosystems and Human Well-Being: Synthesis*; Island Press: Washington, DC, USA, 1997.
37. Cowling, R.M.; Egoh, B.; Knight, A.T.; O'Farrell, P.J.; Reyers, B.; Rouget, M.; Wilhelm-Rechman, A. An operational model for mainstreaming ecosystem services for implementation. *Proc. Natl. Acad. Sci. USA* **2008**, *105*, 9483–9488. [CrossRef]
38. Gould, R.K.; Klain, S.C.; Ardoin, N.M.; Satterfield, T.; Woodside, U.; Hannahs, N.; Chan, K.M.A. A protocol for eliciting nonmaterial values through a cultural ecosystem services frame. *Conserv. Biol.* **2015**, *29*, 575–586. [CrossRef] [PubMed]
39. Kumar, M.; Kumar, P. Valuation of the ecosystem services: A psycho-cultural perspective. *Ecol. Econ.* **2008**, *4*, 808–819. [CrossRef]
40. Raymond, C.M.; Bryan, B.A.; MacDonald, D.H.; Cast, A.; Strathearn, S.; Grandgirard, A.; Kalivas, T. Mapping community values for natural capital and ecosystem services. *Ecol. Econ.* **2009**, *68*, 1301–1315. [CrossRef]
41. Russell, R.; Guerry, A.D.; Balvanera, P.; Gould, R.K.; Basurto, X.; Chan, K.M.A.; Tam, J. Humans and Nature: How Knowing and Experiencing Nature Affect Well-Being. *Annu. Rev. Environ. Resour.* **2013**, *38*, 473–502. [CrossRef]
42. Sherrouse, B.C.; Semmens, D.J. Validating a method for transferring social values of ecosystem services between public lands in the Rocky Mountain region. *Ecosyst. Serv.* **2014**, *8*, 166–177. [CrossRef]
43. Brown, G. Mapping Spatial Attributes in Survey Research for Natural Resource Management: Methods and Applications. *Soc. Nat. Resour.* **2005**, *18*, 17–39. [CrossRef]
44. Maes, J.; Paracchini, M.L.; Zulian, G.; Dunbar, M.B.; Alkemade, R. Synergies and trade-offs between ecosystem service supply, biodiversity, and habitat conservation status in Europe. *Biol. Conserv.* **2012**, *155*, 1–12. [CrossRef]
45. Nelson, E.J.; Daily, G.C. Modelling ecosystem services in terrestrial systems. *F1000 Biol. Rep.* **2010**, *2*, 53. [CrossRef]
46. Petter, M.; Mooney, S.; Maynard, S.M.; Davidson, A.; Cox, M.; Horosak, I. A methodology to map ecosystem functions to support ecosystem services assessments. *Ecol. Soc.* **2013**, *18*, 31. [CrossRef]
47. Sherrouse, B.C.; Semmens, D.J.; Clement, J.M. An application of Social Values for Ecosystem Services (SolVES) to three national forests in Colorado and Wyoming. *Ecol. Indic.* **2014**, *36*, 68–79. [CrossRef]
48. Sherrouse, B.C.; Clement, J.M.; Semmens, D.J. A GIS application for assessing, mapping, and quantifying the social values of ecosystem services. *Appl. Geogr.* **2011**, *31*, 748–760. [CrossRef]
49. Turnhout, E.; Bloomfield, B.; Hulme, M.; Vogel, J.; Wynne, B. Listen to the voices of experience. *Nature* **2012**, *488*, 454–455. [CrossRef] [PubMed]
50. Alessa, L.; Kliskey, A.; Brown, G. Social-ecological hotspots mapping: A spatial approach for identifying coupled social-ecological space. *Landsc. Urban Plan.* **2008**, *85*, 27–39. [CrossRef]
51. Beckmann-Wübbelt, A.; Fricke, A.; Sebesvari, Z.; Yakouchenkova, I.A.; Fröhlich, K.; Saha, S. High public appreciation for the cultural ecosystem services of urban and peri-urban forests during the COVID-19 pandemic. *Sustain. Cities Soc.* **2021**, *74*, 103240. [CrossRef]
52. Bing, Z.; Qiu, Y.; Huang, H.; Chen, T.; Zhong, W.; Jiang, H. Spatial distribution of cultural ecosystem services demand and supply in urban and suburban areas: A case study from Shanghai, China. *Ecol. Indic.* **2021**, *127*, 107720. [CrossRef]
53. Brown, G.; Raymond, C. The relationship between place attachment and landscape values: Toward mapping place attachment. *Appl. Geogr.* **2007**, *27*, 89–111. [CrossRef]
54. Bryan, B.A.; Grandgirard, A.; Ward, J.R. Quantifying and exploring strategic regional priorities for managing natural capital and ecosystem services given multiple stakeholder perspectives. *Ecosystems* **2010**, *13*, 539–555. [CrossRef]
55. Van Berkel, D.B.; Verburg, P.H. Spatial quantification and valuation of cultural ecosystem services in an agricultural landscape. *Ecol. Indic.* **2014**, *37*, 163–174. [CrossRef]
56. Pinheiro, R.O.; Triest, L.; Lopes, P.F.M. Cultural ecosystem services: Linking landscape and social attributes to ecotourism in protected areas. *Ecosyst. Serv.* **2021**, *50*, 101340. [CrossRef]
57. Fisher, B.; Turner, R.K.; Morling, P. Defining and classifying ecosystem services for decision making. *Ecol. Econ.* **2009**, *68*, 643–653. [CrossRef]
58. de Groot, R.S.; Alkemade, R.; Braat, L.; Hein, L.; Willemen, L. Challenges in integrating the concept of ecosystem services and values in landscape planning, management and decision making. *Ecol. Complex.* **2010**, *7*, 260–272. [CrossRef]
59. Flood, K.; Mahon, M.; McDonagh, J. Assigning value to cultural ecosystem services: The significance of memory and imagination in the conservation of Irish peatlands. *Ecosyst. Serv.* **2021**, *50*, 101326. [CrossRef]
60. Hein, L.; van Koppen, K.; de Groot, R.S.; van Ierland, E.C. Spatial scales, stakeholders and the valuation of ecosystem services. *Ecol. Econ.* **2006**, *57*, 209–228. [CrossRef]
61. Ishihara, H. Relational values from a cultural valuation perspective: How can sociology contribute to the evaluation of ecosystem services? *Curr. Opin. Environ. Sustain.* **2018**, *35*, 61–68. [CrossRef]
62. Pascual, U.; Balvanera, P.; Díaz, S.; Pataki, G.; Roth, E.; Stenseke, M.; Yagi, N. Valuing nature's contributions to people: The IPBES approach. *Curr. Opin. Environ. Sustain.* **2017**, *26*, 7–16. [CrossRef]
63. Chan, K.M.A.; Gould, R.K.; Pascual, U. Editorial overview: Relational values: What are they, and what's the fuss about? *Curr. Opin. Environ. Sustain.* **2018**, *35*, A1–A7. [CrossRef]

64. Zafra-Calvo, N.; Balvanera, P.; Pascual, U.; Merçon, J.; Martín-López, B.; van Noordwijk, M.; Cabrol, D. Plural valuation of nature for equity and sustainability: Insights from the Global South. *Glob. Environ. Change* **2020**, *63*, 102115. [CrossRef]
65. Arias-Arévalo, P.; Martín-López, B.; Gómez-Baggethun, E. Exploring intrinsic, instrumental, and relational values for sustainable management of social-ecological systems. *Ecol. Soc.* **2017**, *22*, 43. [CrossRef]
66. Braito, M.T.; Böck, K.; Flint, C.; Muhar, A.; Muhar, S.; Penker, M. Human-Nature Relationships and Linkages to Environmental Behaviour. *Environ. Values* **2017**, *26*, 365–389. [CrossRef]
67. Klain, S.C.; Olmsted, P.; Chan, K.M.A.; Satterfield, T. Relational values resonate broadly and differently than intrinsic or instrumental values, or the New Ecological Paradigm. *PLoS ONE* **2017**, *12*, e0183962. [CrossRef]
68. Arias-Arévalo, P.; Gómez-Baggethun, E.; Martín-López, B.; Pérez-Rincón, M. Widening the Evaluative Space for Ecosystem Services: A Taxonomy of Plural Values and Valuation Methods. *Environ. Values* **2018**, *27*, 29–53. [CrossRef]
69. Ellis, E.C.; Pascual, U.; Mertz, O. Ecosystem services and nature's contribution to people: Negotiating diverse values and trade-offs in land systems. *Curr. Opin. Environ. Sustain.* **2019**, *38*, 86–94. [CrossRef]
70. See, S.C.; Shaikh, S.F.E.A.; Carrasco, W.J.L.R. Are relational values different in practice to instrumental values? *Ecosyst. Serv.* **2020**, *44*, 101132. [CrossRef]
71. Intergovernmental Science-Policy Platform on Biodiversity and Ecosystem Services (IPBES). *Preliminary Guide Regarding Diverse Conceptualization of Multiple Values of Nature and Its Benefits, Including Biodiversity and Ecosystem Functions and Services (Deliverable 3 (d))*; IPBES Secretariat: Bonn, Germany, 2015; p. 95.
72. Himes, A.; Muraca, B. Relational values: The key to pluralistic valuation of ecosystem services. *Curr. Opin. Environ. Sustain.* **2018**, *35*, 1–7. [CrossRef]
73. De Vos, A.; Bezerra, J.C.; Roux, D. Relational values about nature in protected area research. *Curr. Opin. Environ. Sustain.* **2018**, *35*, 89–99. [CrossRef]
74. De Vreese, R.; Van Herzele, A.; Dendoncker, N.; Fontaine, C.M.; Leys, M. Are stakeholders' social representations of nature and landscape compatible with the ecosystem service concept? *Ecosyst. Serv.* **2019**, *37*, 100911. [CrossRef]
75. Lau, J.D.; Hicks, C.C.; Gurney, G.G.; Cinner, J.E. What matters to whom and why? Understanding the importance of coastal ecosystem services in developing coastal communities. *Ecosyst. Serv.* **2019**, *35*, 219–230. [CrossRef]
76. Forsyth, G.G.; van Wilgen, B.W. The recent fire history of the Table Mountain National Park and implications for fire management. *Koedoe* **2008**, *50*, 3–9. [CrossRef]
77. Daitz, D.; Myrdal, B. Table Mountain National Park. In *Evolution and Innovation in Wildlife Conservation. Parks and Game Ranches to Transfrontier Conservation Areas*; Suich, H., Child, B., Eds.; Routledge: Cape Town, South Africa, 2009; pp. 325–339.
78. Standish, B.; Boting, A.; van Zyl, H.; Leiman, T.; Turpie, J. *The Economic Contribution of Table Mountain National Park*; The Graduate School of Business, University of Cape Town: Cape Town, South Africa, 2004; p. 46.
79. Ferreira, S.L. Balancing people and park: Towards a symbiotic relationship between Cape Town and Table Mountain National Park. *Curr. Issues Tour.* **2011**, *14*, 275–293. [CrossRef]
80. Nel, J.; Colvin, C.; Le Maitre, D.; Smith, J.; Haines, I. *South Africa's Strategic Water Source Areas*; CSIR: Stellenbosch, South Africa, 2013.
81. Statistics South Africa. City of Cape Town. Available online: https://www.statssa.gov.za/?page_id=1021&id=city-of-cape-town-municipality (accessed on 6 April 2022).
82. Turok, I. Deconstructing density: Strategic dilemmas confronting the post-apartheid city. *Cities* **2011**, *28*, 470–477. [CrossRef]
83. Lemanski, C. Global Cities in the South: Deepening social and spatial polarisation in Cape Town. *Cities* **2007**, *24*, 448–461. [CrossRef]
84. Davis, G. Biodiversity conservation as a social bridge in the urban context: Cape Town's sense of "urban imperative" to protect its biodiversity and empower the people. In *The Urban Imperative*; Trzyna, T., Ed.; California Institute of Public Affairs: Sacramento, CA, USA, 2005; p. 168.
85. Plieninger, T.; Dijks, S.; Oteros-Rozas, E.; Bieling, C. Assessing, mapping, and quantifying cultural ecosystem services at community level. *Land Use Policy* **2013**, *33*, 118–129. [CrossRef]
86. Vidal, D.G.; Dias, R.C.; Oliveira, G.M.; Dinis, M.A.P.; Filho, W.L.; Fernandes, C.O.; Barros, N.; Maia, R.L. A Review on the Cultural Ecosystem Services Provision of Urban Green Spaces: Perception, Use and Health Benefits. In *Sustainable Policies and Practices in Energy, Environment and Health Research*; World Sustainability Series; Leal Filho, W., Vidal, D.G., Dinis, M.A.P., Dias, R.C., Eds.; Springer: Cham, Switzerland, 2022. [CrossRef]
87. Paul, S.; Nagendra, H. Factors Influencing Perceptions and Use of Urban Nature: Surveys of Park Visitors in Delhi. *Land* **2017**, *6*, 27. [CrossRef]
88. Giles-Corti, B.; Timperio, A.; Bull, F.; Pikora, T. Understanding Physical Activity Environmental Correlates: Increased Specificity for Ecological. *Exerc. Sport Sci. Rev.* **2005**, *33*, 175–181. [CrossRef] [PubMed]
89. McPhearson, T.; Andersson, T.; Elmqvist, T.; Frantzeskaki, N. Resilience of and through urban ecosystem services. *Ecosyst. Serv.* **2015**, *12*, 152–156. [CrossRef]
90. Quintas-Soriano, C.; Brandt, J.S.; Running, K.; Baxter, C.V.; Gibson, D.M.; Narducci, J.; Castroristina, A.J. Social-Ecological Systems Influence Ecosystem Service Perception: A Programme on Ecosystem Change and Society (PECS) Analysis. *Ecol. Soc.* **2018**, *23*, 15–32. [CrossRef]

91. Pan, H.; Page, J.; Cong, C.; Barthel, S.; Kalantari, Z. How ecosystems services drive urban growth: Integrating nature-based solutions. *Anthropocene* **2021**, *35*, 100297. [CrossRef]
92. de Groot, R. Function-analysis and valuation as a tool to assess land use conflicts in planning for sustainable, multi-functional landscapes. *Landsc. Urban Plan.* **2006**, *75*, 175–186. [CrossRef]
93. Bryan, B.A.; Raymond, C.M.; Crossman, N.D.; Macdonald, D.H. Targeting the management of ecosystem services based on social values: Where, what, and how? *Landsc. Urban Plan.* **2010**, *97*, 111–122. [CrossRef]
94. Tian, T.; Sun, L.; Peng, S.; Sun, F.; Che, Y. Understanding the process from perception to cultural ecosystem services assessment by comparing valuation methods. *Urban For. Urban Green.* **2021**, *57*, 126945. [CrossRef]
95. Zhang, C.; Li, J.; Zhou, Z. Ecosystem service cascade: Concept, review, application and prospect. *Ecol. Indic.* **2022**, *137*, 108766. [CrossRef]
96. Fagerholm, N.; Käyhkö, N.; Ndumbaro, F.; Khamis, M. Community stakeholders' knowledge in landscape assessments—Mapping indicators for landscape services. *Ecol. Indic.* **2012**, *18*, 421–433. [CrossRef]
97. Gobster, P.H. Visions of nature: Conflict and compatibility in urban park restoration. *Landsc. Urban Plan.* **2001**, *56*, 35–51. [CrossRef]
98. Postel, S.; Richter, B. *Rivers for Life: Managing Water for People and Nature*; Island Press: Washington, DC, USA, 2003.
99. Erfurt-Cooper, P. European waterways as a source of leisure and recreation. In *River Tourism*; Prideaux, B., Cooper, M., Eds.; CABI: Wallingford, UK, 2009; pp. 95–116.
100. Andersson, E.; Tengö, M.; McPhearson, T.; Kremer, P. Cultural ecosystem services as a gateway for improving urban sustainability. *Ecosyst. Serv.* **2015**, *12*, 165–168. [CrossRef]
101. O'Farrell, P.J.; Anderson, P.M.L.; Le Maitre, D.C.; Holmes, P.M. Insights and opportunities offered by a rapid ecosystem service assessment in promoting a conservation agenda in an urban biodiversity hotspot. *Ecol. Soc.* **2012**, *17*, 27. [CrossRef]
102. Kaczynski, A.T.; Potwarka, L.R.; Smale, B.J.A.; Havitz, M.E. Association of Parkland Proximity with Neighborhood and Park-based Physical Activity: Variations by Gender and Age. *Leis. Sci.* **2009**, *31*, 174–191. [CrossRef]
103. Liu, W.; Chen, W.; Dong, C. Spatial decay of recreational services of urban parks: Characteristics and influencing factors. *Urban For. Urban Green.* **2017**, *25*, 130–138. [CrossRef]
104. Wong, K.K. Urban park visiting habits and leisure activities of residents in Hong Kong, China. *Manag. Sport Leis.* **2009**, *14*, 125–140. [CrossRef]
105. Hörnsten, L.; Fredman, P. On the distance to recreational forests in Sweden. *Landsc. Urban Plan.* **2000**, *51*, 1–10. [CrossRef]
106. McCormack, G.; Giles-Corti, B.; Bulsara, M.; Pikora, T. Correlates of distances traveled to use recreational facilities for physical activity behaviors. *Int. J. Behav. Nutr. Phys. Act.* **2006**, *3*, 18. [CrossRef]
107. Năstase, I.I.; Pătru-Stupariu, I.; Kienast, F. Landscape Preferences and Distance Decay Analysis for Mapping the Recreational Potential of an Urban Area. *Sustainability* **2019**, *11*, 3620. [CrossRef]
108. Łaszkiewicz, E.; Heyman, A.; Chen, X.; Cimburova, Z.; Nowell, M.; Barton, D.N. Valuing access to urban greenspace using non-linear distance decay in hedonic property pricing. *Ecosyst. Serv.* **2022**, *53*, 101394. [CrossRef]
109. Wagner, I.; Krauze, K.; Zalewski, M. Blue aspects of green infrastructure. *Sustain. Dev. Appl.* **2013**, *4*, 145–155.
110. Tinsley, H.E.A.; Tinsley, D.J.; Croskeys, C.E. Park Usage, Social Milieu, and Psychosocial Benefits of Park Use Reported by Older Urban Park Users from Four Ethnic Groups. *Leis. Sci.* **2002**, *24*, 199–218. [CrossRef]
111. Reynisdottir, M.; Song, H.; Agrusa, J. Willingness to pay entrance fees to natural attractions: An Icelandic case study. *Tour. Manag.* **2008**, *29*, 1076–1083. [CrossRef]
112. Chung, J.Y.; Kyle, G.T.; Petrick, J.F.; Absher, J.D. Fairness of prices, user fee policy and willingness to pay among visitors to a national forest. *Tour. Manag.* **2011**, *32*, 1038–1046. [CrossRef]
113. Cundill, G.; Bezerra, J.C.; De Vos, A.; Ntingana, N. Beyond benefit sharing: Place attachment and the importance of access to protected areas for surrounding communities. *Ecosyst. Serv.* **2017**, *28*, 140–148. [CrossRef]
114. White, M.; Smith, A.; Humphryes, K.; Pahl, S.; Snelling, D.; Depledge, M. Blue space: The importance of water for preference, affect, and restorativeness ratings of natural and built scenes. *J. Environ. Psychol.* **2010**, *30*, 482–493. [CrossRef]
115. Boller, F.; Hunziker, M.; Conedera, M.; Elsasser, H.; Krebs, P. Fascinating Remoteness: The Dilemma of Hiking Tourism Development in Peripheral Mountain Areas. *Mt. Res. Dev.* **2010**, *30*, 320–331. [CrossRef]
116. Bauer, N.; Vasile, M.; Mondini, M. Attitudes towards nature, wilderness and protected areas: A way to sustainable stewardship in the South-Western Carpathians. *J. Environ. Plan. Manag.* **2018**, *61*, 857–877. [CrossRef]
117. Zhang, Y.; Liu, Y.; Zhang, Y.; Liu, Y.; Zhang, G.; Chen, Y. On the spatial relationship between ecosystem services and urbanization: A case study in Wuhan, China. *Sci. Total Environ.* **2018**, *637–638*, 780–790. [CrossRef] [PubMed]
118. Kim, Y.; Kim, K.-C.; Lee, D.K.; Lee, H.-W.; Andrada, R.T. Quantifying nature-based tourism in protected areas in developing countries by using social big data. *Tour. Manag.* **2019**, *72*, 249–256. [CrossRef]
119. Cebrián-Piqueras, M.A.; Filyushkina, A.; Johnson, D.N.; Lo, V.B.; López-Rodríguez, M.D.; March, H.; Oteros-Rozas, E.; Peppler-Lisbach, C.; Quintas-Soriano, C.; Raymond, C.M.; et al. Scientific and local ecological knowledge, shaping perceptions towards protected areas and related ecosystem services. *Landsc. Ecol.* **2020**, *35*, 2549–2567. [CrossRef]
120. Ndayizeye, G.; Imani, G.; Nkengurutse, J.; Irampagarikiye, R.; Ndihokubwayo, N.; Niyongabo, F.; Cuni-Sanchez, A. Ecosystem services from mountain forests: Local communities' views in Kibira National Park, Burundi. *Ecosyst. Serv.* **2020**, *45*, 101171. [CrossRef]

121. Shishany, S.; Al-Assaf, A.A.; Majdalawi, M.; Tabieh, M.; Tadros, M. Factors influencing Local Communities Relational Values to Forest Protected Areas in Jordan. *J. Sustain. For.* **2020**, 1–19. [CrossRef]
122. Coelho-Junior, M.G.; de Oliveira, A.L.; da Silva-Neto, E.C.; Castor-Neto, T.C.; de Oliveira Tavares, A.A.; Basso, V.M.; Turetta, A.P.D.; Perkins, P.E.; de Carvalho, A.G. Exploring Plural Values of Ecosystem Services: Local Peoples' Perceptions and Implications for Protected Area Management in the Atlantic Forest of Brazil. *Sustainability* **2021**, *13*, 1019. [CrossRef]
123. Jax, K.; Calestani, M.; Chan, K.M.A.; Eser, U.; Keune, H.; Muraca, B.; O'Brien, L.; Potthast, T.; Voget-Kleschin, L.; Wittmer, H. Caring for nature matters: A relational approach for understanding nature's contributions to human well-being. *Curr. Opin. Environ. Sustain.* **2018**, *35*, 22–29. [CrossRef]
124. Tibesigwa, B.; Ntuli, H.; Lokina, R. Valuing recreational ecosystem services in developing cities: The case of urban parks in Dar es Salaam, Tanzania. *Cities* **2020**, *106*, 102853. [CrossRef]
125. Basu, S.; Nagendra, H. Perceptions of park visitors on access to urban parks and benefits of green spaces. *Urban For. Urban Green.* **2021**, *57*, 126959. [CrossRef]
126. Belaidi, N.; Gaudry, K.H.; Landy, F. Categorisation of People and Places, Indigenous Peoples and Urban National Parks: Between Eviction, Instrumentality and Empowerment. In *From Urban National Parks to Natured Cities in the Global South*; Landy, F., Ed.; Springer: Singapore, 2018. [CrossRef]
127. Brill, G.; Anderson, P.; O'Farrell, P. Urban national parks in the global South: Linking management perceptions, policies and practices to water-related ecosystem services. *Ecosyst. Serv.* **2017**, *28*, 185–195. [CrossRef]
128. Bonilla-Bedoya, S.; Estrella, A.; Santos, F.; Herrera, M.A. Forests and urban green areas as tools to address the challenges of sustainability in Latin American urban socio-ecological systems. *Appl. Geogr.* **2020**, *125*, 102343. [CrossRef]
129. Frank, B.; Delano, D.; Caniglia, S. Urban Systems: A Socio-Ecological System Perspective. *Sociol. Int. J.* **2017**, *1*, 1–8. [CrossRef]
130. De la Mora-De la Mora, G.; López-Miguel, C. Challenges in the management of urban natural protected area systems and the conservation of ecosystem services in Guadalajara and Monterrey, Mexico. *Land Use Policy* **2022**, *114*, 105987. [CrossRef]
131. He, S.; Gallagher, L.; Su, Y.; Wang, L.; Cheng, H. Identification and assessment of ecosystem services for protected area planning: A case in rural communities of Wuyishan national park pilot. *Ecosyst. Serv.* **2018**, *31*, 169–180. [CrossRef]
132. Daw, T.; Brown, K.; Rosendo, S.; Pomeroy, R. Applying the ecosystem services concept to poverty alleviation: The need to disaggregate human well-being. *Environ. Conserv.* **2011**, *38*, 370–379. [CrossRef]
133. Anton, C.; Young, J.; Harrison, P.A.; Musche, M.; Bela, G.; Feld, C.K.; Harrington, R.; Haslett, J.R.; Pataki, G.; Rounsevell, M.D.A.; et al. Research needs for incorporating the ecosystem service approach into EU biodiversity conservation policy. *Biodivers. Conserv.* **2010**, *19*, 2979–2994. [CrossRef]
134. Bradford, J.B.; Betancourt, J.L.; Butterfield, B.J.; Munson, S.M.; Wood, T.E. Anticipatory natural resource science and management for a changing future. *Front. Ecol. Environ.* **2018**, *16*, 295–303. [CrossRef]
135. Nassl, M.; Loffler, J. Ecosystem services in coupled social—Ecological systems: Closing the cycle of service provision and societal feedback. *Ambio* **2015**, *44*, 737–749. [CrossRef]
136. Reyers, B.; Biggs, R.; Cumming, G.S.; Elmqvist, T.; Hejnowicz, A.p.; Polasky, S. Getting the measure of ecosystem services: A social–ecological approach. *Front. Ecol. Environ.* **2013**, *11*, 268–273. [CrossRef]
137. Seppelt, R.; Fath, B.; Burkhard, B.; Fisher, J.L.; Grêt-Regamey, A.; Lautenbach, S.; Pert, P.; Hotes, S.; Spangenberg, J.; Verburg, P.H.; et al. Form follows function? Proposing a blueprint for ecosystem service assessments based on reviews and case studies. *Ecol. Indic.* **2012**, *21*, 145–154. [CrossRef]
138. Wilkinson, C.; Sendstad, M.; Parnell, S.; Schewenius, M. Urban governance of biodiversity and ecosystem services. In *Urbanization, Biodiversity and Ecosystem Services: Challenges and Opportunities: A Global Assessment*; Elmqvist, T., Fragkias, M., Goodness, J., Güneralp, B., Marcotullio, P.J., McDonald, R.I., Parnell, S., Schewenius, M., Sendstad, M., Seto, K.C., et al., Eds.; Springer: Dordrecht, The Netherlands, 2013; pp. 539–587. [CrossRef]
139. De Groot, R.S.; Fisher, B.; Christie, M.; Aronson, J.; Braat, L.; Gowdy, J.; Haines-Young, R.; Maltby, E.; Neuville, A.; Polasky, S.; et al. Integrating the ecological and economic dimensions in biodiversity and ecosystem service valuation. In *The Economics of Ecosystems and Biodiversity: The Ecological and Economic Foundations*; Kumar, P., Ed.; Earthscan: London, UK; Earthscan: Washington, DC, USA, 2010; p. 41. [CrossRef]
140. Granek, E.F.; Polasky, S.; Kappel, C.V.; Reed, D.J.; Stoms, D.M.; Koch, E.W.; Kennedy, C.J.; Cramer, L.A.; Hacker, S.D.; Barbier, E.B.; et al. Ecosystem services as a common language for coastal ecosystem-based management. *Conserv. Biol.* **2010**, *24*, 207–216. [CrossRef] [PubMed]
141. Rissman, A.R.; Gillon, S. Where are ecology and biodiversity in social-ecological systems research? A review of research methods and applied recommendations. *Conserv. Lett.* **2016**, *10*, 86–93. [CrossRef]
142. Comberti, C.; Thornton, T.F.; de Echeverria, V.W.; Patterson, T. Ecosystem services or services to ecosystems? Valuing cultivation and reciprocal relationships between humans and ecosystems. *Glob. Environ. Chang.* **2015**, *34*, 247–262. [CrossRef]
143. Zhang, L.; Cong, C.; Pan, H.; Cai, Z.; Cvetkovic, V.; Deal, B. Socioecological informed comparative modeling to promote sustainable urban policy transitions: Case study in Chicago and Stockholm. *J. Clean. Prod.* **2021**, *281*, 125050. [CrossRef]
144. Atkinson, G.; Bateman, I.; Mourato, S. Recent advances in the valuation of ecosystem services and biodiversity. *Oxford Rev. Econ. Policy* **2012**, *28*, 22–47. [CrossRef]
145. Almenar, J.B.; Elliot, T.; Rugani, B.; Philippe, B.; Gutierrez, T.N.; Sonnemann, G.; Geneletti, D. Nexus between nature-based solutions, ecosystem services and urban challenges. *Land Use Policy* **2021**, *100*, 104898. [CrossRef]

146. Castellar, J.A.C.; Popartan, L.A.; Pueyo-Ros, J.; Atanasova, N.; Langergraber, G.; Säumel, I.; Corominas, L.; Comas, J.; Acuna, V. Nature-based solutions in the urban context: Terminology, classification and scoring for urban challenges and ecosystem services. *Sci. Total Environ.* **2021**, *779*, 146237. [CrossRef]
147. Mononen, L.; Auvinen, A.-P.; Ahokumpu, A.-L.; Rönkä, M.; Aarras, N.; Tolvanen, H.; Kamppinen, M.; Viirret, E.; Kumpula, T.; Vihervaara, P. National ecosystem service indicators: Measures of social–ecological sustainability. *Ecol. Indic.* **2016**, *61*, 27–37. [CrossRef]

land

Article

Fostering the Resiliency of Urban Landscape through the Sustainable Spatial Planning of Green Spaces

Donatella Valente [1], María Victoria Marinelli [1,2,*], Erica Maria Lovello [1], Cosimo Gaspare Giannuzzi [3] and Irene Petrosillo [1]

[1] Laboratory of Landscape Ecology, Department of Biological and Environmental Sciences and Technologies, University of Salento, 73100 Lecce, Italy; donatella.valente@unisalento.it (D.V.); ericamaria.lovello@studenti.unisalento.it (E.M.L.); irene.petrosillo@unisalento.it (I.P.)

[2] Institute for Advanced Space Studies "Mario Gulich" (IG), National Commission for Space Activities (CONAE), Córdoba 5186, Argentina

[3] Regional Agency for the Protection of the Environment of the Apulia Region, Scientific Direction U.O.C. Natural Environment Service Regional Sea Center, 70126 Bari, Italy; c.giannuzzi@arpa.puglia.it

* Correspondence: mariavictoria.marinelli@unisalento.it

Abstract: Background: It has been recognized that urban green spaces play a crucial role in providing many landscape services. The research aimed at identifying the main knowledge gaps in this framework and to support urban planning, taking into account the spatial configuration of green areas through a pilot study area, and mapping urban landscape services. Methods: In this research, (1) a systematic review, analyzed through a network analysis; (2) an urban pilot study to map the Urban Green Index and, jointly, the spatial composition and configuration of urban green areas, through the integration of three landscape metrics; and (3) the mapping of Urban Landscape Services Index have been carried out. Results: The 37% of the reviewed articles focused on regulating services, while the network analysis identified four clusters. The total Urban Green Index was 26%, and some districts showed a percentage that surpassed it. The total overall Green Connectivity Index was 21%. Some districts were the best providers of landscape services. Conclusions: This research was in line with the EU Joint Science for Policy Report suggesting giving emphasis to the spatial pattern map of green spaces in European cities to provide spatial data available for decision-makers in relation to GI deployment.

Keywords: urban green areas; green connectivity index; landscape services; urban spatial planning

Citation: Valente, D.; Marinelli, M.V.; Lovello, E.M.; Giannuzzi, C.G.; Petrosillo, I. Fostering the Resiliency of urban Landscape through the Sustainable Spatial Planning of Green Spaces. *Land* **2022**, *11*, 367. https://doi.org/10.3390/land11030367

Academic Editors: Alessio Russo and Giuseppe T. Cirella

Received: 11 January 2022
Accepted: 27 February 2022
Published: 3 March 2022

1. Introduction

It is widely recognized that most of the world's population will live in cities in the near future [1–3]. The problem of urbanization is particularly alarming since it represents a complex process of land use transformation (from rural to urban), which involves a profound change of landscape structure, function, and dynamics [4], causing urban landscape degradation and loss of urban biodiversity. This continuous urbanization of the planet, with effects on citizens' quality of life, has led to the identification of the local (urban) level as the right scale to achieve the 11th UN Sustainable Development Goal (SDG) [5]. This goal is focused on "sustainable cities and communities" and, through innovative urban planning choices, it aims at making cities resilient. The key element for achieving the SDG in cities is certainly represented by urban green areas [6] usually composed of natural or semi-natural components of the urban landscape such as forests, grasslands, street trees, public and private gardens, vertical gardens etc., which ensure crucial urban landscape services [7–9] that depend on the resilience of urban landscapes. Sustainable cities require innovative urban planning strategies, based on new ecological knowledge and perspectives that are sometimes missing at urban scale. In this context, there is an explicit call for more green spaces in cities, and the European policy as well as the US New Urban Agenda ask to include urban green areas into urban planning challenges for improving societal quality of life. It has been recognized that urban green space plays a crucial role in providing many landscape services [10–16]. More in detail, urban green areas provide cultural services, such as aesthetics, recreation, tourism, etc. [17–20], and regulatory services [10,21] such as reduction of air pollution, through carbon storage and sequestration [22–24], reduction of water pollution [25–27], water regulation [17,18,28], noise reduction [19,26], micro-climate regulation [29–31]. Urban green spaces provide habitats for different species and are, therefore, essential for maintaining biodiversity in urban landscapes [26,28,32]. Finally, as recognized by the Millennium Ecosystem Assessment [33] urban landscape services support the quality of life [34,35] and the well-being of citizens [36]. However, the challenge is to design a strategically planned network of green areas able to deliver a wide range of Landscape Urban Services (LS) within a city, through a spatial approach. Very often, the contribution of green areas to the provision of LS is related to their presence and amount, by using green area type (gardens, street trees, urban parks) as proxies of service flow [37,38]. However, the naturalness degree, diversity, size, shape, and above all the spatial pattern of urban green areas also play a crucial role in determining the contribution of green areas to the provision of urban landscape services [39–41]. The last point is of particular importance as urban green spaces are, often, increasingly fragmented because of urban development [42,43] and there is still a lack of research on the role of their spatial pattern in guaranteeing equal social access. Landscape ecology theory has strongly recognized that spatial composition (amount) and configuration (spatial arrangement) of urban green areas are two interplaying landscape components that can affect urban landscape heterogeneity, given by the complexity and variability of the properties of a landscape in space and time [44–52]. Therefore, both landscape properties influence the way landscape services are provided [53–56], since they depend on the abundance and variety of green area types, and on their spatial distribution in terms of spatial connectivity and/or fragmentation [57–60]. Even if most green spaces are public, access depends on distance [61]. Therefore, their spatial configuration plays an important role in the equal distribution of environmental benefits among people living in different suburban areas. In this context, the aims of this research are:

1. a systematic review through a network analysis approach to identify the main research items and knowledge gaps related to urban green areas and landscape services, and to analyze how these concepts are interrelated to each other and to the spatial configuration of green spaces;
2. to better focus the research on urban green area planning: (a) a pilot study has been carried out in the municipality of Lecce to analyze the amount of urban green areas at urban and suburban (district) scale through the use of a simple Urban Green Index; and (b) the joint analysis of the spatial composition and configuration of urban green spaces has been carried out through the integration of three landscape metrics; and the Urban Landscape Services Index has been estimated and mapped at urban and suburban scale as a support urban green areas planning to foster the resilience of urban landscape.

2. Materials and Methods

2.1. Systematic Review

To carry out the systematic review, Web ISI and Scopus were used to collect potentially relevant papers to analyze the connection among alternative terms: "Green Infrastructure" or "Urban Green Areas" and "Natural Capital" or "Ecosystem Services". These four terms have been selected for two main reasons: (1) to analyze the main temporal trend of publications focused on these interlinked concepts as well as the typology of landscape services (supporting, provisioning, regulating, and cultural) taken into account, using descriptive statistics. (2) to investigate the role played by the spatial connotations, the keywords collected from the systematic review have been analyzed in terms of co-occurrence network, underlying the possible relationships among them. Network analysis has been carried out using the open-source VOSviewer software. The size of the labels and nodes is determined by "total link strength", the number of documents linking the keywords, and the thickness of the lines connecting the nodes depends on "link strength" between two keywords [30].

The output of the analysis allowed the identification of possible clusters, whose number usually depends on the resolution parameter: the higher its value, the higher the level of detail [30]. In this study, 9 was the set as the minimum cluster size, therefore the minimum cluster is made up of at least 9 keywords.

2.2. Spatial Analysis of Urban Green Spaces and Assessment of the Landscape Services Index (LSI)

Focusing on urban green area planning, a pilot study was carried out in the urbanized area of the municipality of Lecce (Apulia Region—southern Italy) (Figure 1) to analyze the amount of urban green areas at urban and suburban (district) scale through the use of a simple Urban Green Index, and then a spatial configuration was carried out.

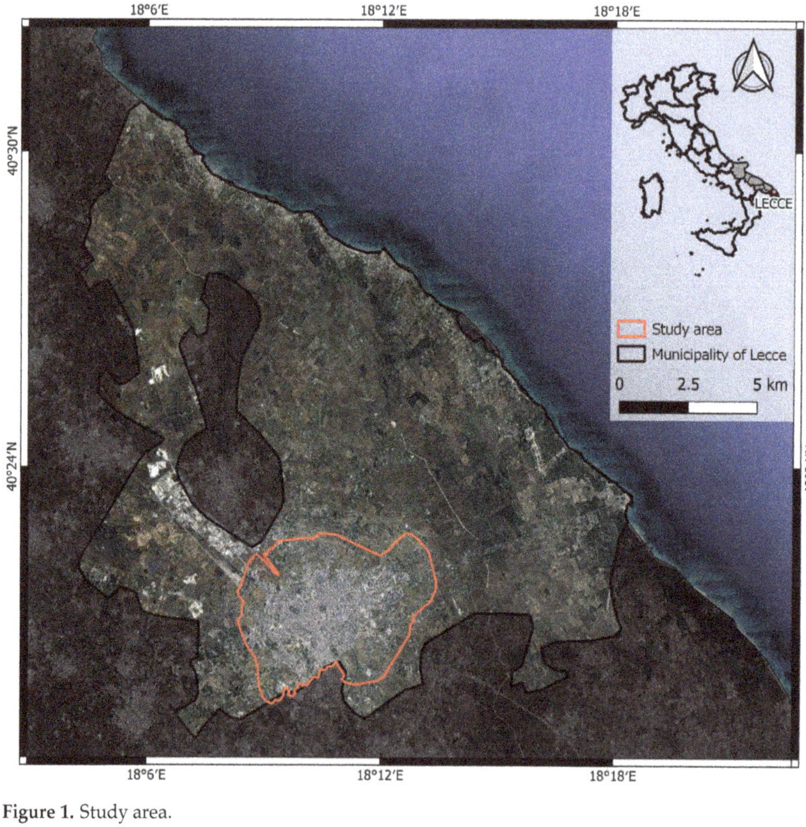

Figure 1. Study area.

More in detail, the study area has an extension of 30.16 km^2 over the total area of the municipality of Lecce 238.82 km^2 and presents the number of inhabitants as 95,037. Urban green planning is a priority for the municipal administration of the city of Lecce with four priority actions included in the purposes of green plan drafting: the census of the green areas, the proposal of a green area regulation, a programmatic framework for urban forestry interventions and the elaboration of the tree balance. In this perspective, interventions are planned to increase the presence of new green areas, and to improve the maintenance of the existing ones. This pilot study helps in testing some landscape indices that can support more effective strategies for implementing green areas to enhance their connectivity.

In the study area, spatial analysis was carried out through the following steps:

- identifying and mapping the public green areas using high-spatial-resolution satellite imagery using the QuickMapServices in QGIS 3.20.2 software.
- measuring and mapping the Urban Green Index (UGI) through the PLAND landscape metric, using the FRAGSTATS 4.2 software. This index measures the amount of green area over the whole urban landscape under study, and the districts' green index.
- measuring and mapping the Green Connectivity Index (GCI) through the integration of three landscape metrics: Class Area (CA), Aggregation Index, and COHESION index [62–64]. More specifically, Class Area, given by the number of green patches in the study area, has been used to quantify the spatial composition of green areas. On the other side, Aggregation Index, given indications on the spatial aggregation among green patches, and COHESION, quantifying the connectivity among green patches in an urban landscape, have been innovatively integrated to measure the spatial configuration of green areas in an urban landscape. Their use has been tested in the pilot study area, taking into account that CA can assume values > 0, Aggregation Index and COHESION range from 0 and 100. Thus, the GCI allows analysis of the urban green landscape taking into consideration not only the quantity of green areas (amount) but also their spatial aggregation and connectivity. The GCI is always > 0 and is given by:

$$Green \quad Connectivity \quad Index \quad (GCI) = CA \times AI \times COHESION \tag{1}$$

- measuring and mapping the urban Landscape Service Index (LSI), though a new classification of the urban green areas of the study area into two sub-classes—Forest and Non-Forest—and by considering three main urban landscape services associated with these sub-classes: carbon sequestration, temperature regulation, and runoff regulation. The LSI has been calculated as follows:

$$Landscape \quad Services \quad Index \quad (LSI) = aGCI_{Forest} + bGCI_{Non-Forest} \tag{2}$$

  where "a" and "b" are the weighting factors in terms of broad contribution of each class to each selected landscape service: green areas classified as Forest contribute more to each landscape services than green areas classified as Non-Forest, with the exception of runoff regulation (Table 1), while GCI_{Forest} is the green connectivity index for class Forest, and $GCI_{Non-Forest}$ is the green connectivity index for class Non-Forest.
- Finally, the LSI has been normalized.

Table 1. Weighting factors in terms of contribution of each sub-class to each selected landscape service (adapted from [65]).

Weighting Factors	Carbon Sequestration	Temperature Regulation	Runoff Regulation
a (Forest)	1	1	1
b (Non-Forest)	0.5	0.5	1

3. Results

3.1. Systematic Review

The systematic review resulted in the collection of 671 articles published over a period ranging from 2006 to 2020. The trend of the number of articles per year has highlighted the growing interest of the scientific literature towards the interlinked topics "Green Infrastructure" and "Urban Green Areas", with "Natural Capital" and "Ecosystem Services" (Figure 2a).

The articles were analyzed by their content to understand which categories of urban landscape services have been mainly associated with green infrastructures/areas. The results showed that some articles did not deal with one or more specific categories of landscape services, while others have been more focused on specific landscape services that in 37% of the articles have been represented by regulating services followed by cultural services (32%) (Figure 2b).

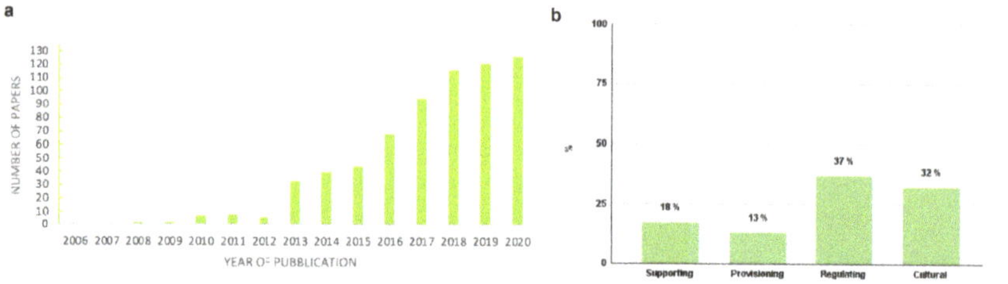

Figure 2. (**a**) Temporal trend of the papers collected in the systematic review. (**b**) The landscape services most investigated in the Review.

Most of the study areas were in Europe, China, and the United States. The results of the Network Analysis are shown in Figure 3. The sample used in the Network Analysis consisted of 1895 keywords.

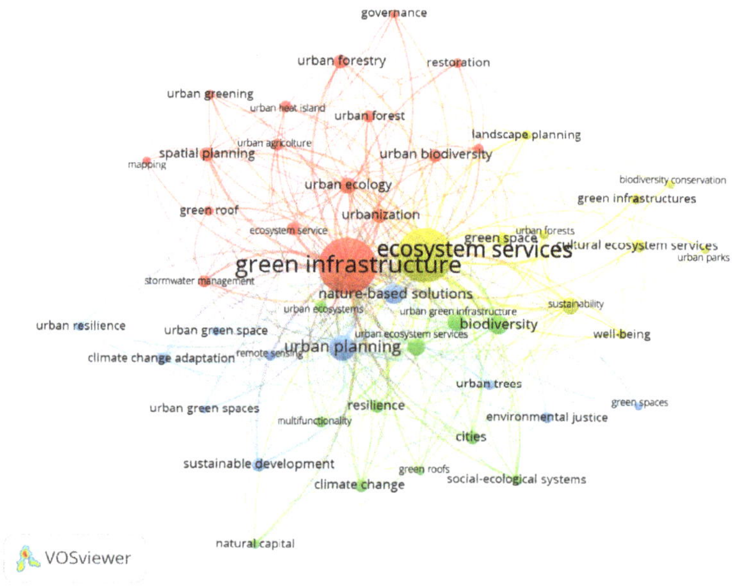

Figure 3. Network Analysis Map.

The size of the label and node of a keyword was determined by its weight: the greater its weight, the larger its label and node. The color of each keyword was dependent on belonging to a specific cluster. The lines among the keywords, on the other hand, represent the interlinks among them [66].

The results showed that four clusters were identified with the following central nodes: green infrastructures, ecosystem services, urban planning, and biodiversity. The most frequent keyword was "green infrastructures" (red cluster), which was, together with "ecosystem services", one of the central topics of the systematic review. It showed a great weight, evident from the size of the node, and the greatest number of elements (keywords) nearby (Figure 3). The terms "green space" and "nature-based solutions" in adjacent positions to "green infrastructures", was highlighted as all these terms are often used in an interchangeable way to indicate urban green spaces.

Another important keyword is "ecosystem services" (yellow cluster), a transversal topic associated with several research items such as well-being, landscape planning, and green spaces (Figure 3).

Finally, "urban planning" and "biodiversity" represents the last two clusters, less important than "green infrastructures" and "ecosystem services". In particular, "urban planning" (blue cluster) was more related to "urban green spaces", "climate change adaptation", "nature-based solutions", and "environmental justice" at urban scale, while "biodiversity" (green cluster) was more related to "resilience", "climate change", and "natural capital".

3.2. Spatial Analysis of Urban Green Areas

For the spatial analysis of urban green areas, it has been an expedient to take into consideration the relationship between spatial composition and configuration of urban green areas and the related provision of landscape services. A map of Lecce public urban greenery is shown in Figure 4.

Figure 4. (**a**) Urban greenery map of the municipality of Lecce; (**b**) Districts' Urban Green Index (UGI) Map; (**c**) Districts' Green Connectivity Index (GCI) Map.

It is possible to notice that the green areas are found mainly in the districts around Lecce old town, as well as the areas characterized by more recent urban sprawl. The urban green index (UGI) of the municipality of Lecce is 26% of the total area. When this index is estimated at the district scale (Figure 4b), it is possible to highlight that Districts 1, 6, 7, 9, 10, and 11 show a percentage of UGI higher than 26%.

The high value of UGI in the case of District 6 (historic center) is linked to the presence of two green areas in a district that has shown a very small extent in comparison with others with the same amount of green areas but with a higher extent, as with District 4 and District 3, showing, the lowest UGI. The result is in line with other studies focused on other urban contexts [67,68]. However, these districts are characterized by private urban gardens and green roofs that are not included in the results of the present research.

To take into consideration jointly the spatial composition and configuration of the urban green areas, the Green Connectivity Index (GCI) has been elaborated and mapped (Figure 4c). The GCI for the whole study area is about 21%, and Districts 1, 5, 9, 10, 11, and 13 shows values of GCI higher than 21%. These higher values of GCI could be related to the presence of green areas more aggregated in these districts in comparison with the whole study area.

3.3. Urban Landscape Services

The Landscape Services Index is based on the reclassification and mapping of urban green spaces in "Forest" and "Non-Forest" green areas (Figure 5a), showing a greater extent of "Non-Forest" areas than "Forest" areas (Figure 5b). More specifically, the "Non-Forest" areas are predominant in the peri-urban areas, as a ring outside the most urbanized area, and are more aggregated than "Forest" areas.

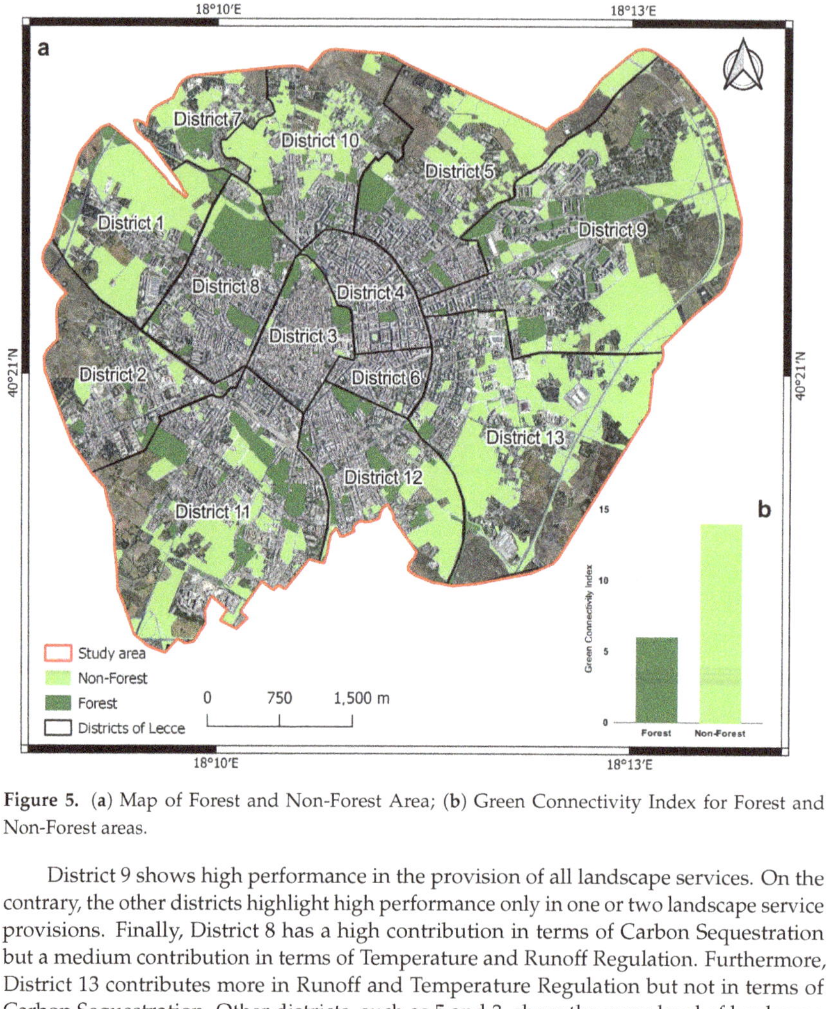

Figure 5. (**a**) Map of Forest and Non-Forest Area; (**b**) Green Connectivity Index for Forest and Non-Forest areas.

District 9 shows high performance in the provision of all landscape services. On the contrary, the other districts highlight high performance only in one or two landscape service provisions. Finally, District 8 has a high contribution in terms of Carbon Sequestration but a medium contribution in terms of Temperature and Runoff Regulation. Furthermore, District 13 contributes more in Runoff and Temperature Regulation but not in terms of Carbon Sequestration. Other districts, such as 5 and 3, show the same level of landscape service provision.

From the analysis of the Landscape Services Index, however, it has emerged that the districts that best provide the three urban landscape services analyzed in this study are 1, 8, 9, 11 and 13 (Figure 6a), while Figure 6b–d show the districts that mostly provide each landscape service (Carbon Sequestration, Temperature Regulation and Runoff Regulation).

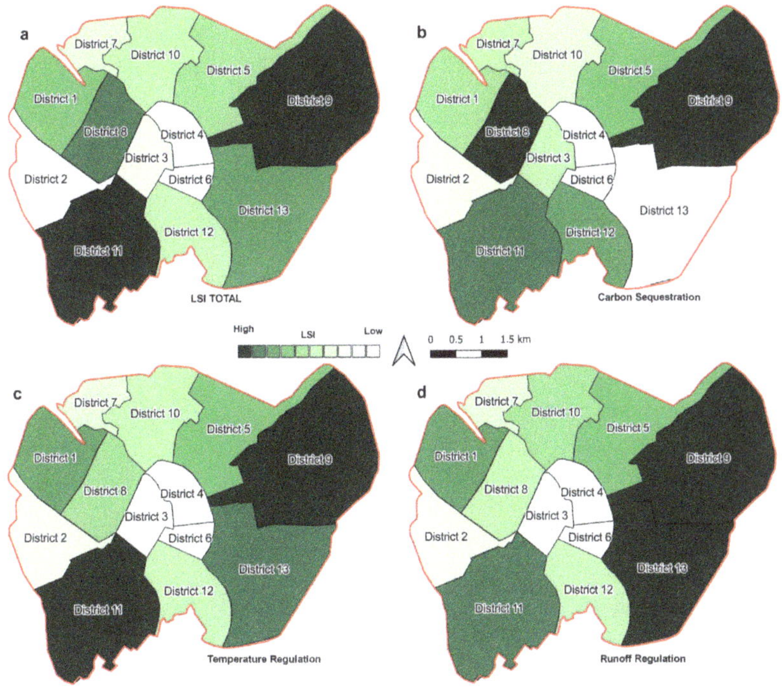

Figure 6. (**a**) Map of the Landscape Services Index (LSI) at the District scale, (**b**) Map of the provision of Carbon sequestration at the District scale, (**c**) Map of the provision of Temperature Regulation at the District scale, and (**d**) Map of Runoff Regulation at the District scale.

4. Discussion

Today, urban areas represent a fundamental tool to support and promote sustainable development within urban contexts [69]. As highlighted by the systematic review, the scientific literature has shown a growing interest in urban green areas from 2006 to 2020 (Figure 2a). Furthermore, the worldwide interest in this theme has also been noticed in a recent review by Chatzimentor and colleagues in 2020 [70] but probably the greatest interest towards green areas in Europe is related to the economic support given to their implementation in European countries.

The regulated landscape services provided by urban green areas have played a key role in building healthy, livable cities that are resilient to natural disasters [71], and this is the reason these services have been the most frequently analyzed services in the investigated literature (Figure 2b). It has been also observed that the regulating services have been followed by cultural services, therefore most of the studies have also been tackled from a social perspective, probably because urban landscape services that foster cultural diversity and social relations comply with the Sustainable Development Goal on Sustainable Cities and Communities (SDG11) [72]. This has been further confirmed in the four clusters resulting from the Network Analysis (Figure 3), where among the main scientific research topics it has been possible to notice that cultural ecosystem services are part of the ecosystem service cluster (yellow cluster). The deep knowledge of urban green areas could represent a useful support for appropriate and sustainable urban planning that supports both natural capital and human well-being. The systematic review has noticed the strong interdisciplinarity of the topic, as demonstrated by the interrelations between red and yellow clusters among the terms planning, biodiversity, well-being, and services. This could mean that urban green planning should be able to integrate ecological and social values. Urban green spaces can therefore provide landscape services which, by affecting the constituents of human well-being, produce several social and cultural benefits. Another interconnection has been highlighted between blue and green clusters, where the terms urban planning, climate change and natural capital are of particular interest. Part of the scientific literature on green spaces has focused on the role of biodiversity and urban tree coverage in supporting the resilience of cities [72,73].

However, the systematic review has highlighted that there are few keywords focused on spatial aspects such as spatial planning and landscape planning. This represents a knowledge gap that it is crucial to fill, since the connectivity of a landscape, intended as a "spatial characteristic of systems that supports the occurrence of specific processes and functions, through adjacency, proximity or connection and functional connection" [74], must be taken into account when planning urban green areas for landscape services. In this perspective, the spatial analysis of green areas in the municipality of Lecce was intended to highlight how crucial it is to analyze the amount of green areas together with their spatial arrangement (configuration) to foster landscape services at urban scale. Planning for improving the spatial connectivity of urban landscapes is therefore fundamental for contrasting the serious consequences related to climate change, by implementing the provision of landscape services as shown in Figure 6, as well as the equal distribution of environmental benefits among people living in different districts. For instance, Districts 3, 4, and 6 need new planning strategies to guarantee the implementation of carbon sequestration, and temperature and runoff regulation. The only consideration of the amount of urban green areas do not help in planning for enhancing urban landscape services. The correct implementation of urban greenery will move cities towards the enhancement of human well-being and urban sustainability [75].

The research needs further development implemented through the inclusion of private green areas that could modify the results mainly in those districts where built-up areas do not leave available space for the increasing of new urban greenery, as in the case of the old town.

In addition, given the recognized role of green areas in regulating the urban microclimate, this research will be developed by analyzing the effects of green areas on the mitigation of the heat island, by measuring in situ the climate regulation service. All this information will be used to better plan the urban and peri-urban green areas for enhancing the provision of urban landscape services and the resiliency of the city.

5. Conclusions

The role of urban green spaces in producing urban landscape services is fundamental for the resilience of a city [76]. The link between the provision of landscape services and urban green areas has been the focus of several studies, revealing a positive correlation between the amount of green space and the services they provide [22,77,78]. However, the spatial configuration of green areas plays an important role, since it helps in understanding if a single large green area (SL) is better than several small (SS) interlinked green areas or vice versa. The so-called SLOSS debate, very well known in nature conservation theory, can be of relevance also in the case of green area planning. This is a typical ecological concept to discuss if it is better to protect a single large or several small protected areas to enhance biodiversity conservation. In this perspective, the contribution of urban green areas to the provisioning of landscape services should be analyzed in terms of spatial composition and the configuration of green spaces [8].

In the context of ever-increasing urban sprawl, it is necessary to evaluate the impact of this expansion on the configuration of the urban landscape, and on its connectivity. The design of a network of green areas can improve the connectivity of highly urbanized landscapes by building green corridors to protect biodiversity, improve the quality of life, and enhance the resilience of the cities [79]. Characteristics such as the proportion between green areas of the different areas of the city, the complexity of their spatial configuration and, consequently, their connectivity, can affect the role of green areas in urban contexts.

The European Commission, in one of its most recent Joint Science for Policy Reports (2019), suggested giving emphasis to the spatial pattern map of green spaces in European cities to provide spatial data available for decision-makers in relation to GI deployment.

The Green Infrastructure concept is better known and applied in the urban local context, particularly for its benefits in regulating landscape services, as demonstrated in this study. Planning that develops connected green areas can solve several urban challenges and contributes significantly to the creation of resilient future cities that support landscape services and human well-being in facing climate change.

Author Contributions: Conceptualization, D.V., M.V.M., E.M.L. and I.P.; methodology, D.V., M.V.M., E.M.L. and I.P.; software, M.V.M., E.M.L. and C.G.G.; validation, D.V., M.V.M., E.M.L. and I.P.; formal analysis, D.V., M.V.M., E.M.L. and I.P.; data curation, D.V., M.V.M., E.M.L., C.G.G. and I.P.; writing—original draft preparation, D.V., E.M.L. and I.P.; writing—review and editing, D.V., M.V.M., E.M.L. and I.P.; visualization, M.V.M., E.M.L., C.G.G.; supervision, I.P. All authors have read and agreed to the published version of the manuscript.

Funding: The Italian Ministry of Education, Universities and Research (MIUR) has awarded the Department of Biological and Environmental Sciences and Technologies of the University of Salento with a special grant as one of the best departments of Italian universities with the finance five-year development projects (Project code: F85D18000130001).

Institutional Review Board Statement: Not applicable.

Informed Consent Statement: Not applicable.

Data Availability Statement: Data available on request.

Acknowledgments: We strongly thank the anonymous Reviewers for their useful suggestions that have improved the previous version of the manuscript.

Conflicts of Interest: The authors declare no conflict of interest.

References

1. Ritchie, H.; Roser, M. Urbanization. In *Our World in Data*. 2018. Available online: https://ourworldindata.org/urbanization (accessed on 19 December 2020).
2. United Nations Department of Economic and Social Affairs. World Urbanization Prospects: The 2018 Revision, Online Edition. 2018. Available online: https://population.un.org/wup/Publications/Files/WUP2018-Report.pdf (accessed on 19 December 2020).
3. Yu, Z.; Yang, G.; Zuo, S.; Jørgensen, G.; Koga, M.; Vejre, H. Critical review on the cooling effect of urban blue-green space: A threshold-size perspective. *Urban For. Urban Green.* **2020**, *49*, 126630. [CrossRef]
4. Pickett, S.T.; Cadenasso, M.L.; Grove, J.M.; Nilon, C.H.; Pouyat, R.V.; Zipperer, W.C.; Costanza, R. Urban ecological systems: Linking terrestrial ecological, physical, and socioeconomic components of metropolitan areas. *Annu. Rev. Ecol. Syst.* **2001**, *32*, 127–157. [CrossRef]
5. Nations, U. *Transforming our World: The 2030 Agenda for Sustainable Development*; Department of Economic and Social Affairs: New York, NY, USA, 2015.
6. Elgizawy, E. The significance of urban green areas for the Sustainable community. In Proceedings of the Al-Azhar Engineering-Thirteen International Conference, Cairo, Egypt, 23–25 December 2014.
7. Li, F.; Liu, X.; Zhang, X.; Zhao, D.; Liu, H.; Zhou, C.; Wang, R. Urban ecological infrastructure: An integrated network for ecosystem services and sustainable urban systems. *J. Clean. Prod.* **2017**, *163*, S12–S18. [CrossRef]
8. Valente, D.; Pasimeni, M.R.; Petrosillo, I. The role of green infrastructures in Italian cities by linking natural and social capital. *Ecol. Indic.* **2020**, *108*, 105694. [CrossRef]
9. Marinelli, M.V.; Valente, D.; Scavuzzo, C.M.; Petrosillo, I. Landscape service flow dynamics in the metropolitan area of Córdoba (Argentina). *J. Environ. Manag.* **2021**, *280*, 111714. [CrossRef] [PubMed]
10. Bolund, P.; Hunhammar, S. Ecosystem services in urban areas. *Ecol. Econ.* **1999**, *29*, 293–301. [CrossRef]
11. Burkhard, B.; Kroll, F.; Nedkov, S.; Müller, F. Mapping ecosystem service supply, demand and budgets. *Ecol. Indic.* **2012**, *21*, 17–29. [CrossRef]
12. Derkzen, M.L.; van Teeffelen, A.J.; Verburg, P.H. Quantifying urban ecosystem services based on high-resolution data of urban green space: An assessment for Rotterdam, the Netherlands. *J. Appl. Ecol.* **2015**, *52*, 1020–1032. [CrossRef]

13. Gkatsopoulos, P. A methodology for calculating cooling from vegetation evapotranspiration for use in urban space microclimate simulations. *Procedia Environ. Sci.* **2017**, *38*, 477–484. [CrossRef]
14. Gómez-Baggethun, E.; Barton, D.N. Classifying and valuing ecosystem services for urban planning. *Ecol. Econ.* **2013**, *86*, 235–245. [CrossRef]
15. Tratalos, J.; Fuller, R.A.; Warren, P.H.; Davies, R.G.; Gaston, K.J. Urban form, biodiversity potential and ecosystem services. *Landsc. Urban Plan.* **2007**, *83*, 308–317. [CrossRef]
16. Van Oudenhoven, A.P.; Petz, K.; Alkemade, R.; Hein, L.; de Groot, R.S. Framework for systematic indicator selection to assess effects of land management on ecosystem services. *Ecol. Indic.* **2012**, *21*, 110–122. [CrossRef]
17. Zinia, N.J.; McShane, P. Ecosystem services management: An evaluation of green adaptations for urban development in Dhaka, Bangladesh. *Landsc. Urban Plan.* **2018**, *173*, 23–32. [CrossRef]
18. Muthulingam, U.; Thangavel, S. Density, diversity and richness of woody plants in urban green spaces: A case study in Chennai metropolitan city. *Urban For. Urban Green.* **2012**, *11*, 450–459. [CrossRef]
19. Paulin, M.; Remme, R.; de Nijs, T.; Rutgers, M.; Koopman, K.; de Knegt, B.; van der Hoek, D.; Breure, A. Application of the natural capital model to assess changes in ecosystem services from changes in green infrastructure in Amsterdam. *Ecosyst. Serv.* **2020**, *43*, 101114. [CrossRef]
20. Dickinson, D.C.; Hobbs, R.J. Cultural ecosystem services: Characteristics, challenges and lessons for urban green space research. *Ecosyst. Serv.* **2017**, *25*, 179–194. [CrossRef]
21. Lonsdorf, E.V.; Nootenboom, C.; Janke, B.; Horgan, B.P. Assessing urban ecosystem services provided by green infrastructure: Golf courses in the Minneapolis-St. Paul metro area. *Landsc. Urban Plan.* **2021**, *208*, 104022. [CrossRef]
22. Chen, W.Y.; Hu, F.Z.Y. Producing nature for public: Land-based urbanization and provision of public green spaces in China. *Appl. Geogr.* **2015**, *58*, 32–40. [CrossRef]
23. De la Sota, C.; Ruffato-Ferreira, V.; Ruiz-García, L.; Alvarez, S. Urban green infrastructure as a strategy of climate change mitigation. A case study in northern Spain. *Urban For. Urban Green.* **2019**, *40*, 145–151. [CrossRef]
24. De Valck, J.; Beames, A.; Liekens, I.; Bettens, M.; Seuntjens, P.; Broekx, S. Valuing urban ecosystem services in sustainable brownfield redevelopment. *Ecosyst. Serv.* **2019**, *35*, 139–149. [CrossRef]
25. Sutton, P.C.; Anderson, S.J. Holistic valuation of urban ecosystem services in New York City's Central Park. *Ecosyst. Serv.* **2016**, *19*, 87–91. [CrossRef]
26. Mexia, T.; Vieira, J.; Príncipe, A.; Anjos, A.; Silva, P.; Lopes, N.; Freitas, C.; Santos-Reis, M.; Correia, O.; Branquinho, C.; et al. Ecosystem services: Urban parks under a magnifying glass. *Environ. Res.* **2018**, *160*, 469–478. [CrossRef]
27. Ramyar, R.; Saeedi, S.; Bryant, M.; Davatgar, A.; Hedjri, G.M. Ecosystem services mapping for green infrastructure planning—The case of Tehran. *Sci. Total Environ.* **2020**, *703*, 135466. [CrossRef] [PubMed]
28. Chenoweth, J.; Anderson, A.R.; Kumar, P.; Hunt, W.F.; Chimbwandira, S.J.; Moore, T.L. The interrelationship of green infrastructure and natural capital. *Land Use Policy* **2018**, *75*, 137–144. [CrossRef]
29. Bartesaghi-Koc, C.; Osmond, P.; Peters, A. Spatio-temporal patterns in green infrastructure as driver of land surface temperature variability: The case of Sydney. *Int. J. Appl. Earth Obs. Geoinf.* **2019**, *83*, 101903. [CrossRef]
30. Van Eck, N.; Waltman, L. *VOSviewer Manual Version 1.6. 16*; Univeristeit Leiden: Leiden, The Netherlands, 2020; pp. 1–52.
31. Arghavani, S.; Malakooti, H.; Bidokhti, A.A.A.A. Numerical assessment of the urban green space scenarios on urban heat island and thermal comfort level in Tehran Metropolis. *J. Clean. Prod.* **2020**, *261*, 121183. [CrossRef]
32. Mukherjee, M.; Takara, K. Urban green space as a countermeasure to increasing urban risk and the UGS-3CC resilience framework. *Int. J. Disaster Risk Reduct.* **2018**, *28*, 854–861. [CrossRef]
33. Assessment, M.E. *Ecosystems and Human Well-Being: Wetlands and Water*; World Resources Institute: Washington, DC, USA, 2005.
34. Panagopoulos, T.; Duque, J.A.G.; Dan, M.B. Urban planning with respect to environmental quality and human well-being. *Environ. Pollut.* **2016**, *208*, 137–144. [CrossRef]
35. Loures, L.; Santos, R.; Panagopoulos, T. Urban parks and sustainable city planning—The case of Portimão, Portugal. *Population* **2007**, *15*, 171–180.
36. Hartig, T.; Kahn, P.H. Living in cities, naturally. *Science* **2016**, *352*, 938–940. [CrossRef]
37. Seppelt, R.; Dormann, C.F.; Eppink, F.V.; Lautenbach, S.; Schmidt, S. A quantitative review of ecosystem service studies: Approaches, shortcomings and the road ahead. *J. Appl. Ecol.* **2011**, *48*, 630–636. [CrossRef]
38. Martínez-Harms, M.J.; Balvanera, P. Methods for mapping ecosystem service supply: A review. *Int. J. Biodivers. Sci. Ecosyst. Serv. Manag.* **2012**, *8*, 17–25. [CrossRef]
39. Adams, M.; Lükewille, A. *The European Environment—State and Outlook 2010*; European Environment Agency: København, Denmark, 2010.
40. Beninde, J.; Veith, M.; Hochkirch, A. Biodiversity in cities needs space: A meta analysis of factors determining intra urban biodiversity variation. *Ecol. Lett.* **2015**, *18*, 581–592. [CrossRef] [PubMed]
41. Asadolahi, Z.; Salmanmahiny, A.; Sakieh, Y.; Mirkarimi, S.H.; Baral, H.; Azimi, M. Dynamic trade-off analysis of multiple ecosystem services under land use change scenarios: Towards putting ecosystem services into planning in Iran. *Ecol. Complex.* **2018**, *36*, 250–260. [CrossRef]
42. Fahrig, L. Effects of habitat fragmentation on biodiversity. *Annu. Rev. Ecol. Evol. Syst.* **2003**, *34*, 487–515. [CrossRef]

43. McKinney, M.L. Urbanization, Biodiversity, and Conservation: The impacts of urbanization on native species are poorly studied, but educating a highly urbanized human population about these impacts can greatly improve species conservation in all ecosystems. *BioScience* **2002**, *52*, 883–890. [CrossRef]
44. Li, H.; Reynolds, J.F. A new contagion index to quantify spatial patterns of landscapes. *Landsc. Ecol.* **1993**, *8*, 155–162. [CrossRef]
45. Gustafson, E.J. Quantifying landscape spatial pattern: What is the state of the art? *Ecosystems* **1998**, *1*, 143–156. [CrossRef]
46. Riitters, K.; Wickham, J.; O'Neill, R.; Jones, B.; Smith, E. Global-scale patterns of forest fragmentation. *Conserv. Ecol.* **2000**, *4*, 23. [CrossRef]
47. Neel, M.C.; McGarigal, K.; Cushman, S.A. Behavior of class-level landscape metrics across gradients of class aggregation and area. *Landsc. Ecol.* **2004**, *19*, 435–455. [CrossRef]
48. Zurlini, G.; Riitters, K.; Zaccarelli, N.; Petrosillo, I.; Jones, K.B.; Rossi, L. Disturbance patterns in a socio-ecological system at multiple scales. *Ecol. Complex.* **2006**, *3*, 119–128. [CrossRef]
49. Zurlini, G.; Riitters, K.H.; Zaccarelli, N.; Petrosillo, I. Patterns of disturbance at multiple scales in real and simulated landscapes. *Landsc. Ecol.* **2007**, *22*, 705–721. [CrossRef]
50. Zurlini, G.; Petrosillo, I.; Zaccarelli, N. Toward a science of humans-in-nature: The role of pattern in assessing multi-scale vulnerability of natural capital. In Proceedings of the 95th ESA Annual Meeting, Pittsburgh, PA, USA, 1–6 August 2010.
51. Zurlini, G.; Jones, K.B.; Riitters, K.H.; Li, B.L.; Petrosillo, I. Early warning signals of regime shifts from cross-scale connectivity of land-cover patterns. *Ecol. Indic.* **2014**, *45*, 549–560. [CrossRef]
52. Proulx, R.; Fahrig, L. Detecting human-driven deviations from trajectories in landscape composition and configuration. *Landsc. Ecol.* **2010**, *25*, 1479–1487. [CrossRef]
53. Petrosillo, I.; Zaccarelli, N.; Zurlini, G. Multi-scale vulnerability of natural capital in a panarchy of social–ecological landscapes. *Ecol. Complex.* **2010**, *7*, 359–367. [CrossRef]
54. Laterra, P.; Orúe, M.E.; Booman, G.C. Spatial complexity and ecosystem services in rural landscapes. *Agric. Ecosyst. Environ.* **2012**, *154*, 56–67. [CrossRef]
55. Turner, M.G.; Donato, D.C.; Romme, W.H. Consequences of spatial heterogeneity for ecosystem services in changing forest landscapes: Priorities for future research. *Landsc. Ecol.* **2013**, *28*, 1081–1097. [CrossRef]
56. Mitchell, M.G.; Bennett, E.M.; Gonzalez, A. Agricultural landscape structure affects arthropod diversity and arthropod-derived ecosystem services. *Agric. Ecosyst. Environ.* **2014**, *192*, 144–151. [CrossRef]
57. Maimaitiyiming, M.; Ghulam, A.; Tiyip, T.; Pla, F.; Latorre-Carmona, P.; Halik, Ü.; Sawut, M.; Caetano, M. Effects of green space spatial pattern on land surface temperature: Implications for sustainable urban planning and climate change adaptation. *ISPRS J. Photogramm. Remote Sens.* **2014**, *89*, 59–66. [CrossRef]
58. Asgarian, A.; Amiri, B.J.; Sakieh, Y. Assessing the effect of green cover spatial patterns on urban land surface temperature using landscape metrics approach. *Urban Ecosyst.* **2015**, *18*, 209–222. [CrossRef]
59. Aronson, M.F.; Lepczyk, C.A.; Evans, K.L.; Goddard, M.A.; Lerman, S.B.; MacIvor, J.S.; Nilon, C.H.; Vargo, T. Biodiversity in the city: Key challenges for urban green space management. *Front. Ecol. Environ.* **2017**, *15*, 189–196. [CrossRef]
60. Jennings, V.; Floyd, M.F.; Shanahan, D.; Coutts, C.; Sinykin, A. Emerging issues in urban ecology: Implications for research, social justice, human health, and well-being. *Popul. Environ.* **2017**, *39*, 69–86. [CrossRef]
61. Chen, Y.; Ge, Y.; Yang, G.; Wu, Z.; Du, Y.; Mao, F.; Liu, S.; Xu, R.; Qu, Z.; Xu, B.; et al. Inequalities of urban green space area and ecosystem services along urban center-edge gradients. *Landsc. Urban Plan.* **2022**, *217*, 104266. [CrossRef]
62. McGarigal, K.; Cushman, S.A.; Ene, E. FRAGSTATS v4: Spatial Pattern Analysis Program for Categorical and Continuous Maps. Computer Software Program Produced by the Authors at the University of Massachusetts, Amherst. Available online: http://www.umass.edu/landeco/research/fragstats/fragstats.html (accessed on 12 December 2019).
63. Haas, J.; Furberg, D.; Ban, Y. Satellite monitoring of urbanization and environmental impacts—A comparison of Stockholm and Shanghai. *Int. J. Appl. Earth Obs. Geoinf.* **2015**, *38*, 138–149. [CrossRef]
64. Kong, F.; Yin, H.; Nakagoshi, N.; Zong, Y. Urban green space network development for biodiversity conservation: Identification based on graph theory and gravity modeling. *Landsc. Urban Plan.* **2010**, *95*, 16–27. [CrossRef]
65. Maheng, D.; Pathirana, A.; Zevenbergen, C. A preliminary study on the impact of landscape pattern changes due to urbanization: Case study of Jakarta, Indonesia. *Land* **2021**, *10*, 218. [CrossRef]
66. Van Oijstaeijen, W.; Van Passel, S.; Cools, J. Urban green infrastructure: A review on valuation toolkits from an urban planning perspective. *J. Environ. Manag.* **2020**, *267*, 110603. [CrossRef]
67. Zhang, Z.; Martin, K.L.; Stevenson, K.T.; Yao, Y. Equally green? Understanding the distribution of urban green infrastructure across student demographics in four public school districts in North Carolina, USA. *Urban For. Urban Green.* **2022**, *67*, 127434. [CrossRef]
68. Meerow, S.; Newell, J.P. Spatial planning for multifunctional green infrastructure: Growing resilience in Detroit. *Landsc. Urban Plan.* **2017**, *159*, 62–75. [CrossRef]
69. Grunewald, K.; Bastian, O. Maintaining Ecosystem Services to Support Urban Needs. *Sustainability* **2017**, *9*, 1647. [CrossRef]
70. Chatzimentor, A.; Apostolopoulou, E.; Mazaris, A.D. A review of green infrastructure research in Europe: Challenges and opportunities. *Landsc. Urban Plan.* **2020**, *198*, 103775. [CrossRef]
71. Cortinovis, C.; Geneletti, D. A framework to explore the effects of urban planning decisions on regulating ecosystem services in cities. *Ecosyst. Serv.* **2019**, *38*, 100946. [CrossRef]

72. Romero-Duque, L.P.; Trilleras, J.M.; Castellarini, F.; Quijas, S. Ecosystem services in urban ecological infrastructure of Latin America and the Caribbean: How do they contribute to urban planning? *Sci. Total Environ.* **2020**, *728*, 138780. [CrossRef] [PubMed]
73. Barona, C.O.; Devisscher, T.; Dobbs, C.; Aguilar, L.O.; Baptista, M.D.; Navarro, N.M.; da Silva Filho, D.F.; Escobedo, F.J. Trends in urban forestry research in Latin America & the Caribbean: A systematic literature review and synthesis. *Urban For. Urban Green.* **2020**, *47*, 126544.
74. Ahern, J. Greenways in the USA: Theory, trends and prospects. In *Ecological Networks and Greenways, Concept, Design, Implementation*; Cambridge University Press: Cambridge, UK, 2004; pp. 34–55.
75. MacGregor-Fors, I.; García-Arroyo, M.; Marín-Gómez, O.H.; Quesada, J. On the meat scavenging behavior of House Sparrows (*Passer domesticus*). *Wilson J. Ornithol.* **2020**, *132*, 188–191. [CrossRef]
76. Schäffler, A.; Swilling, M. Valuing green infrastructure in an urban environment under pressure—The Johannesburg case. *Ecol. Econ.* **2013**, *86*, 246–257. [CrossRef]
77. Estoque, R.C.; Murayama, Y.; Myint, S.W. Effects of landscape composition and pattern on land surface temperature: An urban heat island study in the megacities of Southeast Asia. *Sci. Total Environ.* **2017**, *577*, 349–359. [CrossRef]
78. Li, X.; Zhou, W.; Ouyang, Z.; Xu, W.; Zheng, H. Spatial pattern of greenspace affects land surface temperature: Evidence from the heavily urbanized Beijing metropolitan area, China. *Landsc. Ecol.* **2012**, *27*, 887–898. [CrossRef]
79. Cameron, R.W.; Blanuša, T.; Taylor, J.E.; Salisbury, A.; Halstead, A.J.; Henricot, B.; Thompson, K. The domestic garden—Its contribution to urban green infrastructure. *Urban For. Urban Green.* **2012**, *11*, 129–137. [CrossRef]

land

Article

Socioeconomic and Environmental Benefits of Expanding Urban Green Areas: A Joint Application of i-Tree and LCA Approaches

Mariana Oliveira [1], Remo Santagata [2,*], Serena Kaiser [1], Yanxin Liu [3], Chiara Vassillo [2], Patrizia Ghisellini [2], Gengyuan Liu [4] and Sergio Ulgiati [4,5]

[1] International PhD Programme/UNESCO Chair "Environment, Resources and Sustainable Development", Department of Science and Technology, Parthenope University of Naples, Centro Direzionale, Isola C4, 80143 Naples, Italy
[2] Department of Engineering, Parthenope University of Naples, Centro Direzionale, Isola C4, 80143 Naples, Italy
[3] School of Management and Engineering, Capital University of Economics and Business, Beijing 100070, China
[4] State Key Joint Laboratory of Environment Simulation and Pollution Control, School of Environment, Beijing Normal University, Beijing 100875, China
[5] Department of Science and Technology, Parthenope University of Naples, Centro Direzionale, Isola C4, 80143 Naples, Italy
* Correspondence: remo.santagata@assegnista.uniparthenope.it

Abstract: Green infrastructures deliver countless functions for counteracting climate change, air pollution, floods, and heat islands, contributing at the same time to water and carbon recycling as well as to renewable energies and feedstock provisioning. Properly addressing such environmental problems would require huge investments that could be decreased thanks to the further implementation of urban forests. Local administrations are designing participative projects to improve territories and their living conditions. The i-Tree Canopy modelling tool and the life cycle assessment method are jointly applied to evaluate the potential benefits of increasing tree coverage within the boundaries of the Metropolitan City of Naples, Southern Italy. Results highlighted that tree coverage could increase by about 2.4 million trees, thus generating 51% more benefits in pollutants removal, carbon sequestration and stormwater management. The benefits are also explored and confirmed by means of the life cycle assessment method. The potential tree cover is expected to provide a total annual economic benefit of USD 55 million, purchasing power parity value adjusted, representing USD 18 per citizen and USD 99,117 per square kilometre of implemented urban forest. These results can support a potential replication elsewhere and provide a reference for the sustainable improvement of cities by expanding urban green areas.

Keywords: urban forest; green infrastructures; i-Tree tool; life cycle assessment; ecosystem functions; nature shaped circularity

Citation: Oliveira, M.; Santagata, R.; Kaiser, S.; Liu, Y.; Vassillo, C.; Ghisellini, P.; Liu, G.; Ulgiati, S. Socioeconomic and Environmental Benefits of Expanding Urban Green Areas: A Joint Application of i-Tree and LCA Approaches. *Land* **2022**, *11*, 2106. https://doi.org/10.3390/land11122106

Academic Editors: Alessio Russo and Giuseppe T. Cirella

Received: 23 September 2022
Accepted: 18 November 2022
Published: 22 November 2022

1. Introduction

In the last decades, the dramatic rise of urbanization rates and consequent degradation of urban environments has directed society's attention towards the natural environment, urban forests ecosystems and other urban green infrastructures [1–5]. Such attention has increased awareness about the role of urban trees and green spaces as nature-based solutions [6,7] to improve the quality of life and wellbeing of the current generations and those to come [8], creating more resilient cities [6,9,10] in contrast to the pressing urban challenges [11]. Proper planning, management, and conservation in policy agendas represent conditions for maximizing their beneficial role [11,12] and making cities more resilient to future shocks [3].

Wellbeing depends on the local geography, culture, and ecological circumstances, being affected by the availability of basic materials for a good life and the presence of healthcare services, job opportunities, and relationships within urban communities [13]. Therefore, urban green spaces can fulfil some specific functions and address other global and local issues, such as climate change mitigation and adaptation, food dependency and flood protection, high standards of health and wellbeing, employment, and income needs [3,10,14–17]. Urban forests also deliver ecosystem functions related to carbon storage/sequestration, air quality, stormwater management, energy, habitat, noise, and microclimate. Most of these functions translate into social, economic, health, visual, and aesthetic benefits for urban dwellers [5]. Moreover, urban agriculture improves the quantity and quality of food, providing higher income and employment and promoting community development by intensifying social relationships capable of breaking down cross-generational barriers and distances [18,19].

Linear economy models and categories can mainly describe the monetary transactions related to these relationships. On the contrary, the concept of commons arises by focusing on the benefits of urban reforestation measured in terms of non-profit-oriented categories. The main aspect of commons is their irreducibility to private and public property and the capability to provide socioeconomic and territorial cohesion. However, besides the need for a collective perspective aiming at protection and preservation, in many situations, green areas belong to public institutions or private owners. Hence, civil society should abandon the owner-based perspective in favour of a rights-based active involvement, putting the collective dimension in a central position [20]. The collective dimension represents the physical and theoretical space to create and utilize commons, and the development of human societies creates new commons.

Consequently, the political struggle created by social movements represents a perfect field for shaping and growing commons since their substance is not ontological but originated from their relevance in specific contexts [21], which explains why commons should always be free from profit [22]. The boundaries of commons are often identified with the boundaries of natural goods. Moreover, commons can be born and enjoyed in urban environments, since a more comprehensive set of new commons is recognized within urban frameworks [23]. Beyond the dichotomy between natural and urban commons concepts, green areas are desirable and should be included in urban environments: On the one hand, cities represent aggression against the environment and ecosystems. On the other, urban residents present an urgent need to interact with nature since perceived as an essential need for health and wellbeing. Indeed, the growing contact with nature represents an increment of environmental awareness in people's lives [24,25].

At the global level, several countries (e.g., China, Ghana, Ethiopia) and cities (e.g., New York, Melbourne, and Berlin) have developed urban forestry projects to redesign their forestry policies and increase their green stocks [3,26]. In Italy, the government approved the "Climate Bill", aiming at expanding the national forest heritage, among other actions [27]. Additionally, several cities (e.g., Milan, Modena, Ferrara, Prato, Naples) are planning urban trees expansion programs [12]. In 2013, Italy established Law n. 10 "Standards for the development of public parks and gardens" to define criteria and guidelines for creating green multifunctional systems. This national urban green strategy is divided into three theme areas: biodiversity and ecosystem services, climate change and heat islands, wellbeing, and quality of life.

Consequently, the Campania Region (southern Italy) also focused on developing urban green areas. Regional Law n. 17, "Establishment of the system of regional interest urban parks", defines the urban green system as a set of spaces with environmental and landscape value or of strategic importance for the ecological balance in territorial contexts with high anthropic impact. In 2019, the Metropolitan City of Naples (MCN), in the Campania region, released the "*Ossigeno Bene Comune*" project (OBC Project)—"Oxygen as a Common" in English—to counteract climate change, excessive urbanization and land consumption and regenerate the MCN landscape through nature-based solutions, by planting three million trees—one per inhabitant—throughout its territory [28].

Another important aspect to be addressed when dealing with contemporary urban systems is what we can name "nature-shaped circularity" or "nature-based circular economy pattern". Circular economy (CE) is an evolving concept embodying internal complexities and multiple definitions. As pointed out by many authors [29–33], CE is an economic framework aiming to minimize resource use and waste generation by making the most out of available resources. The "nature-shaped circularity" pattern (through ecosystem services) provides a way forward to operationalize CE designing for green infrastructure planning. In so doing, it enhances the impact of CE on policies and related practices. For instance, accounting for the ecosystem services of urban green infrastructures can (i) provide a more accurate picture of the composition of the urban energy mix (including renewable energy provision from local biomass), (ii) reveal the impacts of green infrastructure on the amount of energy use (including mitigation of energy demand in buildings), and (iii) affect the dynamics of biogeochemical processes in cities (microclimate regulation and carbon sequestration by plants and soils) [34]. Thus, a circular urban system (including resource input, waste generation, emissions, and energy leakage) can be redesigned by nature-based slowing, closing, and narrowing energy and material loops, which represents a change of paradigm towards effective circularity.

Sustainable city development corresponds to a rise in the quality and quantity of all stocks considered as a source of wealth for all countries: natural, cultural, human, and manufactured [35–37], moving away from the fossil fuels-based society and linear economy towards renewable sources of energy and circular economy [30,38,39]. Therefore, besides monitoring the economic and financial assets and their flows, such as GDP [37,40], monitoring the ecosystem services stocks and flows as natural capital assets is also mandatory. However, evaluating the ecosystem functions flows is a challenge since every individual can enjoy them without any monetary payment: this suggests that the evaluation of ecosystem services should not be performed by employing a capital-oriented analysis [22]. From this perspective, ecosystem functions can be classified as commons, according to the definition given above.

This study aims to investigate and discuss the environmental, social, and economic benefits delivered by the increased presence of trees in a territory and to its inhabitants. The current and the potential tree coverage scenarios and the related benefits are assessed by the integration of the i-Tree Canopy online tool [41] and the life cycle assessment [42,43] method to quantify the ecosystem functions, including an economic perspective and pollution sequestration, to support future policies for urban green areas expansion projects and investments. The i-Tree Canopy tool is fairly used within the scientific literature to assess not just tree canopy cover but also other cover classes, thanks to its efficiency in making land cover assessments relatively easy by using aerial imagery [44–48]. However, very little seems to be present about using the i-Tree Canopy tool together with the LCA method [49], and the simultaneous use of the two frameworks to quantify ecosystem functions and assess the pros and cons of reforestation programs seems to be absent.

2. Materials and Methods

2.1. The Study Area

The Metropolitan City of Naples (MCN) in the Campania region, Southern Italy, represents the former Province of Naples, divided into 92 municipalities (Figure 1). The Municipality of Naples is the administrative centre. The MCN covers about 1200 km^2 and has a population of about 3 million inhabitants. The study area includes some of the highest density of population among Italian municipalities (e.g., Portici and Casavatore have about 12,000 inhabitants/km^2), two volcanic sites (Vesuvius and Phlegraean Fields), a long coastline, and urban areas that extend without interruption also across agricultural fields and green areas, thus presenting several types of land cover. The area falls under a Mediterranean climate zone, therefore characterized by Mediterranean vegetation, with diversified agrarian production (potato, peach, apricot, tomato, fennel, plum, grapes, tomato, cauliflower, broccoli, and strawberry) [50].

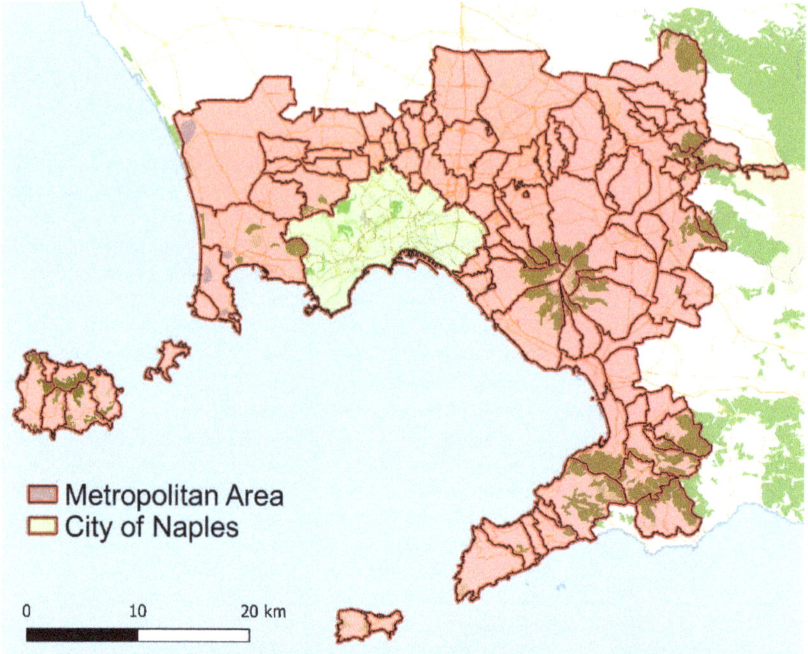

Figure 1. The study area of the Metropolitan City of Naples. The red borders highlight the 91 municipalities of the former Province of Naples (in red) and the City of Naples (in green).

The area accounts for three state-owned forests (Phlegraean Area, Mount Cumae's and Roccarainola's forests), which cover more than 1000 ha mainly with elm, elder, hawthorn, ash, chestnut, alder, and beech tree species [51]. The municipality of Naples includes 53 different public and private parks and green areas, almost entirely managed by municipal services, presenting a tree population of above 60,000 elements (very common *Pinus pinea* to different kinds of citrus trees, *Ginko biloba*, *Olea europaea*, *Cupressus*, and *Ficus*), with a large part of trees aged more than 50 years. Their management within the urban context represents a challenge for administrators [52]. Between 2011 and 2015, the number of new plantations exceeded the number of cuts of trees by 1.58%. However, Naples is 88th among the 105 Italian provincial chief towns, having just six trees per 100 inhabitants [53]. Due to high urbanization, the urban tree stock is minimal.

The most represented cover classes within the MCN 2018 land cover, assessed by CORINE Land Cover [54,55], are the broad-leaved forest (≈21%), the non-irrigated arable land (≈17%), the complex cultivation patterns (≈15%), the fruit trees and berry plantations (≈12%), the continuous and discontinuous urban fabric (respectively ≈12% and ≈8%), and the industrial or commercial units (≈2.5%)—confirming the strong agricultural and naturalistic tendency of the investigated area (65% of the territory).

2.2. i-Tree Canopy

The land cover assessment of the MCN area was performed using i-Tree Canopy tool v.7, using a classification method based on automatically retrieved aerial imagery. The tool is part of the i-Tree Tools software suite that provides urban and rural forestry analysis and benefits assessment tools. The i-Tree Canopy tool estimates tree coverage, as well as other user-decided cover classes, by generating random points within a specified area to be manually classified [41]. In addition, it allows the estimation of the coverage area by providing the most recent aerial Google Maps/Google Earth imagery data (with a resolution of about 0.3 m) in order to perform a statistical analysis of the classification of randomly generated points [48].

The i-Tree Canopy analysis is performed following three steps:

1. Drawing boundaries or importing a file with the boundaries of the investigated area. Boundaries for the MCN were retrieved from a dedicated repository provided by the administration (http://sit.cittametropolitana.na.it/, accessed on 27 August 2019);
2. Naming the cover classes to be assessed. In this work, the cover classes are defined encompassing all the different land cover types present in the studied area (Table 1).
3. Classifying points into the defined cover classes.

Table 1. Cover classes used for the land cover assessment of the Metropolitan City of Naples.

Item	Cover Class	Description
TI	Tree over impervious	Trees over impervious ground
TP	Tree over pervious	Trees over pervious ground
AG	Agricultural land	Pastures, crops or fallows
GH	Grass/herbaceous cover	Yards, parks or fields
IN	Impervious non-plantable	Roads, rails, roofs or monuments
PN	Pervious non-plantable	Dirt road
IP	Impervious partially plantable	Sidewalks, parking lots or plazas
PP	Pervious partially plantable	Bare earth
SB	Shrubs/bushes	Stem based vegetation smaller than trees
WT	Water	Ocean, estuary, river, lake, wetland, etc
OT	Other	Other surfaces

In Table 1, the current tree coverage is represented by TI and TP, and the potential tree coverage is represented by IP (e.g., sidewalks, parking lots, or plazas) and PP (e.g., bare earth).

The accuracy of the final assessment will depend upon the ability of the analyst to correctly classify each randomly generated point into one of the chosen cover classes. The precision of the estimate increases with the increase in assessed points, as the standard error (SE) will decrease. Being "n" the number of points assessed in each class and "N" the total number of assessed points, SE for each class is calculated as in (1):

$$SE = \sqrt{\left(\frac{pq}{N}\right)}, \tag{1}$$

where $p = n/N$ and $q = 1 - p$.

If $n < 10$, the SE is calculated as (2):

$$SE = (\sqrt{n})/N, \tag{2}$$

Endreny et al. (2017) [48] state that 500 random points are adequate to survey megacities. In this study, to produce more accurate results, 803 points were assessed and classified according to the user-defined cover classes reported in Table 1.

The benefits investigated with the i-Tree Canopy tool are based on the percentage of tree coverage and climatic conditions. Table 2 reports the conversion factors for uptaken mass of different polluting flows and for volume of avoided runoff per unit area of trees per year, and the related economic values. The removal rates and monetary values of the considered environmental functions are derived from analyses conducted in the United States using i-Tree Eco, a component tool of the i-Tree software suite, within urban and rural areas and then aggregated at the national level [56]; data are then provided as specific sets related to chosen locations, within the report of tool results. Monetary values for pollutant removal are estimated as the incidence of adverse health effects resulting from changes in pollutants concentrations [56]. The i-Tree Canopy tool, as well as the entire i-Tree suite, is developed by the USDA Forest Service and the Davey Tree Expert company, among others, focusing on USA territories. Thus, the average trees benefits values are calculated, regardless of tree species, based on climatic conditions for USA territories only (and in more recent times for the United Kingdom and Sweden too). In order to apply the tool to regions outside the ones included in the suite, a solution would be to analyse the annual trends and distribution of temperature and precipitations of the investigated location and find the most similar ones in similar climatic areas within the United States. In this study, to overcome geographical data limitation, the Horry County in South Carolina, USA, was selected based on similar MCN climate conditions.

Table 2. Removal rate and monetary value of benefits of tree coverage estimated by i-Tree Canopy.

Annual Air Pollution Removal Benefits			
Item	**Description**	**Removal Rate ($g/m^2/yr$)**	**Monetary Value (USD/t/yr)**
CO	Carbon monoxide	0.03	192.29
NO_2	Nitrogen dioxide	0.22	63.03
O_3	Ozone removed	7.37	549.10
PM_{10}	Particulate matter greater than 2.5 microns and less than 10 microns	1.47	889.02
$PM_{2.5}$	Particulate matter less than 2.5 microns	0.05	38,529.50
SO_2	Sulphur dioxide	0.19	23.54
Annual Hydrological Benefits			
Item	**Description**	**Tree effects ($L/m^2/yr$)**	**Monetary Value ($USD/m^3/yr$)**
AVRO	Avoided runoff	0.60	2.60
E	Evaporation	14.94	
I	Interception	14.95	
T	Transpiration	107.41	
PE	Potential evaporation	545.19	
PET	Potential evapotranspiration	488.30	
Carbon Benefits			
Description		**Carbon Rate (t/ha/yr)**	**Monetary Value (USD/t)**
Carbon sequestered annually		30.60	187.99
Carbon stored in trees' lifetime		768.48	

2.3. Life Cycle Assessment

The life cycle assessment method, standardized by ISO standards and ILCD handbook [42,43,57], assesses the potential environmental burdens of human-dominated processes and systems in a "cradle to grave" perspective, from the extraction of raw materials to the disposal of generated waste, through distribution and use. The results provided different kinds of impact categories, including emissions and resource consumption [58]. It is performed by following a four-step procedure:

(1) Goal and scope definition: in this phase, the objective, the functional unit (FU) and the burdens of the investigated case study are clearly defined. In this work, the LCA method is used to support and integrate the results of the performed i-Tree Canopy study. The chosen FU is the assessment of the impact categories affected by the ecological functions delivered by the current and potential tree cover within the boundaries of the Metropolitan City of Naples.

(2) Inventory analysis: LCA analyses are performed by means of specific inventories listing all relevant input and output flows enabling processes and/or subprocesses within the investigated case studies. Inventories are usually dimensioned considering the chosen FU, and include primary data (directly collected), secondary data (from scientific literature, databases, etc.), and tertiary data (calculations and assumptions). In this work, data for the used inventories come from the i-Tree Canopy tool, listing the pollution and hydrological benefits of tree cover. Hence, the inventory is built considering the annual mass of uptaken flows: particulate <2.5 μm; particulate >2.5 μm and <10 μm; sulphur dioxide; ozone; nitrogen dioxide; carbon monoxide and carbon dioxide; and the annual avoided water runoff in the assessed area in the present and future potential scenarios.

(3) Impact assessment: this step translates the input and output flows in the inventories into potential impacts in different categories by means of specific characterization factors related to distinct impact methods. In this work, the avoided pollution and hydrological effects of the current and potential tree cover are translated into characterized impacts by using the SimaPro software v.9.1.1.1 (https://network.simapro.com, accessed on 15 September 2022), the Ecoinvent database v.3.6 [59], and the ReCiPe midpoint (H) method v.1.04 [60]. The software, the database and the impact method work together to classify the burdens deriving from the assessed inventory into impacts related to specific categories. The various emissions in the different compartments and resource use are then characterized by means of particular factors to express them into the proper units.

(4) Results interpretation: in this phase, the results are carefully checked and evaluated to understand the characteristics of the investigated case study, propose solutions, find the hot spots, etc.

3. Results

This work explored the benefits deriving from the present and the potential tree coverage within the area of the MCN. Among the investigated cover classes, tree over impervious (TI) and tree over pervious (TP) represent the current tree coverage characterized by trees whose canopies stand above cemented sidewalks and plazas and bare earth and meadows, respectively. The potential tree coverage classes impervious partially plantable (IP) and pervious partially plantable (PP) were added to the potential scenario, considering both impervious and pervious soils where trees could be planted without impairing human activities (i.e., parking lots, sidewalks, parks, marginal lands, etc.).

3.1. i-Tree Canopy Results

The MCN cover area, estimated by means of the assessed 803 points, is mainly represented by the 29% trees planted over pervious grounds (TP), 23% impervious area where trees are impractical (IN) and 17% agricultural land (AG) (Table 3).

The area currently covered by trees within the MCN boundaries was identified in 251 points (TP and TI) and was estimated at 367.26 square kilometres. The area for the potential planting of new trees was identified in 128 points (IP and PP), equal to about 187.28 square kilometres (16% of the entire MCN area), which represents a potential increment of 51% in green spaces. The estimated area needed to plant one million is 80 square kilometres of forests [61]. Thus, around 240 square kilometres would be required to achieve the goal of planting three million trees, which is 30% larger than the identified available area. The analysis performed in this work suggests a current availability for plantations equal to around 2.34 million trees.

Table 3. Number of points assessed and estimated area for each cover class.

Cover Class	N° of Points	Area (km^2)	±SE (km^2)	% Cover
TP	230	336.53	29.08	29%
IN	187	273.61	17.50	23%
AG	136	198.99	15.51	17%
IP	70	102.42	11.75	9%
PP	58	84.86	10.70	7%
SB	46	67.31	9.63	6%
GH	30	43.90	7.86	4%
TI	21	30.73	9.64	3%
PN	13	19.02	5.28	2%
WT	9	13.17	4.35	1%
OT	3	4.39	2.61	0%
Total Area	803	1174.93	30	100%

The i-Tree Canopy tool considers a linear correlation between the presence of trees and the functions they provide. In so doing, considering that all available areas will actually be used for tree plantation, the potential scenario presents an estimated 51% increment in terms of benefits related to air pollution absorption (Figure 2a) and carbon sequestrated and stored during trees' lifetime (Figure 2b). A 51% of increment is also estimated for avoided runoff (113 megalitres of water per year) and the hydrological cycle (219,273 megalitres of water per year) of the photosynthesis steps (evaporation, interception, transpiration, potential evaporation, and potential evapotranspiration) (Table 4).

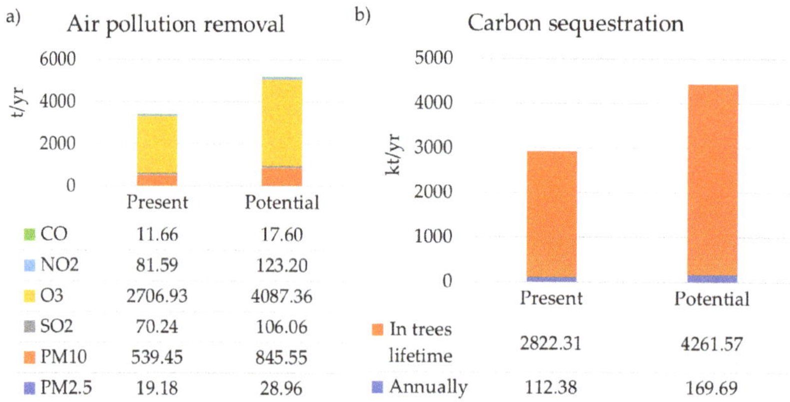

Figure 2. Tree benefits: (**a**) amount of air pollution absorbed expressed in carbon monoxide (CO), nitrogen dioxide (NO_2), Ozone (O_3), particulate matter (PM10 and PM2.5) and sulfuric dioxide (SO_2); (**b**) carbon annually sequestrated and carbon stored in trees' lifetime.

Table 4. Hydrological benefits from tree coverage in megalitres (Ml).

Benefits	Amount (Ml)	
	Present	Potential
AVRO	221.81	334.92
E	5485.22	8282.46
I	5491.00	8291.20
T	39,447.51	59,564.16
PE	200,226.75	302,334.42
PET	179,330.20	270,781.45
Total	430,202.49	649,588.61

The increment of the annual economic benefits is more than USD 12.4 million per year, going from almost USD 24.5 million to nearly USD 36.9 million (Table 5); 87% of this value is due to the carbon sequestrated annually by trees. Considering the amount of carbon stored in trees during their lifetime (4261 kt), the final economic value increases by more than USD 270 million, from about USD 530 million in the present scenario to more than USD 800 million in the potential scenario.

Table 5. Hydrological benefits from tree coverage in Megalitres (Ml).

Benefit	Description	Value (USD)	
		Present	Potential
Carbon sequestration	Sequestered annually in trees	21,126,481	31,900,144
Air pollution removal	CO	2242	3385
	NO_2	5143	7766
	O_3	1,486,373	2,244,364
	SO_2	1654	2497
	PM10	479,582	724,149
	PM2.5	739,064	1,115,957
Hydrological	AVRO	577,302	871,703
Total		24,417,841	36,869,965

3.2. LCA Analysis

The assessed environmental benefits provided by the current and the potential tree cover are configured as an annual sequestration/absorption of different potentially harmful substances in the atmosphere and as avoided water runoff that could cause hydrological damages. These benefits can be translated into LCA-avoided impacts, expressing the annual avoided repercussion in different, specific environmental sectors, thanks to the presence of trees delivering their functions. The LCA analysis of the ecological functions provided by the present and the potential assessed tree cover is reported in Table 6. The different air pollutant substances sequestered by trees and the avoided runoff in the two scenarios affect six ReCiPe midpoint (H) impact categories: global warming; ozone formation, human health; fine particulate matter formation, ozone formation, terrestrial ecosystems; terrestrial acidification; water consumption. The results confirm once again the linear correlation between environmental benefits and the presence of trees proposed by the i-Tree Canopy tool.

Table 6. Characterized ReCiPe midpoint impacts avoided by the present and the possible tree coverage in the Metropolitan City of Naples.

Impact Category	Unit	Present	Potential
Global warming	kg $CO_{2\,eq}$	1.12×10^8	1.70×10^8
Ozone formation, human health	kg $NO_{x\,eq}$	8.16×10^4	1.23×10^5
Fine particulate matter formation	kg $PM_{2.5\,eq}$	4.85×10^4	7.33×10^4
Ozone formation, terrestrial ecosystems	kg $NO_{x\,eq}$	8.16×10^4	1.23×10^5
Terrestrial acidification	kg $SO_{2\,eq}$	9.96×10^4	1.50×10^5
Water consumption	m^3	2.22×10^5	3.35×10^5

Planting new trees in urban environments presents both economic and environmental costs due to the cultivation of tree seedlings in specific nurseries, involving energy, chemicals, and machinery, among other input flows. A careful accounting of the mentioned environmental burdens against the analysis of benefits coming from the presence of trees could facilitate afforestation projects and urban green management plans. Table 7 shows the

ReCiPe midpoint impacts of the production and plantation of 2.34 million tree seedlings, according to the Ecoinvent database used in this study. The comparison between the numbers in Tables 6 and 7 suggests that the impacts of tree production and planting would be entirely compensated by the functions provided by these new trees in a little more than half a year (≈200 days).

Table 7. Characterized ReCiPe midpoint impacts for growing and planting 2.34 million tree seedlings.

Impact Category	Unit	Growing + Planting
Global warming	kg $CO_{2\ eq}$	3.36×10^6
Stratospheric ozone depletion	kg CFC11 $_{eq}$	1.98×10^0
Ionizing radiation	kBq Co-60 $_{eq}$	1.40×10^5
Ozone formation, human health	kg $NO_{x\ eq}$	2.23×10^4
Fine particulate matter formation	kg $PM_{2.5\ eq}$	8.29×10^3
Ozone formation, terrestrial ecosystems	kg $NO_{x\ eq}$	2.28×10^4
Terrestrial acidification	kg $SO_{2\ eq}$	1.57×10^4
Freshwater eutrophication	kg P $_{eq}$	9.03×10^2
Marine eutrophication	kg N $_{eq}$	9.27×10^1
Terrestrial ecotoxicity	kg 1,4-DCB	1.73×10^7
Freshwater ecotoxicity	kg 1,4-DCB	2.65×10^5
Marine ecotoxicity	kg 1,4-DCB	3.43×10^5
Human carcinogenic toxicity	kg 1,4-DCB	2.02×10^5
Human non-carcinogenic toxicity	kg 1,4-DCB	1.22×10^7
Land use	m^2 a crop $_{eq}$	7.68×10^6
Mineral resource scarcity	kg Cu $_{eq}$	5.50×10^4
Fossil resource scarcity	kg oil $_{eq}$	9.44×10^5
Water consumption	m^3	1.47×10^4

4. Discussion

The economic benefit of the current tree coverage, calculated by the i-Tree Canopy tool, is estimated at about USD 24 million annually (≈USD 8 per citizen of the MCN per year). If the assessed potentially plantable area was used for tree plantations, the total benefits would be equal to ≈USD 37 million/year (≈USD 12 per citizen). At the urban level, further economic savings that are not currently included within the perspective of the tool, which, as seen, operated a linear correlation among tree cover and ecological functions provided, would also be found. For instance, tree shadings would affect the local temperature, inducing a reduction in the expenses for electricity that would vary from USD 4 to USD 166 per tree annually due to reduced use of cooling and heating systems [62].

In order to undertake the OBC Project, which originated the idea behind the research presented in this work, the MCN administration committed to funding USD 969 million, within which planting trees is only one of the actions planned.

The total annual savings of the present and potential trees land coverage are estimated at ≈USD 55 million/year purchasing power parity value (PPP) adjusted (considering the value of 0.6708 for Italy in 2019 [63]). It is equivalent to ≈USD 18 per citizen of the MCN per year and ≈USD 99,117 per square kilometre of tree coverage. Endreny et al. (2017) investigated 10 of the biggest megacities worldwide, with an average of 1.8×10^7 citizens, including among the tree benefits also the cost of the avoided kWh of electricity and the avoided Mbtu of energy. The median value of the additional tree coverage in the investigated megacities was estimated at 17.8%, providing an average USD 967,000/km^2 of tree coverage and USD 32/capita, PPP adjusted. Rogers et al. (2015) assessed the tree population for the Greater London Area, calculating an average benefit of about USD 891,000/km^2 and USD 23/capita (PPP adjusted) for a canopy cover equal to 21% of the territory [64].

Scaling up the economic benefits calculated in this work to the projected number of three million trees planted, it would be reasonable to expect a payback time reduced to 24 years from the entire tree coverage within the investigated area, considering the total investment by the MCN administration and disregarding other benefits provided from the presence of urban forests. Therefore, the payback time could be further reduced in a more accurate and detailed evaluation. Further studies should include logistics and implementation costs, project duration, and job opportunities creation to provide a comprehensive evaluation of the investments and provide a complete management report.

The three million trees objective would require the implementation of significant actions to be fulfilled. Among these, the plantation of trees within the areas assessed in this work as IP—impervious partially plantable—might represent a challenge, as this cover class is largely characterized as spaces within the highly populated urban areas of the MCN (sidewalks, plazas, and parking lots represent 9% of the territory—around 102 km^2). In these places, trees are planted individually and spaced, improving the population approach to some ecological functions, such as local temperature and humidity control, runoff, infiltration, and flooding regulation, and creating areas for recreation and leisure. However, this might represent a significant economic cost for the city administration due to the regular maintenance required by street trees (pruning, irrigation, crown thinning, and removal). Even more efforts would be required to convert a significant fraction of the land cover assessed as "not plantable" (e.g., industrial areas, marginal lands, unused deposits, etc.) into entire "urban forests", freeing the soil from old cement and giving back life to the ground underneath. In this case, the bare ground would become reforested wood, offering tree benefits on a regional scale. This strategy also includes other benefits such as aquifer recharge, soil formation, fertility and biodiversity maintenance, recreative and touristic places creation [65], and wellbeing. However, this would require a huge initial effort from the metropolitan city administration, compensated by lesser needs on management and maintenance compared to single-planted street trees.

Nastran et al. (2022) analysed the perceived connections between ecosystem services and green infrastructures in urban ecosystems and their impacts on wellbeing, suggesting that urban forests are the most influencing among the types of infrastructures considered [66]. Valeri et al. (2021) investigated the significance of the selection of target plant species in peri-urban and agricultural areas, as well as in many areas of the MCN [67]. Evans et al. (2022) reviewed scientific literature about green infrastructures and services delivery, in particular describing how their delivery is partly modulated by the kind of spaces where they are assessed [68]. Shao and Kim (2022) investigated the urban heat islands mitigation potential of green infrastructures, at the same time dealing with climate change and providing different functions promoting sustainable development and wellbeing in urban systems [69]. García-Pardo et al. (2022) reviewed remote sensing techniques for ecosystem services analysis, pointing out the importance of the sensors used, the geographical scale and image resolution, and the need for more information and a transdisciplinary framework for the assessment of the ecosystem services [70].

The results and insights obtained in this work, together with the presented joint implementation of the i-Tree tool suite and the LCA assessment method, could be useful for administrators, policymakers, urban planners, and organizations of different natures on successfully planning and managing urban greening projects and interventions, receiving countless environmental functions and benefits in return. The application of both the i-Tree and LCA tools made it possible to analyse the specific polluting substances and hydrological features investigated in this work and characterize them into specific impact categories in different environmental sectors. This allowed a wider understanding of the investigated case study, enabling the possibility of looking at it from multiple perspectives and calculating the time needed to compensate the related environmental and economic investments. However, the i-Tree Canopy tool might adopt a too simplistic point of view when establishing only a linear correlation between the presence of trees and the provided functions. This might be true in some cases but untrue in others, where the synergies among trees and the inclusions of more functions could indicate different types of correlation (e.g., exponential). Thus, the kind of analysis performed in this work could benefit from a deeper and wider investigation of the ecological functions related to trees. This kind of analysis, together with stronger efforts for the conversion of already identified areas, incentives for private sector initiatives, and ecological management for agriculture, can provide further benefits to biodiversity, crops and pastures productivity [71].

The findings of this work then suggest the importance of ecosystems functions and the need for proper planning and management in order to correctly account the balance between investments and advantages. As the kind and the number of functions can be affected by local conditions and by the viewpoint adopted in their analysis, the simultaneous application of different methods, as the performed joint application of LCA and i-Tree Canopy, can provide a very beneficial, multiple perspective to proper planning in worldwide urban environments. Moreover, there seems to be a lack in scientific literature of joint i-Tree Canopy/LCA applications, reinforcing the novelty of the presented work and proposing a framework for benefit/constraint analyses of designed and implemented projects for green infrastructures.

Practical Implications: Social, Environmental, and Economic Benefits

The results of this work show opportunities to improve people's quality of life, reduce environmental damage, and bring economic development by the provision of ecological functions. The increment of 51% in new green areas reduces 51% of harmful hydrological events, air pollution, and greenhouse gases. This is also confirmed by the avoided emission highlighted by the LCA analysis.

Greenhouse gas emissions and air pollutants are harmful to crops' productivity, human health, and, consequently, the economy [72]. Air pollutants are responsible for pulmonary diseases, cancer, and respiratory infections, increasing death rates [73] and public health costs. Additionally, the O_3 exposure reduces plant transpiration and heat fluxes, diminishing rainfall and increasing air temperature. All these factors negatively impact the crop yield, from 18% to 45%, representing approximately 35% of loss related to gross primary productivity (GPP) [74].

> *"The interaction with nature is a fundamental pillar for individuals' wellbeing, regardless of their geographical, cultural, or socioeconomic background".* [24]

Many city residents have no daily contact with local nature and biodiversity, which also impacts individual wellbeing. Different studies show the health benefits of living nearby green areas, improving human wellbeing by reducing blood pressure, heart rate, and muscle tension [75,76]. Unfortunately, too many individuals are far from natural areas in their daily life, and this distance might impact their perception of natural environments and local biodiversity. However, it is also possible to notice that personal perceptions are affected by social, cultural, and psychological backgrounds that should be addressed and investigated in local communities.

The *New York Times* newspaper, on 21 September 1966, reported the former US President Ronald Reagan's words talking about trees as sources of raw materials: "A tree's a tree. How many more redwoods do you need to look at? If you've seen one, you've seen them all". In 1973, Martin Krieger published an article in *Nature* questioning the utility of urban trees: the use of plastic trees could replicate the recreational and shading functions of natural ones; he argued that the definition of natural environment depends on culture and society, and that "preservationism" is more expensive than pragmatic conservationism [77]. The argument resurfaced in 2007 when the journal *Nature* published a blog post discussing Krieger's idea, concluding that living in a Disneyfied world costs more than the direct monetary value of fake wilderness, being the actual cost is the loss of ecosystem value [78].

Nature and urban green spaces as elements of the urban landscape are fundamental to improving the quality of life of citizens. Therefore, they must be considered in urban planning policies to reflect the needs, economic possibilities, and customs of the city's inhabitants. It is, in any case, an interpretation that takes its inspiration from the hope that biopolitics can have the possibility of winning against biopower [79].

Urban nature is vital in all its manifestations, from private gardens, tree plantations, and city parks to land used for recreational purposes within or nearby the city.

5. Conclusions

This study aimed to evaluate the socioeconomic and environmental benefits of planting trees in urban environments. The area within the boundaries of the Metropolitan City of Naples that is available for planting new trees is enough for 2.3 million new plantations on impervious ground (sidewalks, plazas, parking lots, etc., 9% of the territory—around 102 km^2) and pervious ground (7% of the region—about 85 km^2). The i-Tree Canopy tool is a useful web tool for helping policymakers to understand the environmental and economic value of the benefits offered to society and the interest in carrying out a public project entailing urban trees, evaluating specific scenarios to meet requirements and cost-effective solutions. Its combined application with the LCA method delivers a wider understanding of the ecological sectors affected by impacts and benefits of reforestation activities. In so doing, the two instruments showed that careful planning and different strategies, considering investments, costs, and benefits in a long-term perspective, are needed. As public funds available to administrators are often very scarce, public authorities also need to understand the socioeconomic improvements of carrying out urban forestry projects compared to other alternatives. Furthermore, urban forestry projects have several positive social and cultural side-effects: the creation of new green areas for recreational and cultural meetings and the aesthetical valorisation of spaces for cultural progress and civil society involvement. However, further research and collaborations among different stakeholders and the integrated use of diverse assessment methods are needed to evaluate projects under the three pillars (environmental, economic, and social) of sustainable development and the need to accelerate the transition in the EU towards a more regenerative and circular economy at the EU level.

Author Contributions: M.O.: conceptualization, methodology, formal analysis, investigation, data curation, writing—original draft, writing—review and editing; R.S.: conceptualization, methodology, formal analysis, investigation, data curation, writing—original draft, writing—review & editing; S.K.: conceptualization, investigation, writing—original draft, writing—review and editing; Y.L.: conceptualization, investigation, writing—original draft; C.V.: conceptualization, investigation, writing—original draft; P.G.: conceptualization, methodology, writing- original draft; writing—review and editing; G.L.: conceptualization, writing—original draft, funding acquisition; S.U.: conceptualization, methodology, supervision, project administration, funding acquisition. All authors have read and agreed to the published version of the manuscript.

Funding: This study received funding from: the European Union's Horizon 2020 research and innovation programme under the Marie Skłodowska-Curie European Training Networks (grant agreement ReTraCE No. 814247), the Italian Ministry of Foreign Affairs and International Cooperation (MAECI, High Relevance Bilateral Projects, Grant No. PGR00954), the Sino-Italian Cooperation of NSFC Natural Science Foundation of China (Grant No. 71861137001), within the project "Analysis on the metabolic process of urban agglomeration and the cooperative strategy of circular economy" (2018–2020), the European Union's Horizon 2020 research and innovation programme (grant agreement JUST2CE No. 101003491), and the Fundamental Research Funds of Capital University of Economics and Business (Grant No. XRZ2022028).

Data Availability Statement: Data sharing not applicable.

Conflicts of Interest: The funders had no role in the design of the study; in the collection, analyses, or interpretation of data; in the writing of the manuscript; or in the decision to publish the results. The authors declare no conflict of interest.

References

1. Dong, X.; Yang, W.; Ulgiati, S.; Yan, M.; Zhang, X. The Impact of Human Activities on Natural Capital and Ecosystem Services of Natural Pastures in North Xinjiang, China. *Ecol. Model.* **2012**, *225*, 28–39. [CrossRef]
2. EEA Urban Green Infrastructure—European Environment Agency. 2017. Available online: https://www.eea.europa.eu/themes/sustainability-transitions/urban-environment/urban-green-infrastructure (accessed on 1 June 2021).
3. Food and Agriculture Organization. *Guidelines on Urban and Peri-Urban Forestry—Forestry Paper No. 178*; Food and Agriculture Organization: Rome, Italy, 2016; ISBN 978-92-5-109442-6.

4. Ramakrishna, S. Looking through the COVID-19 Lens for a Sustainable New-Modern Society; Springer Nature—Sustainability Community: 2020. Available online: https://sustainabilitycommunity.springernature.com/posts/looking-through-covid-19-lens-for-a-sustainable-new-modern-society (accessed on 1 June 2021).
5. Roy, S.; Byrne, J.; Pickering, C. A Systematic Quantitative Review of Urban Tree Benefits, Costs, and Assessment Methods across Cities in Different Climatic Zones. *Urban For. Urban Green.* **2012**, *11*, 351–363. [CrossRef]
6. Calfapietra, C.; Cherubini, L. Green Infrastructure: Nature-Based Solutions for Sustainable and Resilient Cities. *Urban For. Urban Green.* **2019**, *37*, 1–2. [CrossRef]
7. Nowak, D.J.; Crane, D.E.; Stevens, J.C. Air Pollution Removal by Urban Trees and Shrubs in the United States. *Urban For. Urban Green.* **2006**, *4*, 115–123. [CrossRef]
8. Obama, B. Presidential Memorandum—America's Great Outdoors. 2010. Available online: https://obamawhitehouse.archives.gov/the-press-office/presidential-memorandum-americas-great-outdoors (accessed on 1 June 2021).
9. Mancuso, S. *La Nazione Delle Piante*; Laterza: Bari, Italy, 2019; ISBN 9788858135815.
10. Roeland, S.; Moretti, M.; Amorim, J.H.; Branquinho, C.; Fares, S.; Morelli, F.; Niinemets, Ü.; Paoletti, E.; Pinho, P.; Sgrigna, G.; et al. Towards an Integrative Approach to Evaluate the Environmental Ecosystem Services Provided by Urban Forest. *J. For. Res. (Harbin)* **2019**, *30*, 1981–1996. [CrossRef]
11. Haaland, C.; van den Bosch, C.K. Challenges and Strategies for Urban Green-Space Planning in Cities Undergoing Densification: A Review. *Urban For. Urban Green.* **2015**, *14*, 760–771. [CrossRef]
12. World Forum on Urban Forests Greener, Healthier and Happier Cities for All—A Call for Action. 2018. Available online: https://www.worldforumonurbanforests.org/component/phocadownload/category/38-wfuf-call-for-action.html?download=155:call-for-action (accessed on 1 June 2021).
13. Millenium Ecosystem Assessment. *Ecosystems and Human Well-Being: Opportunities and Challenges for Business and Industry*; World Resources Institute: Washington, DC, USA, 2005.
14. Dall'Ara, E.; Maino, E.; Gatta, G.; Torreggiani, D.; Tassinari, P. Green Mobility Infrastructures. A Landscape Approach for Roundabouts' Gardens Applied to an Italian Case Study. *Urban For. Urban Green.* **2019**, *37*, 109–125. [CrossRef]
15. Escobedo, F.J.; Giannico, V.; Jim, C.Y.; Sanesi, G.; Lafortezza, R. Urban Forests, Ecosystem Services, Green Infrastructure and Nature-Based Solutions: Nexus or Evolving Metaphors? *Urban For. Urban Green.* **2019**, *37*, 3–12. [CrossRef]
16. Sanesi, G.; Colangelo, G.; Lafortezza, R.; Calvo, E.; Davies, C. Urban Green Infrastructure and Urban Forests: A Case Study of the Metropolitan Area of Milan. *Landsc. Res.* **2017**, *42*, 164–175. [CrossRef]
17. Shah, A.M.; Liu, G.; Huo, Z.; Yang, Q.; Zhang, W.; Meng, F.; Yao, L.; Ulgiati, S. Assessing Environmental Services and Disservices of Urban Street Trees. An Application of the Emergy Accounting. *Resour. Conserv. Recycl.* **2022**, *186*, 106563. [CrossRef]
18. Ackerman, K.; Conard, M.; Culligan, P.; Plunz, R.; Sutto, M.-P.; Whittinghill, L. Sustainable Food Systems for Future Cities: The Potential of Urban Agriculture. *Econ. Soc. Rev. (Irel)* **2014**, *45*, 189–206.
19. Ghisellini, P.; Casazza, M. Evaluating the Energy Sustainability of Urban Agriculture Towards More Resilient Urban Systems. *J. Environ. Account. Manag.* **2016**, *4*, 175–193. [CrossRef]
20. Lucarelli, A. Note Minime per Una Teoria Giuridica Dei Beni Comuni. *Espaço Jurídico J. Law* **2011**, *12*, 11–20.
21. Mattei, U. *Beni Comuni. Un Manifesto*; Laterza: Bari, Italy, 2011.
22. Shiva, V. *Water Wars; Privatization, Pollution, and Profit*; South End Press: Cambridge, MA, USA, 2002.
23. Dellenbaugh, M.; Kip, M.; Bieniok, M.; Müller, A.; Schwegmann, M. *Urban Commons: Moving Beyond State and Market*; Birkhäuser: Basel, Switzerland, 2015; ISBN 3038214957.
24. Priego, C.; Breuste, J.H.; Rojas, J. Perception and Value of Nature in Urban Landscapes: A Comparative Analysis of Cities in Germany, Chile and Spain. *Landsc. Online* **2008**, *7*, 1–22. [CrossRef]
25. Rohde, C.L.E.; Kendle, A.D. *Human Well-Being, Natural Landscapes and Wildlife in Urban Areas: A Review*; English Nature; Northminster House: Peterborough, UK, 1994; ISBN 1857161556.
26. Food and Agriculture Organization. *Forests and Sustainable Cities—Inspiring Stories from around the World*; Food and Agriculture Organization: Rome, Italy, 2018; ISBN 978-92-5-130417-4.
27. Salbitano, F.; Sanesi, G. Cambiamento Climatico. In Italia Le Foreste Aumentano. L'emergenza Verde è Nelle Città. 2019. Available online: https://www.avvenire.it/opinioni/pagine/in-italia-le-foreste-aumentano-lemergenza-verde-nelle-citt (accessed on 1 June 2021).
28. Città Metropolitana di Napoli Ossigeno Bene Comune—Dettaglio News—Città Metropolitana Di Napoli. Available online: https://www.cittametropolitana.na.it/en_US/-/ossigeno-bene-comu-1 (accessed on 15 September 2022).
29. Calzolari, T.; Genovese, A.; Brint, A. Circular Economy Indicators for Supply Chains: A Systematic Literature Review. *Environ. Sustain. Indic.* **2022**, *13*, 100160. [CrossRef]
30. Ellen MacArthur Foundation. *Towards the Circular Economy Vol. 1—An Economic and Business Rationale for an Accelerated Transition*; Ellen MacArthur Foundation: Chicago, IL, USA, 2012.
31. Ghisellini, P.; Cialani, C.; Ulgiati, S. A Review on Circular Economy: The Expected Transition to a Balanced Interplay of Environmental and Economic Systems. *J. Clean. Prod.* **2016**, *114*, 11–32. [CrossRef]
32. Langergraber, G.; Pucher, B.; Simperler, L.; Kisser, J.; Katsou, E.; Buehler, D.; Mateo, M.C.G.; Atanasova, N. Implementing Nature-Based Solutions for Creating a Resourceful Circular City. *Blue-Green Syst.* **2020**, *2*, 173–185. [CrossRef]

33. Santagata, R.; Ripa, M.; Genovese, A.; Ulgiati, S. Food Waste Recovery Pathways: Challenges and Opportunities for an Emerging Bio-Based Circular Economy. A Systematic Review and an Assessment. *J. Clean. Prod.* **2021**, *286*, 125490. [CrossRef]
34. Perrotti, D.; Stremke, S. Can Urban Metabolism Models Advance Green Infrastructure Planning? Insights from Ecosystem Services Research. *Environ. Plan. B Urban Anal. City Sci.* **2020**, *47*, 678–694. [CrossRef]
35. Comitato per il Capitale Naturale. *Primo Rapporto Sullo Stato Del Capitale Naturale in Italia*; Comitato per il Capitale Naturale: Rome, Italy, 2017.
36. Ghisellini, P.; Oliveira, M.; Santagata, R.; Ulgiati, S. Dossier Informativo Sul Valore Ambientale Ed Economico Di Progetti Di Investimento in Foreste Urbane e Infrastrutture Verdi (in Italian). 2019. Available online: https://www.researchgate.net/publication/340254697 (accessed on 15 September 2022).
37. Stahel, W.R. The Circular Economy. *Nature* **2016**, *531*, 435–438. [CrossRef]
38. D'Amato, D.; Droste, N.; Allen, B.; Kettunen, M.; Lähtinen, K.; Korhonen, J.; Leskinen, P.; Matthies, B.D.; Toppinen, A. Green, Circular, Bio Economy: A Comparative Analysis of Sustainability Avenues. *J. Clean. Prod.* **2017**, *168*, 716–734. [CrossRef]
39. Oliveira, M.; Miguel, M.; van Langen, S.K.; Ncube, A.; Zucaro, A.; Fiorentino, G.; Passaro, R.; Santagata, R.; Coleman, N.; Lowe, B.H.; et al. Circular Economy and the Transition to a Sustainable Society: Integrated Assessment Methods for a New Paradigm. *Circ. Econ. Sustain.* **2021**, *1*, 99–113. [CrossRef]
40. Hirvilammi, T. The Virtuous Circle of Sustainable Welfare as a Transformative Policy Idea. *Sustainability* **2020**, *12*, 391. [CrossRef]
41. i-Tree i-Tree Canopy—Refereces. 2006. Available online: https://canopy.itreetools.org/references (accessed on 1 June 2021).
42. *ISO UNI EN ISO 14040*; Environmental Management—Life Cycle Assessment—Principles and Framework. International Organization for Standardization: Geneva, Switzerland, 2006.
43. *ISO UNI EN ISO 14044*; Life Cycle Assessment—Requirements and Guidelines. International Organization for Standardization: Geneva, Switzerland, 2006.
44. Parmehr, E.G.; Amati, M.; Taylor, E.J.; Livesley, S.J. Estimation of Urban Tree Canopy Cover Using Random Point Sampling and Remote Sensing Methods. *Urban For. Urban Green.* **2016**, *20*, 160–171. [CrossRef]
45. Omodior, O.; Eze, P.; Anderson, K.R. Using I-Tree Canopy Vegetation Cover Subtype Classification to Predict Peri-Domestic Tick Presence. *Ticks Tick Borne Dis.* **2021**, *12*, 101684. [CrossRef] [PubMed]
46. Puplampu, D.A.; Boafo, Y.A. Exploring the Impacts of Urban Expansion on Green Spaces Availability and Delivery of Ecosystem Services in the Accra Metropolis. *Environ. Chall.* **2021**, *5*, 100283. [CrossRef]
47. Cristiano, S.; Ghisellini, P.; D'Ambrosio, G.; Xue, J.; Nesticò, A.; Gonella, F.; Ulgiati, S. Construction and Demolition Waste in the Metropolitan City of Naples, Italy: State of the Art, Circular Design, and Sustainable Planning Opportunities. *J. Clean. Prod.* **2021**, *293*, 125856. [CrossRef]
48. Endreny, T.; Santagata, R.; Perna, A.; de Stefano, C.; Rallo, R.F.; Ulgiati, S. Implementing and Managing Urban Forests: A Much Needed Conservation Strategy to Increase Ecosystem Services and Urban Wellbeing. *Ecol. Modell.* **2017**, *360*, 328–335. [CrossRef]
49. Liu, X.; Bakshi, B.R.; Rugani, B.; de Souza, D.M.; Bare, J.; Johnston, J.M.; Laurent, A.; Verones, F. Quantification and Valuation of Ecosystem Services in Life Cycle Assessment: Application of the Cascade Framework to Rice Farming Systems. *Sci. Total Environ.* **2020**, *747*, 141278. [CrossRef]
50. ISTAT. Crops: Areas and Production—Overall Data—Provinces. Available online: http://dati.istat.it/Index.aspx?QueryId=37850&lang=en (accessed on 15 September 2022).
51. Regione Campania. *Campania Un Mare Di Foreste*; Assessorato Agricoltura, Foreste, Caccia e Pesca: Naples, Italy, 2011.
52. Comune di Napoli. *Bilancio Arboreo e Gestione Del Verde Della Città Di Napoli 2011–2015*; Comune di Napoli: Naples, Italy, 2016.
53. Legambiente 2019—Ecosistema Urbano—Ecosistema Scuola. Available online: https://ecosistemi.legambiente.it/risultati2019/ (accessed on 15 September 2022).
54. Buttner, G.; Kosztra, B.; Soukup, T.; Sousa, A.; Langanke, T. *CLC2018 Technical Guidelines*; Environment Agency: Wien, Austria, 2017.
55. Heymann, Y.; Steenmans, C.; Croissille, G.; Bossard, M. *CORINE Land Cover. Technical Guide*; European Commission: Luxembourg City, Luxembourg, 1994.
56. Hirabayashi, S. I-Tree Canopy Air Pollutant Removal and Monetary Value Model Descriptions; 2014. Available online: https://www.itreetools.org/documents/560/i-Tree_Canopy_Air_Pollutant_Removal_and_Monetary_Value_Model_Descriptions.pdf (accessed on 1 June 2021).
57. ILCD. *General Guide for Life Cycle Assessment—Detailed Guidance*; Joint Research Centre: Ispra, Italy, 2010; ISBN 978-92-79-19092-6.
58. Pennington, D.W.; Potting, J.; Finnveden, G.; Lindeijer, E.; Jolliet, O.; Rydberg, T.; Rebitzer, G. Life Cycle Assessment Part 2: Current Impact Assessment Practice. *Environ. Int.* **2004**, *30*, 721–739. [CrossRef]
59. Wernet, G.; Bauer, C.; Steubing, B.; Reinhard, J.; Moreno-Ruiz, E.; Weidema, B. The Ecoinvent Database Version 3 (Part I): Overview and Methodology. *Int. J. Life Cycle Assess.* **2016**, *21*, 1218–1230. [CrossRef]
60. Huijbregts, M.A.J.; Steinmann, Z.J.N.; Elshout, P.M.F.; Stam, G.; Verones, F.; Vieira, M.; Zijp, M.; Hollander, A.; van Zelm, R. ReCiPe2016: A Harmonised Life Cycle Impact Assessment Method at Midpoint and Endpoint Level. *Int. J. Life Cycle Assess.* **2017**, *22*, 138–147. [CrossRef]
61. Maclaren, J.P. *Radiata Pine Growers' Manual*; FRI Bullettin No. 184; Scion: Rotorua, New Zealand, 1993.
62. Song, X.P.; Tan, P.Y.; Edwards, P.; Richards, D. The Economic Benefits and Costs of Trees in Urban Forest Stewardship: A Systematic Review. *Urban For. Urban Green.* **2018**, *29*, 162–170. [CrossRef]

63. OECD. Purchasing Power Parities (PPP) (Indicator). 2020. Available online: https://data.oecd.org/conversion/purchasing-power-parities-ppp.htm (accessed on 15 September 2022).
64. Rogers, K.; Sacre, K.; Goodenough, J.; Doick, K. *Valuing London's Urban Forest—Results of the London I-Tree Eco Project*; Treeconomics London: London, UK, 2015.
65. Jack Ruitenbeek, H. Functions of Nature: Evaluation of Nature in Environmental Planning, Management and Decision Making. *Ecol. Econ.* **1995**, *14*, 211–213. [CrossRef]
66. Nastran, M.; Pintar, M.; Železnikar, Š.; Cvejić, R. Stakeholders' Perceptions on the Role of Urban Green Infrastructure in Providing Ecosystem Services for Human Well-Being. *Land* **2022**, *11*, 299. [CrossRef]
67. Valeri, S.; Zavattero, L.; Capotorti, G. Ecological Connectivity in Agricultural Green Infrastructure: Suggested Criteria for Fine Scale Assessment and Planning. *Land* **2021**, *10*, 807. [CrossRef]
68. Evans, D.L.; Falagán, N.; Hardman, C.A.; Kourmpetli, S.; Liu, L.; Mead, B.R.; Davies, J.A.C. Ecosystem Service Delivery by Urban Agriculture and Green Infrastructure—A Systematic Review. *Ecosyst. Serv.* **2022**, *54*, 101405. [CrossRef]
69. Shao, H.; Kim, G. A Comprehensive Review of Different Types of Green Infrastructure to Mitigate Urban Heat Islands: Progress, Functions, and Benefits. *Land* **2022**, *11*, 1792. [CrossRef]
70. García-Pardo, K.A.; Moreno-Rangel, D.; Domínguez-Amarillo, S.; García-Chávez, J.R. Remote Sensing for the Assessment of Ecosystem Services Provided by Urban Vegetation: A Review of the Methods Applied. *Urban For. Urban Green.* **2022**, *74*, 127636. [CrossRef]
71. Primavesi, A.M. *Manejo Ecológico De Pastagens Em Regiões Tropicais E Subtropicais*; Expressão Popular: Sao Paulo, Brazil, 1985; ISBN 9788521303077.
72. Guerreiro, C.B.B.; Foltescu, V.; de Leeuw, F. Air Quality Status and Trends in Europe. *Atmos. Environ.* **2014**, *98*, 376–384. [CrossRef]
73. Cohen, A.J.; Anderson, H.R.; Ostro, B.; Pandey, K.D.; Krzyzanowski, M.; Künzli, N.; Gutschmidt, K.; Pope, A.; Romieu, I.; Samet, J.M.; et al. The Global Burden of Disease Due to Outdoor Air Pollution. *J. Toxicol. Environ. Health A* **2005**, *68*, 1301–1307. [CrossRef]
74. Li, J.; Mahalov, A.; Hyde, P. Simulating the Effects of Chronic Ozone Exposure on Hydrometeorology and Crop Productivity Using a Fully Coupled Crop, Meteorology and Air Quality Modeling System. *Agric. For. Meteorol.* **2018**, *260–261*, 287–299. [CrossRef]
75. Hamer, M.; Stamatakis, E.; Steptoe, A. Dose-Response Relationship between Physical Activity and Mental Health: The Scottish Health Survey. *Br. J. Sports Med.* **2009**, *43*, 1111–1114. [CrossRef]
76. Mitchell, R.; Astell-Burt, T.; Richardson, E.A. A Comparison of Green Space Indicators for Epidemiological Research. *J. Epidemiol. Community Health (1978)* **2011**, *65*, 853–858. [CrossRef]
77. Krieger, M.H. What's Wrong with Plastic Trees? *Science (1979)* **1973**, *179*, 446–455. [CrossRef]
78. Ecological Society of America What's Wrong with Plastic Trees? In the Field. 2007. Available online: https://blogs.nature.com/inthefield/2007/08/esa_whats_wrong_with_plastic_t.html (accessed on 1 June 2021).
79. Foucault, M.; Davidson, A.I.; Burchell, G. *The Birth of Biopolitics: Lectures at the Collège de France, 1978–1979*; Springer: Berlin/Heidelberg, Germany, 2008; ISBN 0230594182.

land

MDPI

Article

A Combined Methods of Senile Trees Inventory in Sustainable Urban Greenery Management on the Example of the City of Sandomierz (Poland)

Wojciech Durlak [1], Margot Dudkiewicz [2,*] and Małgorzata Milecka [2]

[1] Horticultural Production Institute, Faculty of Horticulture and Landscape Architecture, University of Life Sciences in Lublin, 28 Głęboka St., 20-612 Lublin, Poland
[2] Department of Landscape Architecture, Faculty of Horticulture and Landscape Architecture, University of Life Sciences in Lublin, 28 Głęboka St., 20-612 Lublin, Poland
* Correspondence: margot.dudkiewicz@up.lublin.pl; Tel.: +48-81-531-96-70

Abstract: The sustainable management of urban greenery consists, among others, of the inventory, valuation, and protection of trees of monumental size. This article presents the results of the inspection of 13 large trees growing in the city of Sandomierz, located in south-eastern Poland. The examined specimens belong to five species: Norway maple (*Acer platanoides* L.), common ash (*Fraxinus excelsior* L.), white poplar (*Populus alba* L.), English oak (*Quercus robur* L.), and small-leaved lime (*Tilia cordata* Mill.). The health condition of the trees was assessed using acoustic and electrical tomography, as well as chlorophyll fluorescence tests. Diagnostics employing sound waves and electrical resistivity were crucial in assessing tree health. The data based on chlorophyll fluorescence confirmed the results obtained during tomographic examinations. It was an innovative combination of three non-invasive methods of examining the health condition of trees and their valuation. Economic valuation allows us to reduce to common denominator issues that are often difficult to decide due to different perspectives—expressing the economic value of trees. Calculating the value of trees allowed us to show the city's inhabitants the value of trees that are of monumental size. Thanks to the cooperation of scientists with the city authorities, an economic plan for trees of monumental size was created, distinguished by an individualized and holistic approach to each specimen covered by the study. The database prepared has a chance to become an effective management instrument used by environmental protection authorities and a source of knowledge and education for the city's inhabitants.

Keywords: sustainable management; urban greenery; senile trees; monumental trees; Picus Sonic Tomograph 3; Picus TreeTronic; fluorometer; Sandomierz; Poland

Citation: Durlak, W.; Dudkiewicz, M.; Milecka, M. A Combined Methods of Senile Trees Inventory in Sustainable Urban Greenery Management on the Example of the City of Sandomierz (Poland). *Land* **2022**, *11*, 1914. https://doi.org/10.3390/land11111914

Academic Editors: Alessio Russo and Giuseppe T. Cirella

Received: 16 September 2022
Accepted: 24 October 2022
Published: 27 October 2022

1. Introduction

Finding a balance between social, ecological, and economic aspects is one of the most significant challenges in sustainable city management [1–3]. An important problem is managing and shaping properly functioning urban green areas with the preservation of trees of monumental size. The most important management procedure in taking care of trees of monumental size, which are commonly referred to as veteran trees as they usually are old as well, is to think of how to improve their living conditions, e.g., by improving and protecting their rooting area, or gradually improving their light acquisition [4,5]. Senile and veteran trees should be given adequate space; they need enough light and room for the trunk and crown and ample below-ground rooting and soil volume. They should have the opportunity to continue to grow old and be left alone where possible, in favorable surroundings. Sometimes, the care of trees of monumental size includes pruning to remove dead branches, cutting crowns to reduce their weight and volume, and as bracing and other safeguards. Monitoring and control should be carried out not only through visual assessment (VTA) but also with the use of modern technologies. Likewise, protecting historic trees includes entering them into the register of natural monuments [6,7].

Hazardous trees regularly lead to the death or injury of arborists and property owners. It is worth mentioning in this context that the ISA has created the Tree Risk Assessment Qualification (TRAQ). TRAQ promotes the safety of people and property by providing a standardized and systematic process for assessing tree risk. The results of a tree risk assessment can provide tree owners and risk managers with information to help make informed decisions to enhance tree benefits, health, and longevity [8].

This study applies to an important theme in the urban environment, the management of senile trees, which are very valuable for biodiversity, human well-being, and cultural history. The objective of this study was to formulate recommendations for senile trees based on the results of their condition assessment obtained with the use of three non-invasive methods. Studying the condition of the trees is the first step of making a management plan to preserve these trees safely in the environment. The manuscript studied different methods for evaluating the condition of the trees and a method for calculating the tree replacement value.

Need to Protect Older Trees in Cities

In difficult urban conditions, a significant number of trees die or are removed before they reach full maturity [9,10]. When trees grow old, they are considered to be 'senile'. The decline of trees starts with a sparse appearance, yellowing and other types of foliage symptoms, undergrowth, and sickly appearance and dried-up top growth. The branches of trees start to die from the top downwards. External features include the spiral grain in a tree's trunk, thin or balding bark, the loss of apical dominance, crown dieback, and crowns with a few, large limbs, among others [11–13]. Pederson described six external characteristics of potentially old angiosperm trees. These characteristics are found in the stems and crowns of trees and include: (1) smooth bark; (2) low stem taper; (3) high stem sinuosity; (4) crowns comprised of few, thick, twisting limbs; (5) low crown volume; and (6) a low ratio of leaf area to trunk volume. Admittedly, characteristics (4) and (5) are often correlated; a crown with low volume often has only a few twisting limbs. The greater the number of these characteristics in one tree, the greater the likelihood that the tree is old [14].

Senile trees are precious for a place's nature, landscape, culture, and tradition. Unfortunately, more and more often, it turns out that, due to the rapidly progressing urbanization of the landscape, they grow in collision with the urban infrastructure. As a result of which, they are often mechanically damaged or deformed as a result of improperly carried out works by municipal services dealing with infrastructure, and, as a result, they enter the dieback phase too early with a distinct dryness in the crown. Here, it should be emphasized that, most often, measures of crown dieback or chlorosis are used. As a result, they are vulnerable to disposal due to the safety risk to people and property. Old trees, however, are significant witnesses to the history of towns and villages [14,15]. They are a practical component of the cultural landscape—they are often painted, photographed, and described. Moreover, they are also an essential element of the natural world. They are the habitat of rich biological and microbial life. According to British data, over 2000 species of invertebrates (6% of British invertebrate fauna) depend on the habitats of senile trees; therefore, the removal of such trees seriously affects the biodiversity of towns and villages [16,17].

Urban green spaces regulate regional microclimate via shading, evapotranspiration, boosting air movements, and increasing heat exchange, which mitigates urban heat island (UHI) effects at a city scale [18–20]. Furthermore, vegetation in green spaces can not only intercept rainfall and reduce rainfall runoff, but it also brings more rainfall infiltration, which decreases the frequency of urban floods and stormwater treatment costs and damages [21]. Most studies at regional or city scales show a modest modeled reduction in pollution concentration of less than 5% resulting from urban vegetation [22,23]. Trees increase both the surface roughness (slowing air flow thus enhancing deposition and absorption pollutant removal processes) and the area of the ground surface that atmospheric pollutants come into contact with (acting as biological filters, enhanced by the properties of their surfaces) [24]. The introduction of trees within a street canyon also has the potential to significantly alter the soundscape by generating sounds associated with the rustling of leaves in response to wind and attracting bird wildlife sounds that would be rated more positively than a street canyon dominated by road traffic noise [25]. Moreover, trees growing in cities are an element that humanizes space [26].

Several aspects of physical health have been shown to be correlated with aspects of urban "greenery", such as mortality, longevity, heart rates, and weight changes [27,28]. There are also numerous studies relating aspects of mental health to the prevalence of vegetation [29,30]. Responses of human health to actual and perceived biodiversity have been generally categorized as those that cause or reduce harm and that restore or build the capacity for physical and mental health [31]. Some urban residents are highly deprived of virtually any access to nature, such that even modest additions have been shown to have measurable positive effects [32,33]. However, it should be noted that different people may have divergent views. The owners of the property may have an entirely different opinion regarding the same tree, seeing only the inconvenience in the development and maintenance of the property caused by its shading by the crown of the tree or falling leaves or appreciating the beauty of the specimen and its environmental value. Still, other qualities are noticed by someone who looks at a given tree from the street or a window of an apartment in a nearby building and sees it, above all, as an indispensable and permanent element of space, a sentimental landscape. Such a subjective perception of a tree can be related to its value, understood in social, psychological, and economic categories [34–36]. Thanks to such a view of nature, it is easier to reach the imagination of, for example, politicians or city authorities, or, it may be the only way most people will realize that a functioning environment is more valuable than a destroyed one. It is also worth making society aware that the landscape we live in is about ourselves and our culture.

Sustainable landscape management needs to take into account both the natural values of the landscape (e.g., trees) and the human values. The combination of visual inspection (VTA), acoustic and electrical tomography, and stress testing with a fluorometer is a practical approach to measuring the health of various tree species. Thanks to the use of sound tomography, the threats can be quickly diagnosed and restoration work can be carried out. However, the decision-making process should be accompanied by a process of social participation. Social awareness of the value of trees is needed to manage senile trees in the city space and to support various decisions of officials and arborists (e.g., entering into the register of monuments or tree maintenance procedure).

2. Materials and Methods

The research presented in this article shows a combination of several non-invasive methods for detecting damage and cavities inside trunks and the resistance of the trunk to the fracture of city trees. The combination of visual inspection (VTA), acoustic tomography, electrical tomography, and stress testing with a fluorometer is an effective instrument to show the health status of various tree species. It is worth noting here that the evaluation of the statics and condition of trees are two different but important issues. The developed procedures enable the analysis of endangered trees and can improve the risk assessment process near the tree or the risk caused by it. In the presented study, the economic value of trees was also estimated.

The study covered 13 trees (12 nature trees of monumental size, and one tree is a candidate tree) growing in Sandomierz, belonging to five species: white poplar (*Populus alba* L.)(4), pedunculate oak (*Quercus robur* L.) (4), small-leaved lime (*Tilia cordata* Mill.) (3), ash(*Fraxinus excelsior* L.) (1 pc.), and Norway maple (*Acer platanoides* L.)(1 pc.).The trees were selected for research due to the fact that they are all natural monuments of the city of Sandomierz. They grow in various spatial situations: parks, sacred areas, school areas, and as street greenery. The research was carried out in 2021. The study used quantitative and qualitative data processing methods. The following numerical parameters characterizing the tested trees were determined: circumference at the height of 130 cm (cm), the range of the crown (m), and the height of the tree (m). The crown's reach was measured in two directions, N-S and E-W, and the obtained results were averaged. The height of the trees was measured with a Nikon Forestry Pro laser rangefinder. The location of trees was determined based on GPS positioning. When describing the trees, particular attention was paid to the health condition of the trunk (possible necrotic strips, cavities, rot, traces of insect feeding, mushroom fruiting bodies, trunk tilt) and crowns (drought, broken branches, asymmetry) and to the prevailing habitat conditions. The condition of the immediate vicinity of the tree was documented, such as the type of surface, the proximity of buildings, and the intensity of pedestrian and vehicle traffic. The historical and cultural valuesand the role of trees in a location were assessed, e.g., certain individual trees have significance in relation to specific persons or historical events, or some specimens of trees are remnants of old gardens and parks that no longer exist.

2.1. *Sandomierz—General Information and Location of Research Tress*

Sandomierz is a historic city in south-eastern Poland, situated on the Vistula River, with 23,000 residents. Numerous tourist routes run through the city, including the Cistercian route, St. Jakub, Via Jagiellonica, Wooden Architecture, and the Eastern Bicycle Trail "Green Velo." There are many monuments in Sandomierz, and the Old Town is a well-known historical urban, architectural, and landscape complex.

In the area of Sandomierz, there are 12 specimens of old trees under protection in the form of nature monuments (Figure 1). In Poland, one of the formal attributes of a monument tree is the circumference. The trees assessed grow in different parts of the city (Figure 2). There are three natural monuments in Piszczele Park, one Norway maple and two white poplars. The Norway maple (*Acer platanoides* L.) (1) grows in high densities on a slope among other trees on Emmendingen Avenue. Two magnificent white poplars (*Populus alba* L.) located near the main pedestrian and bicycle path leading through the park are also monumental (2 and 3). In Saski Park, the largest in Sandomierz, located near the Opatowska Gate, on the side of J. Słowacki Street, there is a monumental pedunculate oak (*Quercus robur* L.) (4). The remaining trees of monumental sizeare located near historic sacred buildings, e.g., the old small-leaved lime (*Tilia cordata* Mill.) grows at the entrance to the Dominicans' property from the side of St. James, in the corner of the brick fence (5). The second of the limes (6) is located at the very entrance to the St. James church. On the other hand, at the bell tower of the church of St. Paul the Apostle, the common ash (*Fraxinus excelsior* L.) grows (7), and in the vicinity of the presbytery buildings, there is an imposing English oak (*Quercus robur* L.) (8). Within the Old Town of Sandomierz, there is another natural monument, a small-leaved lime (*Tilia cordata* Mill.) (9) growing near the Collegium Gostomianum building (now High School No. 1 in Sandomierz) at J. Długosz Street. However, at Milbert Street, a magnificent pedunculate oak (*Quercus robur* L.) grows (10). The following two monuments of nature are the white poplars (*Populus alba* L.) (11 and 12) growing in the right-bank part of Sandomierz on the playground at K. K. Baczyński Street. A specimen that meets the conditions of a massive tree was also studied, an English oak (*Quercus robur* L.) (13) from Opatowska Street, which was proposed for protection as a natural monument.

Figure 1. Trees of monumental size in Sandomierz. (**1**) *Acer platanoides* in Park Piszczele; (**2**) and (**3**) two trees of the species *Populus alba* in Park Piszczele; (**4**) *Quercus robur* in Saski Park, (**5**) and (**6**) two trees of the *Tilia cordata* species at the church of St. James; (**7**) *Fraxinus excelsior* at the church of St. Paul; (**8**) *Quercus robur* at the rectory of the Church of St. Paul; (**9**) *Tilia cordata* at Collegium Gostomianum; (**10**) *Quercus robur* at Milbert Street; (**11**,**12**) two *Populus alba* trees at Baczynski Street; (**13**) *Quercus robur*, acandidate for a natural monument, Opatowska Street (photo by M. Dudkiewicz, 2021).

Figure 2. Location of trees of monumental size in the city of Sandomierz (made by M.Milecka).

2.2. Measures

2.2.1. Tree Age Assessment

Several methods were used to calculate the age of trees: age tables by Majdecki [37]; the standard formula for assessing the age of trees (trunk circumference 130 cm/2.5 = tree age); age tables of Mydłowska [38]; and a tree age calculator [39]. It should be noted that the age of the trees is visually difficult to estimate. The significant emerging discrepancies in assessment result mainly from different calculation methods. Therefore, it is best to consider similar results as the most reliable. Trees from the same planting period may often differ in the circumference of the trunks. This state of affairs can be influenced by many factors, such as habitat conditions, soil fertility, soil moisture, growth rate, or genetic factors.

2.2.2. Acoustic Tomography

In the case of old trees, diebacks are the leading cause of their fall or breakage. Structurally weak specimens are more susceptible to strong wind or heavy snow. As weather anomalies resulting from climate change occur more frequently, the number of incidents and financial damage caused by falling trees in cities is increasing. Internal flaws are difficult to assess by visual inspection alone, which makes managing and preparing for the appropriate risks difficult [40]. The method of risk assessment VTA—based on the visual assessment of an expert—is often insufficient to provide a reasonable opinion [41–43].

Generally, two types of methods are used to diagnose internal defects in tree trunks or branches: invasive and non-invasive. The first one involves the use of devices that interfere with the internal tissues of the tree, which include *Presler auger*. The second, however, does not damage the wood structures and uses acoustic or electromagnetic waves. In the case of old trees, especially those designated as protected trees, the physical damage to trees should be minimized, and non-invasive diagnostic methods should be used. Therefore, diagnostic examinations of tree trunk health conditions were carried out using the Picus Caliper and Picus Tomograph (sonic and electric) of the German company Argus Electronic GmbH (Figures 3 and 4).

Figure 3. Stem geometry measurement with the electronic PiCUS Caliper (photo by M. Dudkiewicz, 2021).

Figure 4. PiCUS sonic tomograph central unit and tree trunk measurement (photo by M. Dudkiewicz, 2021).

Acoustic tomography enables the non-invasive measurement of the internal structures of a tree trunk in terms of detecting rot or decay using sound waves, without the need to make drills that are harmful to the tree [44,45]. It is worth remarking that a tomograph study considers the condition of wood only at the very point of cross section of the measurement, not any higher or upper (of course, it can be used to conclude the condition of the trunk in general).Acoustic tomography consists of a central unit, sensors placed around the tree trunk and connected with pins driven shallowly into the trunk, and specialized software. Tomographic examinations were carried out at a height of 50 to 160 cm above the ground, depending on the tree. The optimal test height was determined on the basis of tapping the trunk with a rubber mallet at different levels beforehand. Then, after interpreting the appearing noises, a decision was made on the level of distribution of the measurement sensors on the tree trunk. Sensors measure the time of the sound wave propagation in the wood caused by the impact of an electronic hammer. The distances between the sensors are measured using a special PiCUS Caliper operating in a Bluetooth system, which allows for a detailed mapping of the shape of the tree trunk. The obtained and processed results are a colored tomogram showing the changes inside the tree trunk. At this point, it is worth noting that a tomogram taken at a single time shows the conditions at the time of measurement. Determining a change in stem characteristics requires repeated measurements over time. When analyzing the measurement results, particular attention should be paid to the color of the obtained image, which determines the so-called wood density map. Different colors mean different speeds of sound propagation inside the trunk, depending on the elasticity and density of the wood. The light brown to black color corresponds to the range of the speed of sound from 60 to 100% of the highest speed of sound, which means a living and healthy wood tissue. Various shades of green color correspond to the speed of the sound wave from 40 to 60% of the speed of sound at an average level, which is equivalent to a slight deterioration of the structure of the wood. The color pink means the propagation of sound from 20 to 40%, and a color from blue to white is in the range of 0–20% (the slowest speed of sound). Therefore, these are the areas with the weakest structure, where there is damage and intense decay of wood. The yellow lines on the cross-section of the trunk suggest the appearance of internal cracks. Internal cracks are not automatically dangerous. It depends on their location and the structure and condition of the tree/trunk/branch [46,47].

To obtain more precise results, sonic tomography with electrical tomography is suggested. As both research methods complement each other, the result is a detailed image of the inside of the tree trunk [48,49].

2.2.3. Electric Tomography

The PiCUS TreeTronic (Figure 5) is a device that uses an electrical voltage to define areas within a tree trunk of different electrical resistivity. The result of such a study is a two-dimensional map of the impedance of the wood called the electrical resistivity tomogram. The electrical resistivity of wood depends on various factors. These include the tissue's water content, the concentration of elements (ions), or the structure of cells (the so-called reaction wood has a different resistivity than regular wood). As in the case of sonic tomography, the obtained tomogram is in the form of a colored map of the trunk's cross-section, on which the marked colors correspond to a specific value of electrical resistivity. Blue indicates low-resistivity areas characterized by high water content in the tissues. Green and yellow indicate increased resistivity, and red corresponds to high resistivity (low water content, etc.). To analyze the electrical resistivity tomogram, three types of resistivity in trees are considered: Type 1, Type 2, and Type 3.

Type 1 is the low resistivity (high conductivity = blue colors) of the areas facing outwards and the high resistivity (lower conductivity = red colors) of the areas towards the inside of the tree trunk. Each species has a typical wet and dry distribution. Most European tree species, i.e., maple, chestnut, birch, chestnut, beech, ash, poplar, willow, rowan, linden, oak, elm, larch, pine, spruce, and many others, are classified as Type 1. Such trees usually have lower sapwood (blue) resistivity on the edges and higher resistivity in heartwood (red). Chestnut trees or poplars often develop inner wet wood, even when young. Wet wood changes the resistivity of the heartwood. However, it should be considered that the detected early stages of wet wood do not threaten the trunk's stability. Type 2 is high resistivity (low conductivity = red) to the outside and low resistivity (high conductivity = blue) towards the inside of the tree trunk. Type 3 shows a ring system, where high resistivity (low conductivity = red) covers the middle part of the trunk and low resistivity (high conductivity = blue) is located in the area facing the outside of the trunk and its core part. It is a typical arrangement for healthy trees, e.g., English oak [50].

Figure 5. Tree measurement with the Picus TreeTronic electric tomograph (photo by M. Dudkiewicz, 2021).

2.2.4. Measure the Photosynthesis Process

Using the OS5p+ fluorometer produced by the American company Opti Sciences, the stress level in the examined trees was measured, which allowed for assessing their condition. The measurement of chlorophyll fluorescence and the analysis of this phenomenon is a recognized method that allows you to quickly and non-invasively check the photosynthesis process and determine the health condition and the occurrence of stress phenomena in plants. Clips with a blackout tab were placed on the leaves in different parts of the crown of each tree. After 30 min, chlorophyll fluorescence was measured.

The fluorometer is a device that allows quick and non-invasive measurements of the photosynthesis process (Figures 6–8). The measurement of chlorophyll fluorescence is particularly useful for assessing the health condition of plants and in situations where various environmental factors affect them [51–53]. The clips were attached, 5 pieces per tree, at equal intervals around the crown, at crown base height (CBH).The measurement procedure is based on placing a unique clip on the leaf blade, closed for about 20–30 min (dark time needed for the measurement). Then, the measurement is made by inserting the probe tip into the clip and opening the window. The display shows the results defining the level of the respective coefficients. One of them (F_V/F_M determinant, maximum photochemical efficiency) is commonly used as a reliable diagnostic indicator, among others, to evaluate the impact of stressful conditions on plants [54]. In conditions of total growth and the absence of stress factors in plants, this index, depending on the species, ranges from about 0.83 to 0.85 relative units [55]. The action of biotic and abiotic stressors reduces the quantum efficiency of the photosystem II (PSII) and its value. Very low indices of this parameter—at the level of 0.20–0.30—indicate the occurrence of irreversible changes in the PSII structure [56]. It should be pointed out that fluorescence measurements mostly consider the functioning of the photosynthetic system (especially PSII) in the leaves and indicate the effect of some stresses but not all, e.g., it does not consider the condition of the trunk or branches, i.e., about the structural stability at all. There may be internal decay in the trunk, but the outer layers of conductive tissues are perfectly functioning and the photosynthetic system is working fine, and although the chlorophyll measurements are not good, this should not be a cause for concern.

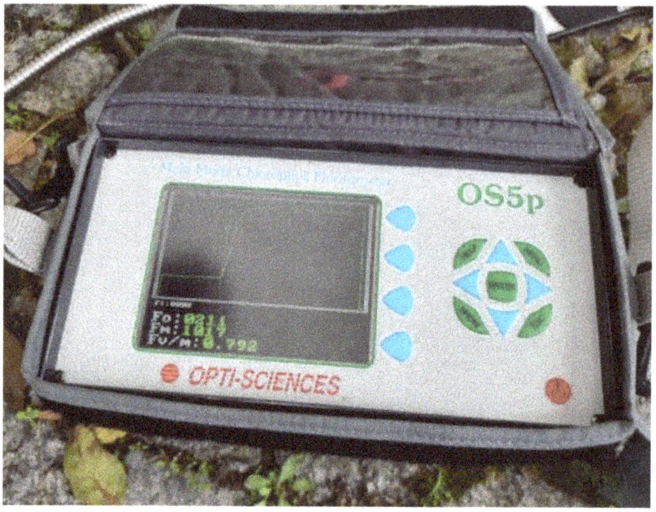

Figure 6. Advanced Fluorometer OS5p+ to measure the photosynthesis process (photo by M. Dudkiewicz, 2021).

Figure 7. Measurement clip (photo by M. Dudkiewicz, 2021).

Figure 8. The moment of taking the measurement (photo by M. Dudkiewicz, 2021).

2.2.5. Economic Valuation of Trees

Economic valuation methods are used around the world to give an economic value to trees. Many regions and countries in Europe have produced their own economic valuation model, e.g., CAVAT in the UK, VAT03 in Denmark, Koch's method in Germany, and Norma Grenada in Spain, and the same trend can be seen outside of Europe, such as with CTLA in the USA and the Revised Burnley Method in Australia. The valuation of the value of trees was performed based on the method developed by the team of authors led by Szczepanowska [57,58]. In this case, the tree valuation method used in Poland is based on comparative analyses of foreign methods, including in the EU countries, and on domestic analyses. It uses American solutions promoted by the CTLA to a large extent, taking into account the conditions prevailing in Poland. However, it should also be pointed out that this method does not take into consideration the value of the ecosystem services of the tree nor the biodiversity effect, but it is purely a tree replacement cost. Thus, in the case of veteran trees with high biodiversity value, it unfortunately underestimates the actual value of the tree.

The method assumes that growing trees constitute a permanent value (this is confirmed by the content of Article 48 of the Civil Code, which states that trees constitute a component of the land from the moment of planting or sowing). The basis for the valuation of trees in the method is the cost of tree replacement. The method distinguishes between the following two situations: when a damaged tree can be physically replaced by a comparable tree available on the market (then the basis of the valuation are the costs of a tree analogous to a growing tree) and when the tree cannot be physically restored (in such a case, the replacement cost is considered as the basis for the valuation of trees available as nursery material, additionally enlarged using increment factors). The coefficients are determined for a given tree size, considering the tree's species value, growth rate, and adaptation to urban conditions.

When assessing the value of the stand, the following formula was used:

$$RWD_{>25} = W \times G \times P \times K \times L$$

where:

$RWD_{>25}$—actual value of a tree with a trunk circumference of more than 25 cm;

W—base value of the tree;

G—tree species value (tabular list of species, the value of the tree is higher for e.g., oak and lower for poplars);

P—tree growth rate coefficient (tabular list of species);

K—tree health factor;

L—tree location factor [58–60].

The obtained results concerning the actual value of trees are presented in Table 1.

Table 1. Results of the dendrological survey of trees of monumental size in Sandomierz (by authors).

No.	Species Name	The Circumference of the Trunk at the Height of 1.3 m (cm)	Height (m)	Crown Reach (m)	Localization	Age	Mean Age	Value (Euro) *
1	Norway maple (*Acer platanoides* L.)	275	13.2	EW 18.6m NS 15 m Mean: 16.8 m	Piszczele Park 50°40′56.9″ N 21°44′40.1″ E	157 [1] 110 [2] 157 [3] 158 [4]	145	6820.50
2	White poplar (*Populus alba* L.)	577	36.6	EW 18.8 m NS 23.7 m Mean: 21.2 m	Piszczele Park 50°40′57.8″ N 21°44′25.0″ E	152 [1] 230 [2] 152 [3] 260 [4]	198	11,674.03
3	White poplar (*Populus alba* L.)	472	35	EW 22.2 m NS 24 m Mean: 23.1 m	Piszczele Park 50°40′57.6″ N 21°44′26.0″ E	124 [1] 188 [2] 124 [3] 213 [4]	162	8284.30
4	English oak (*Quercus robur* L.)	397	17.6	EW 25.5 m NS 24.5 m Mean: 25 m	Saski Park 50°40′56.3″ N 21°44′38.1″ E	275 [1] 158 [2] 275 [3] 313 [4]	255	21,723.34
5	Small-leaved linden (*Tilia cordata* Mill.)	390	10.8	EW 12.8 m NS 11.7 m Mean: 12.2 m	At the entrance to the vineyard and the Church of St. James 50°40′37.1″ N 21°44′41.0″ E	162 [1] 156 [2] 162 [3] 268 [4]	187	13,536.18
6	Small-leaved linden (*Tilia cordata* Mill.)	384	8.4	EW 10 m NS 9 m Mean: 9.5 m	At the entrance to the Church of St. James 50°40′37.8″ N 21°44′39.3″ E	159 [1] 153 [2] 159 [3] 264 [4]	184	9475.32
7	Common ash (*Fraxinus excelsior* L.)	310	22.2	EW 21 m NS 21 m Mean: 21 m	At the Church of St. Paul 50°40′57.6″ N 21°44′26.0″ E	165 [1] 124 [2] 164 [3] 175 [4]	157	16,547.62
8	English oak (*Quercus robur* L.)	525	18.4	EW 27 m NS 28 m Mean 27.5 m	At the rectory of the Church of St. Paul 50°40′40.2″ N 21°44′22.9″ E	364 [1] 210 [2] 364 [3] 414 [4]	338	49,692.52
9	Small-leaved linden (*Tilia cordata* Mill.)	419	13.8	EW 14.2 m NS 18.9 m Mean: 33.1 m	At the Collegium Gostomianum 50°40′39.2″ N 21°45′6.2″ E	174 [1] 167 [2] 174 [3] 288 [4]	201	16,976.20
10	English oak (*Quercus robur* L.)	532	18.4	EW 20.7 m NS 18.5 m Mean: 19.6 m	Milberta St. 50°42′3.8″ N 21°45′0.4″ E	369 [1] 212 [2] 369 [3] 384 [4]	333	23,378.06
11	White poplar (*Populus alba* L.)	465	18.6	EW 17.7 m NS 17 m Mean: 17.3 m	Baczyńskiego St. 50°39′41.4″ N 21°44′51.9″ E	122 [1] 186 [2] 123 [3] 210 [4]	160	12,051.42
12	White poplar (*Populus alba* L.)	430	17.8	EW 18.3 m NS 15 m Mean: 16.6 m	Baczyńskiego St. 50°39′41.1″ N 21°44′52.2″ E	113 [1] 172 [2] 113 [3] 194 [4]	148	10,804.72
13	English oak (*Quercus robur* L.)	357	16.4	EW 25.3 m NS 26.5 m Mean: 25.9 m	Opatowska St. 50°40′51.6″ N 21°45′0.1″ E	247 [1] 142 [2] 247 [3] 281 [4]	229	44,574.54

* converted from PLN to EUR at the rate of 1 Euro = 4.70 PLN of 20.07.2022.; [1] according to Majdecki's age table; [2] according to the formula, circumference/2.5 = age of the tree; [3] according to Mydłowska's tables; [4] http://www.tree-guide.com/tree-age-calculator (accessed on 26 October 2022).

3. Results

3.1. Trees and Examination

The first step in the research was to assess each tree to visualize signs of structural instability, mechanical damage, or fungi and prepare the necessary photographic documentation.

The results of the dendrometric tests are presented in Table 1.

3.2. Results of Computed Tomography

Since chemical properties change earlier than physical properties, combining acoustic and electrical tomography enabled a more advanced assessment of the internal tree trunk structures. In numerous instances, the measurement result can be used to identify the type of rot or determine whether the inside of the trunk is moldy or infected with bacteria. In addition, the result of the electrical test also provides information on the degree of moisture in the wood. A low impedance value indicates a high water content. The distribution of areas with different resistivity (different humidity) can also be used to analyze the efficiency of the root system and water transport inside the tree.

To find out more about these results and to further explain them or create a clearer image, the collected data are presented in the Supplementary File S1. Table 2 summarizes the results of tomographic examinations in the form of generated tomograms.

Table 2. Results of the computed tomographyof senile trees in Sandomierz(by authors).

No.	Species Name and Locatization	Picus Sonic Tomograph 3	PicusTreeTronic
1.	*Acer platanoides* L. Piszczele Park	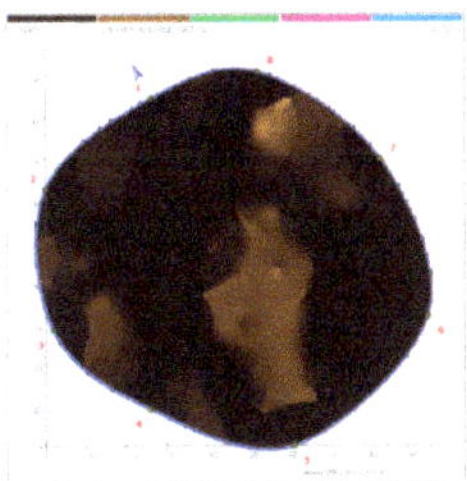	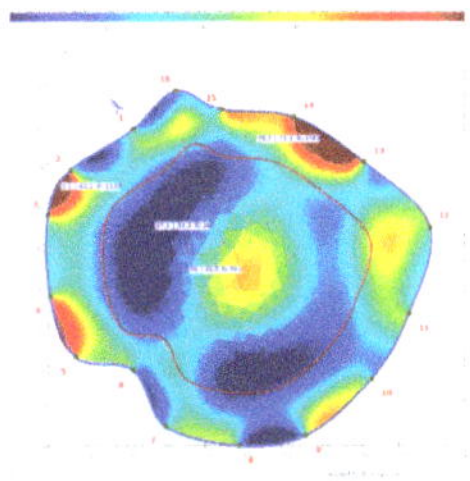
2.	*Populus alba* L. Piszczele Park	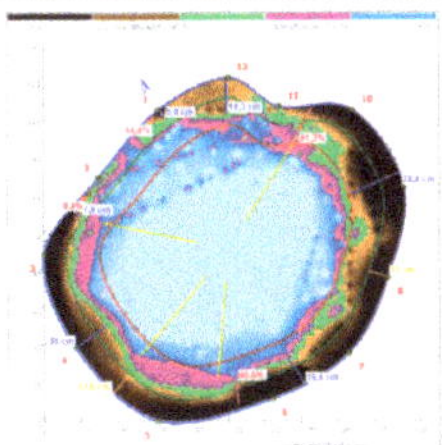	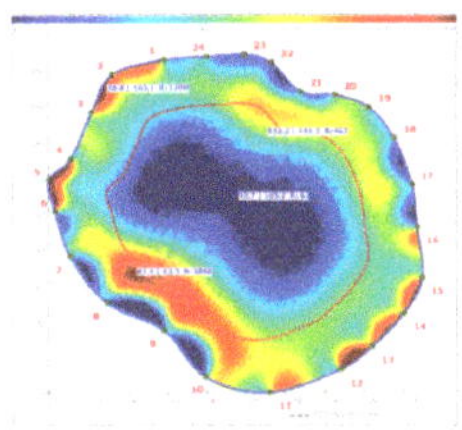
3.	*Populus alba* L. Piszczele Park	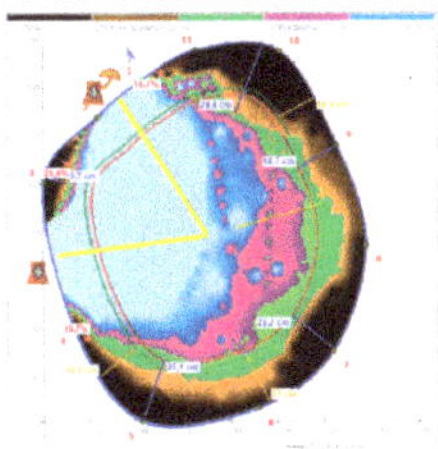	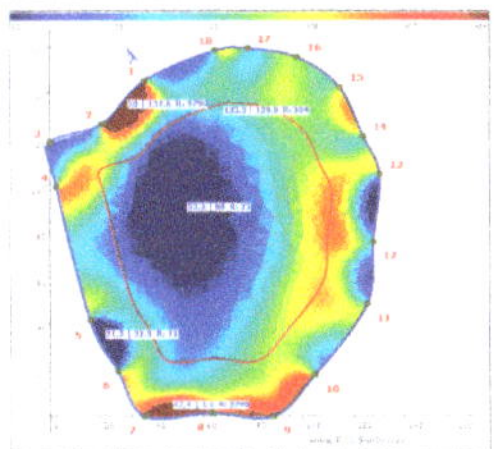

Table 2. *Cont.*

No.	Species Name and Locatization	Picus Sonic Tomograph 3	PicusTreeTronic
4.	*Quercus robur* L. Saski Park	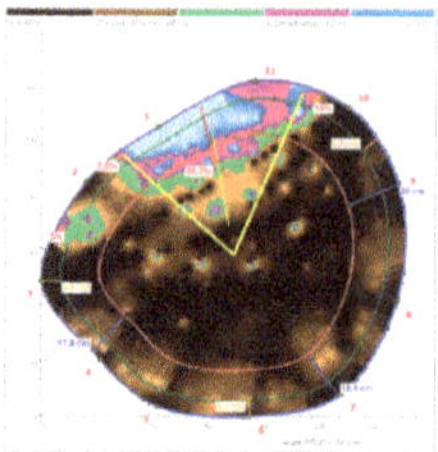	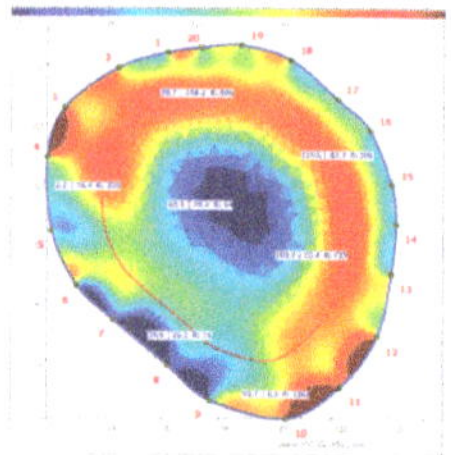
5.	*Tilia cordata* Mill. At the entrance to the vineyard and the Church of St. James	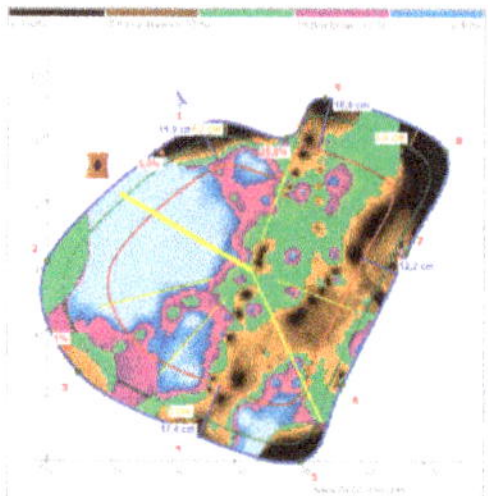	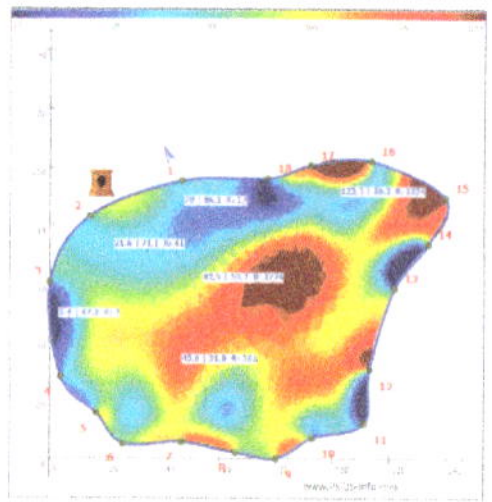
6.	*Tilia cordata* Mill. At the entrance to the Church of St. James	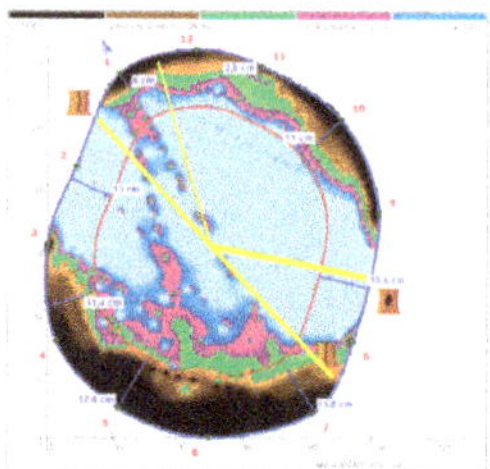	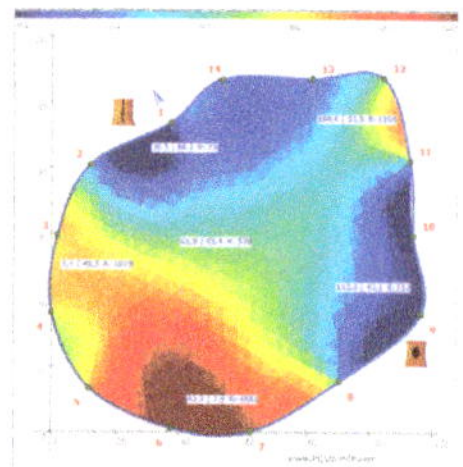
7.	*Fraxinus excelsior* L. At the church of St. Paul	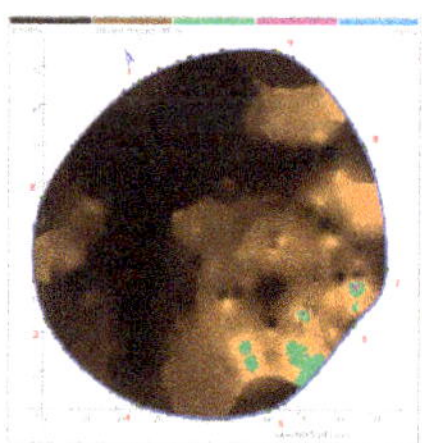	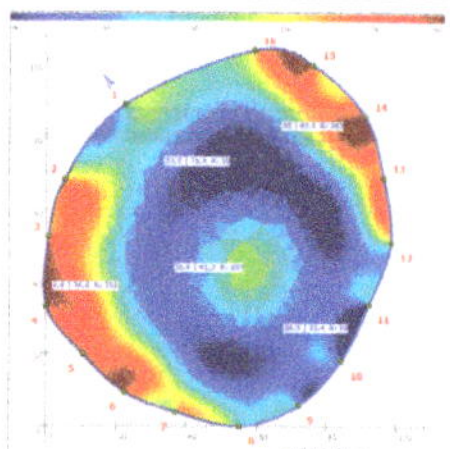
8.	*Quercus robur* L. At the rectory of the Church of St. Paul	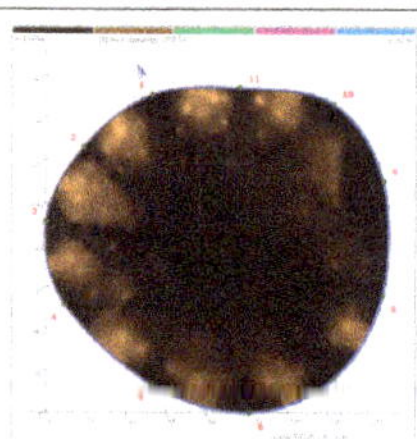	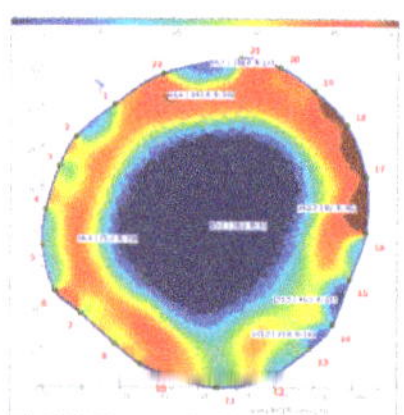

Table 2. *Cont.*

No.	Species Name and Locatization	Picus Sonic Tomograph 3	PicusTreeTronic
9.	*Tilia cordata* Mill. At the Collegium Gostomianum	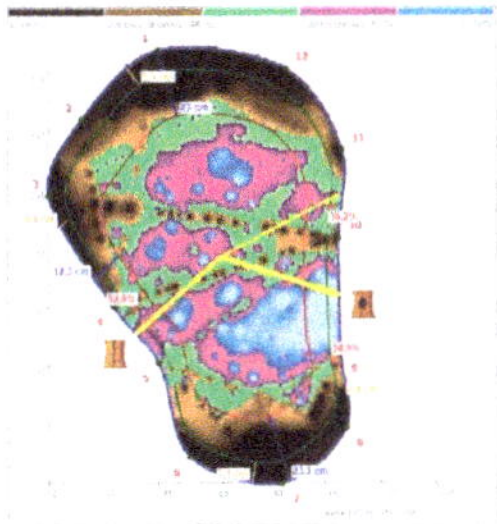	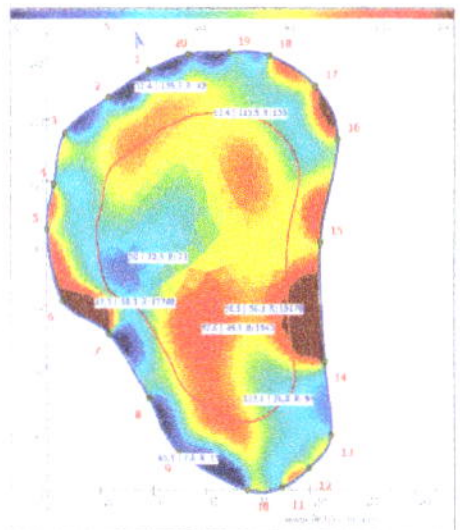
10.	*Quercus robur* L. Milberta St.	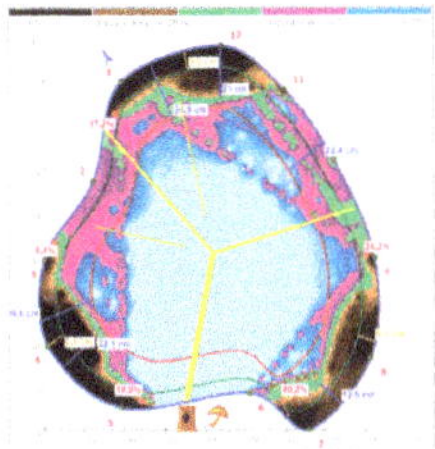	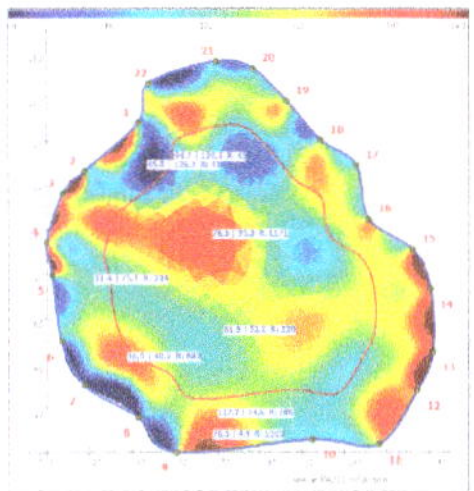
11.	*Populus alba* L. Baczyńskiego St.	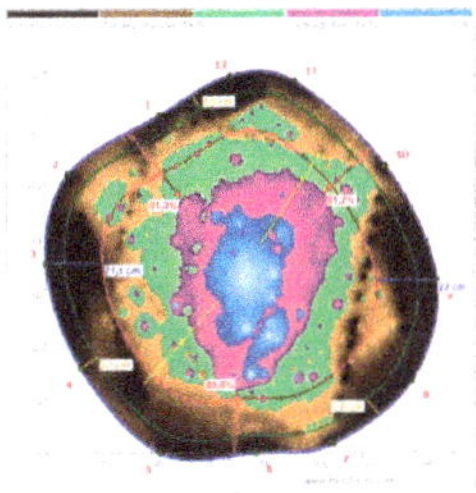	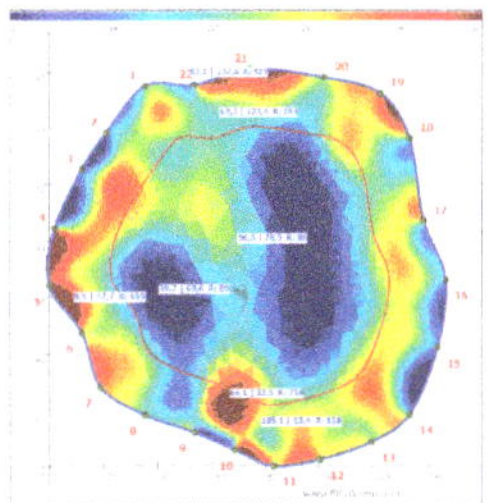
12.	*Populus alba* L. Baczyńskiego St.	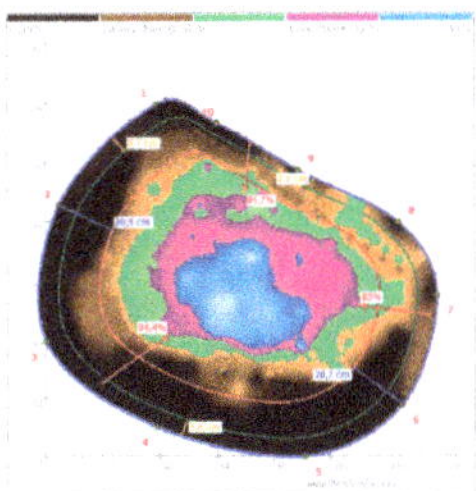	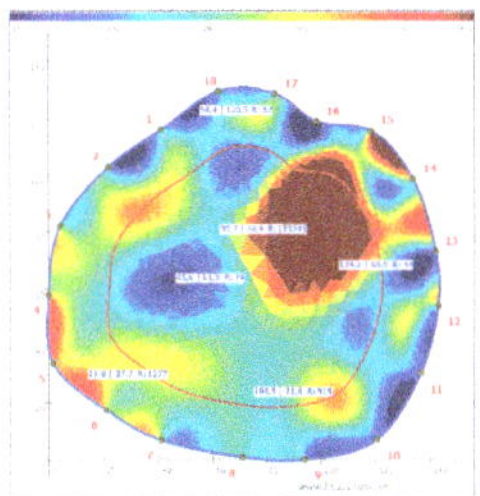
13.	*Quercus robur* L. Opatowska St.	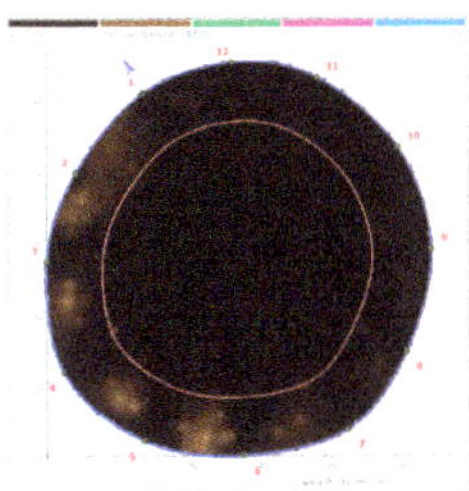	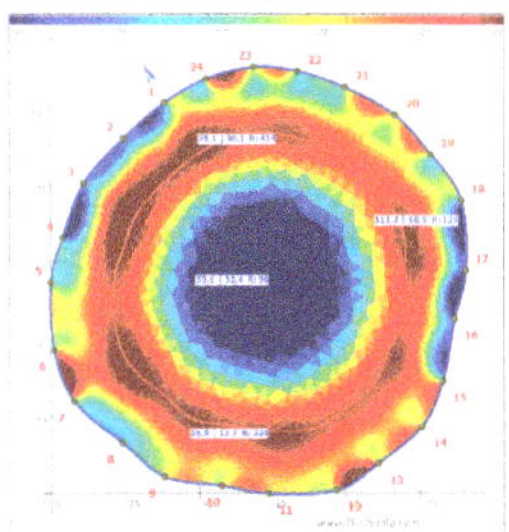

The Effect of Hollowness on the Statics of Senile Trees

Scientists have developed a large body of literature about tree biomechanics. In the 1960s, arborists began to describe trees as engineered structures, using equations and terms such as modulus of rupture, applied bending moment, and lever arm. Biomechanical experiments are designed to quantify the forces imposed on trees and their ability to support the load or fail [61,62].

A hollowing trunk is a natural process, and it is not necessarily a sign of an ailing tree. The center of the tree is deadwood, which is slowly decayed by fungi. Hollow tubes have the advantage of resisting bending and torsional moments with a relatively lower weight per unit length than solid cylinders of the same weight [63]. These hollow structures are ubiquitous in nature, such as bamboo stems, cereal stalks, decayed hollow tree trunks [64], and animal bones [65], as well as in artificial structures, such as buildings, bridge frames, athletic halls, and aircraft fuselages. Hollow-bearing trees are being removed at a faster rate than hollows form in many parts of the world (urban and rural) [66]. There are ongoing declines in the abundance of fauna dependent on hollow trees for survival. Nowadays, many hollow trees are cut down from fear that they may fall at some point in the future and injure someone. Yet, many hollow trees are clearly structurally sound, as they remain in place for decades. Tree trunks can be up to 70% hollow before the probability of failure suddenly increases [67].

The safety assessment of hollow trees has always fascinated arborists, and the criteria to be employed have led to severe public discussions in the professional scene in Europe. Based on Mattheck et al. (1995), many tree consultants state that the required thickness of the residual wall should not be below a t/R ratio of 0.3 to prevent shell-buckling and cross-sectional flattening [68]. On the other hand, Wessolly and Erb (1998) published an opposite theory, by which much lower thickness are often calculated and accepted [69]. Their methods (SIA and Elastomethod) are based on the bending theory of the hollow beam. The degree of hollowness or different wood strengths were only two quite subordinate aspects of the statics triangle, which consists of load, shape, and material. The size of the actual cavity provides no information on the safety of the tree; even a safe tree may be nearly entirely hollow. There is no fixed boundary value such as 0.3, for example. The importance of the cavity can only be assessed in comparison with the static cushion of the tree. This requires a load analysis, simplified by DIN 1055 with the appropriate cw-value or, even better, matched to trees by a static integrated tree assessment (SIA) [70].

Sometimes, strong wind can significantly affect a tree's stability. Wind–tree interactions take place at a wide range of temporal and spatial scales [71]. Aerodynamic drag at all surfaces of the aerial parts of trees—from individual leaves to whole tree crowns [72]—perturbs the airflow inside forest canopies [73]. Wind load acts on the tree crown and causes large bending moments on the trunk and root plate. These moments are the main sources of trunk failure and root overturning. The breaking of thick-walled hollow tree trunks subjected to gale gusts at the crown usually results from bending stresses. Failure begins with fiber buckling on the compression side, followed by fiber tearing on the tension side. It has been indicated that tangential cracking due to cross-sectional flattening, followed by longitudinal splitting, is dominant for hollow trunks. Growth stresses are ubiquitous in trees and may affect the failure mechanism. The longitudinal growth stress of normal wood is tensile on the surface of the trunk and is compressive inside [74], whereas tangential growth stresses are always compressive on the surface but tensile inside. Because the longitudinal compression strength of wood is only about one-third of its tensile strength, longitudinal tensile growth stresses can partially compensate for the compressive stresses on the leeward side of a trunk induced by the bending moment of wind loads acting on the crown. The compressive surface growth stress in the tangential direction can partially resist longitudinal splitting [75]. Furthermore, as the branches move around in the wind, they dissipate wind energy, which reduces the load transferred to the trunk and increases the mechanical stability of the tree [67]. These features may be recognized as the self-optimized design of trees as a result of evolution.

Old hollow trees with severe mechanical faults and old lapsed pollards that are prone to structural collapse require priority treatment. Where tree stability is already heavily compromised, reduction needs to be sufficient to reduce the crown load center and lever arm to an acceptable level [76].

At this point, it is worth paying attention to the fact that decay in the trees is a normal process (not an illness) and belongs to the life of a tree, and there is and should be decaying parts in old trees. Decaying wood is one of the most important parts of an old tree and is what supports other species and biodiversity. Decay is not always a risk, and a professional must conclude on the basis of VTA and by looking carefully at the structure and studying decay with these non-invasive methods or with microdrill resistance (which may be required in studying large branches in the canopy, where the use of a tomograph is very difficult) whether there are structural weaknesses and what to do about them.

3.3. Chlorophyll Fluorescence Results

The results of the conducted research are the results of chlorophyll fluorescence measurements presented with appropriate coefficients (Table 3). The F_V/F_M determinant, commonly used in assessing the impact of stress conditions on plants, is considered a reliable diagnostic indicator. The F_V/F_M ratio not lower than 0.83–0.85 relative units is the optimal value for most vascular plants, indicating their good condition. The influence of various stress factors, both biotic and abiotic, usually contributes to the reduction of this index. The index values at 0.20–0.30 relative units are considered borderline. The F_V/F_M ratio values were lower than the reference ones, especially in trees with visible damage or infected with pathogens. It may suggest a reduced and less efficient efficiency of the photosynthetic apparatus compared to healthy trees. In a stressful situation, the energy reserves of such trees decrease to such an extent that the condition deteriorates and, as a result, weakens the tree, which in extreme cases leads to its dieback. However, the obtained results did not come close to the lower limit, where photosystem II changes to(PSII)'s structure becomes irreversible.

Table 3. Results of the chlorophyll fluorescence of trees of senile size in Sandomierz (by authors).

No.	Species Name	The Obtained Values of the Chlorophyll Fluorescence Index (F_V/F_M) in Successive Replications					Average Value of the F_V/F_M Ratio	Place of Growth
1	Norway maple (*Acer platanoides* L.)	0.825	0.825	0.821	0.826	0.822	0.823	Piszczele Park
2	White poplar (*Populus alba* L.)	0.813	0.818	0.815	0.815	0.803	0.812	Piszczele Park
3	White poplar (*Populus alba* L.)	0.811	0.804	0.807	0.811	0.803	0.807	Piszczele Park
4	English oak (*Quercus robur* L.)	0.821	0.823	0.816	0.81	0.826	0.819	Saski Park
5	Small-leaved linden (*Tilia cordata* Mill.)	0.804	0.805	0.809	0.800	0.808	0.805	At the entrance to the vineyard and the Church of St. James
6	Small-leaved linden (*Tilia cordata* Mill.)	0.779	0.755	0.708	0.769	0.780	0.758	At the entrance to the Church of St. James
7	Common ash (*Fraxinus excelsior* L.)	0.825	0.835	0.829	0.821	0.822	0.826	At the church of St. Paul
8	English oak (*Quercus robur* L.)	0.827	0.839	0.833	0.832	0.836	0.833	At the rectory of the Church of St. Paul
9	Small-leaved linden (*Tilia cordata* Mill.)	0.804	0.802	0.783	0.789	0.798	0.795	At the Collegium Gostomianum
10	English oak (*Quercus robur* L.)	0.759	0.729	0.754	0.722	0.740	0.740	Milberta St.
11	White poplar (*Populus alba* L.)	0.821	0.815	0.821	0.820	0.822	0.819	Baczyńskiego St.
12	White poplar (*Populus alba* L.)	0.807	0.808	0.818	0.809	0.829	0.814	Baczyńskiego St.
13	English oak (*Quercus robur* L.)	0.824	0.826	0.826	0.830	0.826	0.826	Opatowska St.

The lowest F_V/F_M indexes determining the physiological condition of trees were recorded in the pedunculate oak (inv. No. 10) growing at Milbert Street (0.740), the small-leaved lime growing at Collegium Gostomianum (0.795) (inv. No. 9) and the entrance to the Vineyard of the Dominican Monastery (inv. No. 5) (0.805), and the white poplar (inv. No. 3) from Park Piszczele (0.807) (Table 3). This indicates that the mentioned trees grow under stressful conditions or are subjected to factors that cause them. Their condition is not entirely satisfactory. In the case of oak No. 10, a large (over 30%) branch deadwood, there is leaf infection by powdery oak mildew, the progressive decay of the inside of the trunk caused by yellow sulfur mucus, as well as numerous deep open and surface cavities in which wood decay increases stress, which is reflected in the health of the tree. Among all the examined specimens, this one is the weakest in terms of physiology.

In the small-leaved lime (inv. No. 9) at Collegium Gostomianum, its condition is also significantly impaired due to a massive open loss of the trunk and deep damage to the limbs. It is also caused by a fungal infection visible superficially on some fragments of the boughs and the weakening of the main stem, which was tied several decades ago with a rigid through the bond. The numerous deadwood branches also indicate the weakened vitality of the tree from the physiological point of view.

Lime No. 5, growing in extreme habitat conditions (pressed into a brick fence), is characterized by a similarly weakened vitality. The deterioration of the physiological condition of the tree may also be influenced by a significant open loss of a trunk with a rotting interior and weaker nutrition of the crown, which results in smaller chlorotic leaves, especially in the upper parts of the crown, and 20% deadwood.

Among trees with visible damage, the white poplar (inv. No. 3) from Park Piszczele is in poor condition, especially the trunk. In this tree's case, its physiological state is undoubtedly influenced by the significant activity of saprophytic and parasitic fungi, including peat and sulfurous gall. They contribute to the successive and slow decomposition of the wood, weakening the tree's statics.

Reasonably good health characterizes the remaining stand despite various types of damage to some trees, such as broken branches or boughs, cavities, deadwood, or the activity of insects and fungi.

Pedunculate oaks with inv. No. 8 and 13 (at Opatowska Street and the presbytery of the church of St. Paul), common maple with inv. No. 1 (in the Piszczele Park), and the ash tree amounted to inv. No. 7 (at the church of St. Paul) are in the best condition. Based on the measurements, the chlorophyll fluorescence coefficients in the leaves of these trees were, respectively, 0.833; 0.826; 0.823, and 0.826 relative units, corresponding to a value close to the optimal one, which guarantees the excellent condition of the plant.

3.4. Valuation of Trees

Of all the trees included in the analysis, oaks had the highest financial value, regardless of their health condition. It should be said, however, that the items in the best condition are valued the highest and their value, after considering all factors, exceeds EUR 40,000. Those that are damaged to a different extent are also valuable trees, and their monetary value exceeds the level of EUR 20,000. This is mainly related to their dimensions, age, and location. The remaining trees belonging to the faster-growing species (linden, poplars, maple, and ash) are characterized by a slightly lower value than the majestic oaks. Nevertheless, depending on the state of conservation, species, and place of growth, their monetary value is also not small. It ranges from less than EUR 7000 (Norway maple) to almost EUR 17,000 (small-leaved lime).

Disclosing the economic value of the assessed trees to the public may significantly contribute to a greater awareness of the importance of such trees for the environment, increasing the value of the area on which they grow, as well as appropriate care for their health condition. It can also force decision-makers to use effective management methods to maintain such a valuable tree stand in the best condition for an extended time. Tree value calculations are included in the available inventory table (Table 1).

3.5. Practical Conclusions

The aim of the study was to formulate recommendations for senile trees. Considering the above, it should be noted that the vast majority of trees assessed in this study are in good health despite visible defects or old age.

A comparison of the results from the tomography and chlorophyll fluorescence is presented in Table 4. Particularly noteworthy are the common ash (inv. No. 7), pedunculate oak (inv. No. 8) and the pedunculate oak (inv. No. 13). The trees mentioned above are in excellent health, which was shown by tests carried out with the use of sonic and electrical tomography. The inside of the trunk of the studied trees is 99–100%of sound wood, without any symptoms of damage or fungal infections. Their physiological state, as confirmed by the fluorescence of chlorophyll, is satisfactory. The obtained results are close to the optimal values for trees growing in good, stress-free habitat conditions.

Table 4. A comparison of the results from the tomography and chlorophyll fluorescence of senile trees in Sandomierz (by authors).

No.	Species Name	Picus Sonic Tomograph 3		Picus TreeTronic	Fluorometer OS5p+
		Type of Wood	Percent		Average Value of the F_V/F_M Ratio
1	Norway maple (*Acer platanoides* L.)	technically efficient wood	100%	56 Ω·m–362 Ω·m	0.823
2	White poplar (*Populus alba* L.)	damaged wood	63%	63 Ω·m	0.812
		transitional wood	7%	1860 Ω·m	
		technically efficient wood	30%	2200 Ω·m	
3	White poplar (*Populus alba* L.)	damaged wood	56%	11–33 Ω·m	0.807
		transitional wood	13%	2790 Ω·m	
		technically efficient wood	31%	9780 Ω·m	
4	English oak (*Quercus robur* L.)	damaged wood	12%	64 Ω·m	0.819
		transitional wood	4%	725 Ω·m	
		technically efficient wood	84%	2360 Ω·m; resistance values ranged from in the middle of the trunk from 64 Ω·m to 206–210 Ω·m in the area facing the outside of the trunk (wood typical of pedunculate oak)	
5	Small-leaved linden (*Tilia cordata* Mill.)	damaged wood	43%	7–41 Ω·m	0.805
		transitional wood	24%	531 Ω·m	
		technically efficient wood	33%	2274–2554 Ω·m	
6	Small-leaved linden (*Tilia cordata* Mill.)	damaged wood	66%	75–112 Ω·m	0.758
		transitional wood	8%	636 Ω·m	
		technically efficient wood	26%	4681 Ω·m	
7	Common ash (*Fraxinus excelsior* L.)	damaged wood	0.7%	41 Ω·m	0.826
		transitional wood	0.3%	100 Ω·m	
		technically efficient wood	99%	251–262 Ω·m; resistance values ranged from in the middle of the trunk cross-section is a ring system (28–39 Ω·m)—suggesting good tissue hydration	
8	English oak (*Quercus robur* L.)	damaged wood	0%	-	0.833
		transitional wood	0%	-	
		technically efficient wood	100%	298–462 Ω·m; resistance values ranged from 61 Ω·m in the middle of the trunk to 153–175 Ω·m—may suggest good tissue hydration.	
		transitional wood	21%	17,740 Ω·m	
		technically efficient wood	44%	15–94 Ω·m	
10	English oak (*Quercus robur* L.)	damaged wood	73%	41–65 Ω·m	0.740
		transitional wood	7%	220 Ω·m	
		technically efficient wood	20%	1171 Ω·m	

Table 4. *Cont.*

No.	Species Name	Picus Sonic Tomograph 3		Picus TreeTronic	Fluorometer OS5p+
		Type of Wood	Percent		Average Value of the F_V/F_M Ratio
11	White poplar (*Populus alba* L.)	damaged wood	24%	80–86 Ω·m	0.819
		transitional wood	20%	200 Ω·m	
		technically efficient wood	56%	700 Ω·m	
12	White poplar (*Populus alba* L.)	damaged wood	23%	122,516 Ω·m	0.814
		transitional wood	15%	9500 Ω·m	
		technically efficient wood	62%	8000 Ω·m	
13	English oak (*Quercus robur* L.)	damaged wood	0%	-	0.826
		transitional wood	0%	-	
		technically efficient wood	100%	338 to 414 Ω·m; resistance values ranged from 90 Ω·m in the middle of the trunk to 129 Ω·m, which is caused by good tissue hydration.	

The most problematic trees in terms of their health and physiological condition are the English oak (inv. No. 10) growing at Milbert Street, the white poplar from Park Piszczele (inv. No. 3), and the two lime trees—one at the entrance to the Dominican Vineyard (inv. No. 5) and the other at Collegium Gostomianum (inv. No. 9). These trees have significant deep cavities, are infected with fungal pathogens, and are characterized by a low F_V/F_M ratio determining the physiological state, which indicates growth under stressful conditions. The oak growing at Milbert Street has significant deep cavities and considerable damage, and the infection by sulfuric gallbladder does not give any optimism about its future. However, it is worth treating this tree with respect and allowing it to grow as long as its vital forces allow it. It is a tree historically associated with the area in which it grows; thus, attempting to minimize the impact of unfavorable factors will allow it to survive for a long time. In the case of the white poplar from Piszczele Park, one should consider the progressive decomposition of the inside of the trunk caused by the sulfuric gallbladder. The tree is still relatively stable due to the produced column that supports and strengthens the trunk from the damaged side. Nevertheless, regular monitoring will allow appropriate steps for its continued existence in the future. The two lime trees mentioned above are already senile and strongly "experienced by life". The rigid pass-through bonds strengthen their trunks, reducing the potential risk of breaking them. The vitality of trees is still significant, which allows us to think positively about the prognosis for the future.

The rest of the assessed trees are in different health conditions but are good enough that, currently, they require only minor maintenance.

Old trees are a precious component of tall greenery for the city, and even despite their advanced age or considerable damage, they are an invaluable element of biodiversity. Each too-hasty decision to remove a tree is a massive loss for the environment, both in terms of image, nature, and history, as well as finances, which is not always remembered.

The management of veteran trees is in the first place improving the safety and the living conditions around the tree and, as last options, touching the tree (e.g., pruning).

The area on which senile trees grow should be equally adequately protected, and, in the event of an excessive threat from them, appropriate steps should be taken to minimize the risk to the environment. Considering the so-called tree stand management, it is advisable to carry out appropriate treatments to keep the trees in reasonably good condition (Table 5).Where soil conditions under a veteran tree appear to be unfavorable—for example, severe soil compaction—the first thing to do is to stop or reduce the causes of those conditions. In the case of trees in Sandomierz, it is suggested to scatter the bark under trees No. 1, 2, and 3 and systematically mow the lawn grass under the others.

Table 5. Management plan forsenile treesin Sandomierz (by authors).

No.	Species Name	Removal of Branches	Crown Correction (Removal of Stumps)	Health Condition Monitoring	Ordering the Tree Surrounding	Other
1	Norway maple (*Acer platanoides* L.)	yes (branch deadwood 25%)	no	yes (every two years)	yes	-
2	White poplar (*Populus alba* L.)	yes (branch deadwood 10–15%)	no	yes (each year)	no	-
3	White poplar (*Populus alba* L.)	yes (branch deadwood 15%)	no	yes (every six months)	no	-
4	English oak (*Quercus robur* L.)	yes (branch deadwood 25%)	no	yes (each year)	no	-
5	Small-leaved linden (*Tilia cordata* Mill.)	yes (branch deadwood 20%)	no	yes (each year)	no	-
6	Small-leaved linden (*Tilia cordata* Mill.)	no (branch deadwood 10–15%)	no	yes (each year)	no	-
7	Common ash (*Fraxinus excelsior* L.)	yes (branch deadwood 20%)	no	yes (every two years)	no	-
8	English oak (*Quercus robur* L.)	yes (branch deadwood 20%)	no	yes (every two years)	no	-
9	Small-leaved linden (*Tilia cordata* Mill.)	yes (branch deadwood 15%)	no	yes (each year)	no	-
10	English oak (*Quercus robur* L.)	yes (branch deadwood 30%)	yes	yes (each year)	yes	-
11	White poplar (*Populus alba* L.)	yes (branch deadwood 10–15%)	yes	yes (each year)	no	-
12	White poplar (*Populus alba* L.)	yes (branch deadwood 10–15%)	yes	yes (each year)	no	-
13	English oak (*Quercus robur* L.)	yes (branch deadwood 5%)	no	yes (every two years)	no	Protected by an entry in the register of monuments

Tree species do differ in their ecological requirements for light, such as for germination and establishment, and in their ability to tolerate some shade [4,5,77]. However, all trees need light, especially trees in, or entering, the senile stage of life. In terms of trees' access to light, no problems were noticed. Only tree No. 1 is located in a thicket of neighboring trees, therefore, bushes clearing is suggested.

4. Discussion—Combination of Four Tree Assessment Methods

The above results confirm that the use of non-invasive methods is a source of practical information about the health condition of trees. The conducted investigations and analyses, as well as the application of the obtained results, can be used to protect senile trees. The presented results are a response to the need for research on trees of monumental size in the field of sustainable urban greenery management and the perception of the economic value of trees by the community. Old trees growing in urban areas are an essentially natural and cultural element but require unique solutions to minimize risk in their surroundings. The survival, growth, and management of street trees in a city pose unique problems, as most street trees grow in a stressful city environment. The negative urban factors affecting trees are soil compaction, variable water-air ratios, water shortages due to surface runoff, high temperature, salt aerosol, or air pollution. High air temperature and its often bad composition (CO_2, O_3, NO_2, etc.) are unfavorable for plants, especially those growing on paved areas, such as squares. The mentioned factors influence the physiology of trees, e.g., photosynthesis or enzymatic activity [78].In addition, poor air quality can cause mechanical changes such as obstruction of the stomata caused by dust suspended in polluted air. Such changes result in the deterioration of gas exchange; thus, they disturb the life processes of trees [79]. Climate change effects, such as heavy rainfall and local tornadoes, also increase the risk of tree damage [80]. Moreover, studies also indicate that communication with the public is an obstacle to effective management, as the public is not sufficiently aware of the benefits of trees [81,82]. Many inhabitants do not perceive the importance of street trees due to unknown economic values [83,84].

The fall of a tree or limb can cause significant damage to public infrastructure, personal property, and even human life [85], and, when discussing research on the biomechanics of trees, it is worth noting the difference between forest trees and trees growing individually in an open field. The latter group includes most of the natural monuments in the Sandomierz. Open-grown trees usually grow with considerable branch mass, and the dynamic response in winds may be different to other tree forms. The shape or morphology of the tree and the distribution of oscillating branch masses become important during dynamic studies [86]. For example, slender forest conifers sway in a relatively simple manner in contrast to open-grown trees, which have many independent and larger branch masses [87]. The dynamic interaction of branches in winds can significantly modify the frequency and damping of a tree [88]. Studies examining tree failure due to winds in urban areas have been undertaken after wind storms, but with only limited correlation to actual wind velocity and gustiness [89,90].There is currently no definitive method to predict the failure of an individual tree. Assessments of trees include visual tree assessment [91], tree risk assessment methodology [92], quantified tree risk assessment [93], and static integrated methods that combine static pulling with dynamic wind load assessment [94]. It is noteworthy that crown thinning was less effective at reducing trunk movement [95]. Branches on thinned trees appeared to move more than branches on other treatments but not in the same direction. This complex branch movement indicates that the dynamic effects of branches may play an important role in acting as a buffer to dampen and reduce motion [88].

Urban trees, distinct from forest trees, are characterized by complex growth environments, a greater vulnerability to damage, and wind tunnel effects caused by high-rise construction [96–98]. Therefore, the arborist control of old trees is essential in the city. It requires precise diagnostic techniques to detect, for example, rot, internal decay, or other structural defects in tree trunks. Visual tree assessment (VTA) is still the starting point for such research. However, internal defects of tree trunks often remain beyond the sight of a specialist [99–101]. However, there are technological tools to assist with tree inspection. These tools provide results that can be used in calculating fall risk, contributing to the decision to keep or remove the tree.

The most appropriate methods have minimal effect on living tissues, e.g., acoustic tomography, giving 80% accuracy of the results [102,103]. Over the past two decades, technologies for non-invasive research and the evaluation of urban trees have attracted the attention of dendrologists worldwide [104–106]. It is interesting, for example, to compare sound tomography and GPR surveys. Research using non-destructive radar techniques (GPR) can be a diagnostic tool to assess the health status of living tree trunks based on the internal distribution of dielectric permittivity. However, it is still a method that requires refinement and is very time-consuming (some hardware modifications are needed to examine the tree). Compared to tomographic diagnostics, it also does not provide easy-to-interpret images and requires, at the same time, extensive experience in reading the obtained results [107,108]. On the other hand, acoustic and electrical tomography is a perfect combination. While acoustic techniques provide detailed information on the quality of the wood, they are ineffective when it comes to differences between decayed wood and bacterial wet wood or decay and wood loss. In contrast, electrical resistivity methods can identify wood decay early.

However, it is worth noting that the resistivity tomography method provides low spatial resolution. The first pioneering works on the electrical resistivity of wood in living trees were published in 1965–1995. The best-known example of an electrical conductivity meter is the Shigometer, which consists of a twisted wire probe and resistivity meter [109]. When operating the device, an electrical probe is placed into a small, pre-drilled hole approximately 3 mm in diameter. The pattern of resistivity of the wood to a pulsed direct current is recorded [110]. Shigo (1991) claimed that, in the region adjacent to wood decay, the concentration of cations in the wood would increase and, therefore, the electrical resistivity would decrease. However, electrical resistivity also decreases if wood is healthy but dry [111] and increases in dry decayed wood and when the probe moves from sapwood to heartwood [112]. Electrical resistivity may also decrease when the probe reaches bacterial wetwood [109]. Several researchers have had very inconsistent results with the Shigometer. Results have been dependent on the tightness of fit of the electrode and the relative moisture and resin content of the timber [110]. Shigo and Shortle (1985) advised that the Shigometer does not function in resin soaked, frozen, or dead wood, and the operator should carefully control the amount of contact the needle electrodes have with the wood [109]. Nicolotti et al. (2003) assessed results from electrical resistivity and ultrasonic tomography and georadar [113]. They reported good results with electric tomography. Electrical tomography was deemed promising by Hagrey (2007), but the results were qualitative rather than quantitative [114]. Problems with the Shigometer in eucalypts may be similar to electrical tomography, as the raw data are the same (electrical resistivity).

The test carried out in the Sandomierz showed that the measurement should be performed with sonic tomography, and then the result should be confirmed with an electric tomograph. The tomogram obtained from the sonic tomograph may show the inside of the trunk in blue, but we do not know if it is void or rot. On the other hand, an electric tomograph confirms one of these variants.

Two or more methods are usually combined to examine the health status of a tree for comparison and the validation of the results [115,116]. In this case, in Sandomierz, additional studies of chlorophyll fluorescence and the economic valuation of trees were carried out to support the effective monitoring and management of trees of considerable size.

The EU ecological policy formulates postulates on the knowledge, inventory, valuation, and protection of ecosystem services. Provisions in this regard are contained in the EU biodiversity strategy [117] and the Seventh Environmental Action Program [118]. In urbanized areas with a small share of natural ecosystems, trees play a unique role in shaping the environment of human life. The role of trees in urban space has been widely discussed and documented in the achievements of science. They provide a range of natural and cultural benefits. At the same time, the awareness of the authorities, administration, and residents about the trees found in a given area, their importance, and their economic value is often limited. Trees are treated as one of the prominent elements of the urban landscape. There is not much attention paid to the role they play in shaping the quality of life in the city.

For many years, economists were only interested in wood's very narrowly understood utility value. Unfortunately, in Poland, the practical approach to nature and the emphasis on its servant role towards humans are still dominant, as well as the perception of activities for the protection of nature and its preservation as obstacles to the socio-economic development of a commune, region, or country. With this perception of nature, a tree growing outside the forest, e.g., in a city, is—in the opinion of many property owners—primarily an obstacle in the development and use of a given area, a source of nuisance in its maintenance, or a threat to the environment, and the value of the tree is determined only by the value of wood cutting [36].

It is worth noting that most tree ecosystem services can also be estimated and the value can be expressed in the form of a specific economic value. One of the most famous examples of the use of tree ecosystem services valuation recently has been a study conducted in New York. Their subject was evaluating costs and benefits related to street trees managed by the city authorities. The study covered almost 600 thousand such trees (but 4.5 million trees growing in parks and private lands were not included in it). The analyzed trees yielded net benefits (after adjusting for costs) of USD122 million per year (USD 209 per tree); for every dollar spent on tree maintenance in New York City, the city benefits USD 5.6. Benefits such as the reduction of energy consumption, the absorption of CO_2 and other pollutants, water retention in the landscape, or the impact on the property value were considered [119]. Experience in evaluating various natural resources in cities on a global scale includes tens, perhaps hundreds of thousands of such studies, but in Poland, such studies have so far been carried out rather sporadically.

Preparing the valuation of trees and including them, e.g., in analyses of the value of the property, should first become a recommended, and ultimately even mandatory, tool for environmental management, as well as property management, especially by municipal governments and other entities managing public space, including the State Treasury [120].

The example of activities presented in the manuscript shows the possibilities of implementing the postulates of ecological policy by local authorities, guided by the need for in-depth economic knowledge about the local natural environment and improving the quality of life of inhabitants of urbanized areas. The economic valuation of trees facilitates the decision-making process regarding tree stand management in urban areas. It allows us to reduce common denominator issues that are often difficult to decide due to the diverse needs, perspectives, and positions of society presented, expressing their value in money, which continually facilitates the message about the value of things and places, including those seemingly immeasurable.

5. Conclusions

This study presented a combination of several non-invasive technologies for detecting and assessing the health status of urban trees. The combination of visual inspection (VTA), acoustic and electrical tomography, and stress testing with a fluorometer is a practical approach to measuring the health of various tree species. The procedures developed by us enabled the analysis of the conservation status of endangered trees and may improve the risk assessment process caused by them.

Non-invasive tools using acoustics and electric waves are still new in Poland, but the authors of this manuscript conducted similar research (one of the first in Poland) in the case of the historic greenery of the Botanical Garden in Lublin. Then, the effectiveness of the use of tomographic tests in the management of greenery in historic gardens was confirmed. The research presented in this manuscript was supplemented with fluorimetry and tree valuation results. The valuation of a tree is an additional protection tool because it contributes to its public approval and sometimes makes the inhabitants aware of the natural value of an old, often problematic specimen. Thanks to the use of combined methods of tree assessment, it is possible to diagnose old specimens, take care of the safety of people and property in their vicinity, and test candidates qualified for legal protection. Combined methods of tree assessment can be an important activity in the sustainable management of urban greenery. This case study of 13 trees growing in Sandomierz contributes to broadening the knowledge about this technique and its application in Poland and Europe.

Supplementary Materials: The following supporting information can be downloaded at: https://www.mdpi.com/article/10.3390/land11111914/s1, Supplementary File S1.

Author Contributions: Conceptualization, M.D., W.D. and M.M.; methodology, M.D. and W.D.; software W.D., formal analysis.; investigation, M.D. and W.D.; data curation, M.D. and W.D.; writing—original draft preparation M.D. and W.D.; writing—review and editing, M.D., W.D. and M.M.; visualization, M.D. and W.D.; supervision, M.D. and W.D. All authors have read and agreed to the published version of the manuscript.

Funding: This research received no external funding.

Data Availability Statement: Not applicable.

Acknowledgments: The research was financed from the own funds of the Department of Landscape Architecture and the Institute of Horticultural Production.

Conflicts of Interest: The authors declare no conflict of interest.

References

1. Khalaim, O.; Zabarna, O.; Kazantsev, T.; Panas, I.; Polishchuk, O. Urban Green Infrastructure Inventory as a Key Prerequisite to Sustainable Cities in Ukraine under Extreme Heat Events. *Sustainability* **2021**, *13*, 2470. [CrossRef]
2. Lăzăroiu, G.; Ionescu, L.; Uță, C.; Hurloiu, I.; Andronie, M.; Dijmărescu, I. Environmentally Responsible Behavior and Sustainability Policy Adoption in Green Public Procurement. *Sustainability* **2020**, *12*, 2110. [CrossRef]
3. Bednarska-Olejniczak, D.; Olejniczak, J.; Svobodová, L. Towards a Smart and Sustainable City with the Involvement of Public Participation—The Case of Wroclaw. *Sustainability* **2019**, *11*, 332. [CrossRef]
4. Understanding and Managing Veteran Trees. An Advanced Course by the VETcert Project. 2019. Available online: https://www.vetcert.eu/sites/default/files/2020-03/VETcert%20course%2010102019_Delegate%20version-compressed.pdf (accessed on 12 October 2022).
5. Gilmartin, E. Ancient and Veteran Trees: An Assessment Guide. The Woodland Trust. 2022. Available online: https://www.woodlandtrust.org.uk/media/51153/ancient-and-veteran-trees-an-assessment-guide.pdf (accessed on 12 October 2022).
6. Dudkiewicz, M.; Durlak, W. Sustainable Management of Very Large Trees with the Use of Acoustic Tomography. *Sustainability* **2021**, *13*, 12315. [CrossRef]
7. Durlak, W.; Dudkiewicz, M.; Pudelska, K.; Dąbski, M. Using Picus® Sonic Tomograph to assess the health state of trees of monumental sizes. *Teka Kom. Archit. Urban. I Studiów Kraj.* **2017**, *13*, 73–82. [CrossRef]
8. The ISA Tree Risk Assessment Qualification (ISA TRAQ). Available online: https://www.isa-arbor.com/Credentials/ISA-Tree-Risk-Assessment-Qualification (accessed on 12 October 2022).
9. Roman, L.A.; Battles, J.J.; McBride, J.R. The balance of planting and mortality in a street tree population. *Urban Ecosyst.* **2013**, *17*, 387–404. [CrossRef]
10. Smith, I.A.; Dearborn, V.K.; Hutyra, L.R. Live fast, die young: Accelerated growth, mortality, and turnover in street trees. *PLoS ONE* **2019**, *14*, e0215846. [CrossRef] [PubMed]
11. Szczepanowska, H.B. *Drzewa W Mieście*; Hortpress: Warszawa, Poland, 2001.
12. Stahle, D.W.; Chaney, P.L. A predictive model for the location of ancient forests. *Nat. Areas J.* **1994**, *14*, 151–158. Available online: https://www.researchgate.net/publication/292690828_A_predictive_model_for_the_location_of_ancient_forests (accessed on 26 October 2022).
13. Huckaby, L.S.; Kaufmann, M.R.; Fornwalt, P.J.; Stoker, J.M.; Dennis, C. Identification and ecology of old ponderosa pine trees in the Colorado Front Range. *Gen. Tech. Rep.* **2003**, *110*, 47. [CrossRef]
14. Pederson, N. External Characteristics of Old Trees in the Eastern Deciduous Forest. *Nat. Areas J.* **2010**, *30*, 396–407. [CrossRef]
15. Judice, A.; Gordon, J.; Abrams, J.; Irwin, K. Community Perceptions of Tree Risk and Management. *Land* **2021**, *10*, 1096. [CrossRef]
16. Alexander, K.N.A. *Revision of the Index of Ecological Vontinuity as Used for Saproxylic Beetles*; English Nature Research Report ENRR574; English Nature: London, UK, 2004.
17. Rosłon-Szeryńska, E.; Fortuna-Antosziewicz, B.; Łukaszkiewicz, J.; Borowski, J. Strategy for preservation of veteran trees in the city-a model of conduct based on plant defense mechanisms and the idea of sustainable development. *Ecol. Eng. Environ. Technol.* **2018**, *19*, 12–21. [CrossRef]
18. O'Malley, C.; Piroozfar, P.; Farr, E.R.P.; Pomponi, F. Urban Heat Island (UHI) mitigating strategies: A case-based comparative analysis. *Sustain. Cities Soc.* **2015**, *19*, 222–235. [CrossRef]
19. Wu, Z.; Zhang, Y. Spatial Variation of Urban Thermal Environment and Its Relation to Green Space Patterns: Implication to Sustainable Landscape Planning. *Sustainability* **2018**, *10*, 2249. [CrossRef]
20. Song, P.; Kim, G.; Mayer, A.; He, R.; Tian, G. Assessing the Ecosystem Services of Various Types of Urban Green Spaces Based on i-Tree Eco. *Sustainability* **2020**, *12*, 1630. [CrossRef]
21. Nowak, D.J.; Dwyer, J.F. Understanding the benefits and costs of urban forest ecosystems. In *Urban and Community Forestry in the Northeast*; Kuser, J., Ed.; Springer: New York, NY, USA, 2007; pp. 25–46.

22. Tiwary, A.; Sinnett, D.; Peachey, C.; Chalabi, Z.; Vardoulakis, S.; Fletcher, T.; Leonardi, G.; Grundy, C.; Azapagic, A.; Hutchings, T.R. An integrated tool to assess the role of new planting in PM10 capture and the human health benefits: A case study in London. *Environ. Pollut.* **2009**, *157*, 2645–2653. [CrossRef] [PubMed]
23. Tallis, M.; Taylor, G.; Sinnett, D.; Freer-Smith, P. Estimating the removal of atmospheric particulate pollution by the urban tree canopy of London, under current and future environments. *Landsc. Urban Plan.* **2011**, *103*, 129–138. [CrossRef]
24. Beckett, K.P.; Freer-Smith, P.H.; Taylor, G. Urban woodlands: Their role in reducing the effects of particulate pollution. *Environ. Pollut.* **1998**, *99*, 347–360. [CrossRef]
25. Salmond, J.A.; Tadaki, M.; Vardoulakis, S.; Arbuthnott, K.; Coutts, A.; Demuzere, M.; Dirks, K.N.; Heaviside, C.; Lim, S.; MacIntyre, H.; et al. Health and climate related ecosystem services provided by street trees in the urban environment. *Environ. Health* **2016**, *15*, 95–111. [CrossRef] [PubMed]
26. Memluk, M.Z. *Designing Urban Squares*; Intech Open: London, UK, 2013. [CrossRef]
27. Nesbitt, L.; Hotte, N.; Barron, S.; Cowan, J.; Sheppard, S.R. The social and economic value of cultural ecosystem services provided by urban forests in North America: A review and suggestions for future research. *Urban For. Urban Green.* **2017**, *25*, 103–111. [CrossRef]
28. Kondo, M.C.; Fluehr, J.M.; McKeon, T.; Branas, C.C. Urban Green Space and Its Impact on Human Health. *Int. J. Environ. Res. Public Health* **2018**, *15*, 445. [CrossRef]
29. Bratman, G.N.; Hamilton, J.P.; Daily, G.C. The impacts of nature experience on human cognitive function and mental health. *Ann. N. Y. Acad. Sci.* **2012**, *1249*, 118–136. [CrossRef] [PubMed]
30. Houlden, V.; Weich, S.; de Albuquerque, J.P.; Jarvis, S.; Rees, K. The relationship between greenspace and the mental wellbeing of adults: A systematic review. *PLoS ONE* **2018**, *13*, e0203000. [CrossRef] [PubMed]
31. Marselle, M.R.; Hartig, T.; Cox, D.T.; de Bell, S.; Knapp, S.; Lindley, S.; Triguero-Mas, M.; Böhning-Gaese, K.; Braubach, M.; Cook, P.A.; et al. Pathways linking biodiversity to human health: A conceptual framework. *Environ. Int.* **2021**, *150*, 106420. [CrossRef]
32. Nadkarni, N.M.; Hasbach, P.H.; Thys, T.; Crockett, E.G.; Schnacker, L. Impacts of nature imagery on people in severely nature-deprived environments. *Front. Ecol. Environ.* **2017**, *15*, 395–403. [CrossRef]
33. Pataki, D.E.; Alberti, M.; Cadenasso, M.L.; Felson, A.J.; McDonnell, M.J.; Pincetl, S.; Pouyat, R.V.; Setälä, H.; Whitlow, T.H. The Benefits and Limits of Urban Tree Planting for Environmental and Human Health. *Front. Ecol. Evol.* **2021**, *9*, 395–403. [CrossRef]
34. Fay, N. Environmental arboriculture, tree ecology and veteran tree management. *Arboric. J.* **2002**, *26*, 213–238. [CrossRef]
35. Green, T. Arborists should have a central role in educating the public about veteran trees. *Arboric. J.* **2002**, *26*, 239–248. [CrossRef]
36. Dorda, A. What is the value of trees–before and after felling? *Przegląd Przyr.* **2017**, *28*, 3–22.
37. Majdecki, L. *Tabela Wiekowa Drzew*; Oddział Architektury Krajobrazu SGGW: Warszawa, Poland, 1980–1986.
38. Mydłowska, A. Protection of greenery in the investment process, taking into account dendrological knowledge. In *Attachment to the Training Materials Fri Visual Diagnosis of Trees Threating Safety-Methods and Ways to Improve the Control of the Urban Tree Stand*; Dendros Poznań: Poznań, Poland, 2014.
39. Available online: http://www.tree-guide.com/tree-age-calculator (accessed on 29 May 2022).
40. Hansen, J.; Sato, M.; Ruedy, R. Perception of climate change. *Proc. Natl. Acad. Sci. USA* **2012**, *109*, E2415–E2423. [CrossRef] [PubMed]
41. Durlak, W.; Lublinie, U.P.W.; Dudkiewicz, M.; Pudelska, K.; Dąbski, M. Diagnozowanie kondycji drzew z wykorzystaniem tomografii komputerowej. *Acta. Sci. Pol. Circumiectus* **2017**, *2*, 71–83. [CrossRef]
42. Linhares, C.S.F.; Gonçalves, R.; Martins, L.M.; Knapic, S. Structural stability of urban trees using visual and instrumental techniques: A review. *Forests* **2021**, *12*, 1752. [CrossRef]
43. Papandrea, S.F.; Cataldo, M.F.; Zimbalatti, G.; Proto, A.R. Comparative evaluation of inspection techniques for decay detection in urban trees. *Sens. Actuators A Phys.* **2022**, *340*, 113544. [CrossRef]
44. Göcke, L.; Rust, S.; Weihs, U.; Günther, T.; Rücker, C. Combining Sonic and Electrical Impedance Tomography for the Nondestructive Testing of Trees. In Proceedings of the 15th International Symposium on Nondestructive Testing of Wood, Duluth, MN, USA, 10–12 September 2007; pp. 31–42.
45. Brazee, N.J.; Marra, R.; Gocke, L.; Van Wassenaer, P. Non-Destructive Assessment of Interna! Decay in Three Hardwood Species of Northeastern North America Rusing Sonic and Electrical Impedance Tomography. *Forestry* **2011**, *84*, 33–39. Available online: https://www.researchgate.net/publication/327778264_Estimating_carbon_loss_due_to_internal_decay_in_living_trees_using_tomography_Implications_for_forest_carbon_budgets (accessed on 29 May 2022). [CrossRef]
46. Available online: https://www.argus-electronic.de/en/tree-inspection/support/download/manuals (accessed on 16 August 2022).
47. Chomicz, E. Non-Invasive Diagnosis of the Condition of Historic Trees with the Use of Picus Tomographs. *Conserv. Cour.* **2010**, *8*, 29–32. Available online: https://www.nid.pl/upload/iblock/7f6/7f6ac9e8b82416efa9be341f0b30531e.pdf (accessed on 29 May 2022).
48. Rust, S.; Weihs, U.; Günther, T.; Rücker, C.; Goecke, L. Combining Sonic and Electrical Impedance Tomography for the Nondestructive Testing of Trees. Western Arborist. 2008. Available online: https://www.researchgate.net/publication/230882039_Combining_Sonic_and_Electrical_Impedance_Tomography_for_the_Nondestructive_Testing_of_Trees (accessed on 16 August 2022).

49. Dudkiewicz, M.; Durlak, W. Sonic Tomograph as a Tool Supporting the Sustainable Management of Historical Greenery of the UMCS Botanical Garden in Lublin. *Sustainability* **2021**, *13*, 9451. [CrossRef]
50. Available online: https://www.argus-electronic.de (accessed on 16 August 2022).
51. Sepúlveda, P.; Johnstone, D.M. A Novel Way of Assessing Plant Vitality in Urban Trees. *Forests* **2018**, *10*, 2. [CrossRef]
52. Maldonado-Rodriguez, R.; Pavlov, S.; Gonzalez, A.; Oukarroum, A.; Strasser, R.J. Can machines recognise stress in plants? *Environ. Chem. Lett.* **2003**, *1*, 201–205. [CrossRef]
53. Percival, G. Evaluation of physiological tests as predictors of young tree establishment and growth. *Arboric. Urban For.* **2004**, *30*, 80–91. Available online: https://joa.isa-arbor.com/request.asp?JournalID=1&ArticleID=129&Type=2 (accessed on 16 August 2022). [CrossRef]
54. Maxwell, K.; Johnson, G.N. Chlorophyll fluorescence—A practical guide. *J. Exp. Bot.* **2000**, *51*, 659–668. [CrossRef] [PubMed]
55. Angelini, G.; Ragni, P.; Esposito, D.; Giardi, P.; Pompili, M.L.; Moscardelli, R.; Giardi, M.T. A device to study the effect of space radiation on photosynthetic organisms. *Phys. Med.* **2001**, *17* (Suppl. 1), 267–268. Available online: https://www.researchgate.net/publication/11576798_A_device_to_study_the_effect_of_space_radiation_on_photosynthetic_organisms (accessed on 22 October 2022). [PubMed]
56. Cetner, M.D.; Dąbrowski, P.; Samborska, A.; Łukasik, I.; Swoczyna, T.; Pietkiewicz, S.; Bąba, W.; Kalaji, H.M. Zastosowanie pomiarów fluorescencji chlorofilu w badaniach środowiskowych. *Kosmos* **2016**, *65*, 197–205. Available online: http://kosmos.icm.edu.pl/PDF/2016/197.pdf (accessed on 22 October 2022).
57. Economic Valuation Methods and Their Use in Valuing Veteran Trees. 2019. Available online: https://www.vetcert.eu/sites/default/files/2019-11/Economic%20Valuation%20Methods%20.pdf (accessed on 12 October 2022).
58. Szczepanowska, H.B.; Olizar, J.; Borkowski, J.; Sitarski, M.; Suchocka, M.; Szadkowska, E. Work synthesis. In *Development of a New Method for Determining the Value of Trees Together with the Differentiating Indices and the Substantive Justification and Legitimacy of Its Introduction into the Legal Circuit*; Publishing house of the Institute of Spatial Management and Housing: Warsaw, Poland, 2010.
59. Kronenberg, J. Ecosystem Services—A New Look at the Value of the Natural Environment. EkoCity#Environmental. 2016. Available online: https://repozytorium.uni.lodz.pl/xmlui/handle/11089/17980?show=full (accessed on 12 October 2022).
60. Borowski, J.; Pstrągowska, M. *Determination of Species and Incremental Coefficients Used in the Method of Tree Valuation in Urbanized Areas*; SGGW–IGPiM: Warsaw, Poland, 2007; pp. 1–21.
61. James, K.R.; Dahle, G.A.; Grabosky, J.; Kane, B.; Detter, A. Tree biomechanics literature review: Dynamics. *Arboric. Urban For.* **2014**, *40*, 1–15. Available online: https://www.isa-arbor.com/Portals/0/Assets/PDF/research/BiomechanicsAUF.pdf (accessed on 22 October 2022). [CrossRef]
62. Cullen, S. Trees and Wind: A Bibliography for Tree Care Professionals. *Arboric. Urban For.* **2002**, *28*, 41–51. Available online: https://joa.isa-arbor.com/request.asp?JournalID=1&ArticleID=25&Type=2 (accessed on 22 October 2022). [CrossRef]
63. Huang, Y.-S.; Hsu, F.-L.; Lee, C.-M.; Juang, J.-Y. Failure mechanism of hollow tree trunks due to cross-sectional flattening. *R. Soc. Open Sci.* **2017**, *4*, 160972. [CrossRef] [PubMed]
64. Spatz, H.-C.; Niklas, K.J. Modes of failure in tubular plant organs. *Am. J. Bot.* **2013**, *100*, 332–336. [CrossRef]
65. Taylor, D.; Dirks, J.-H. Shape optimization in exoskeletons and endoskeletons: A biomechanics analysis. *J. R. Soc. Interface* **2012**, *9*, 3480–3489. [CrossRef] [PubMed]
66. Kane, B.; Ryan, D.; Bloniarz, D. Comparing Formulae that Assess Strength Loss Due to Decay in Trees. *Arboric. Urban For.* **2001**, *27*, 78–87. [CrossRef]
67. James, K.R. A Dynamic Structural Analysis of Trees Subject to Wind Loading. Ph.D. Thesis, The University of Melbourne, Melbourne, Australia, 2010. Available online: https://minerva-access.unimelb.edu.au/items/e6806505-69c3-5538-bf2a-430a8fda5671 (accessed on 12 October 2022).
68. Mattheck, C.; Bethge, K.; Breloer, H. Allgemeingültigkeit der Regeln zur Bewertung von Risikobäumen. *Patzer Verlag. Das Gart.* **1994**, *6*, 407–412.
69. Wessolly, L.; Erb, M. *Handbuch der Baumstatik und Baumkontrolle*; Berlin Patzer Verlag: Berlin, Germany, 1998.
70. Wessoly, L. How hollow may a tree be? *Neue Landsch.* **1996**, *11*, 847–850. Available online: http://www.historictreecare.com/wp-content/uploads/2015/02/Wessolly-How-hollow-may-a-tree-be.pdf (accessed on 12 October 2022).
71. De Langre, E. Effects of Wind on Plants. *Annu. Rev. Fluid Mech.* **2008**, *40*, 141–168. [CrossRef]
72. Kane, B.; Pavlis, M.; Harris, J.R.; Seiler, J.R. Crown reconfiguration and trunk stress in deciduous trees. *Can. J. For. Res.* **2008**, *38*, 1275–1289. [CrossRef]
73. Turnipseed, A.A.; Anderson, D.E.; Blanken, P.D.; Baugh, W.M.; Monson, R.K. Airflows and turbulent flux measurements in mountainous terrainPart 1. Canopy and local effects. *Agric. For. Meteorol.* **2003**, *119*, 1–21. [CrossRef]
74. Huang, Y.-S.; Chen, S.-S.; Kuo-Huang, L.-L.; Lee, C.-M. Growth strain in the trunk and branches of Chamaecyparis formosensis and its influence on tree form. *Tree Physiol.* **2005**, *25*, 1119–1126. [CrossRef]
75. Schindler, D.; Bauhus, J.; Mayer, H. Wind effects on trees. *Forstwiss. Cent.* **2011**, *131*, 159–163. [CrossRef]
76. Dujesiefken, D.; Fay, N.; de Groot, J.W.; de Berker, N. Trees—A Lifespan Approach. Contributions to arboriculture from European Practitioners. Available online: http://drzewa.org.pl/wp-content/uploads/2018/05/Trees_Lifespan_Approach.pdf (accessed on 22 October 2022).
77. Niinemets, U.; Valladares, F. Tolerance to shade, drought, and waterlogging of temperate northern hemisphere trees and shrubs. *Ecol. Monogr.* **2006**, *76*, 521–547. [CrossRef]

78. Moradi, A.; Abkenar, K.T.; Mohammadian, M.A.; Shabanian, N. Effects of dust on forest tree health in Zagros oak forests. *Environ. Monit. Assess.* **2017**, *189*, 549. [CrossRef]
79. Young, R.F. Mainstreaming urban ecosystem services: A national survey of municipal foresters. *Urban Ecosyst.* **2013**, *16*, 703–722. [CrossRef]
80. Driscoll, A.N.; Ries, P.D.; Tilt, J.H.; Ganio, L.M. Needs and barriers to expanding urban forestry programs: An assessment of community officials and program managers in the Portland—Vancouver metropolitan region. *Urban For. Urban Green.* **2015**, *14*, 48–55. [CrossRef]
81. Grado, S.; Measells, M.; Grebner, D. Revisiting the Status, Needs, and Knowledge Levels of Mississippi's Governmental Entities Relative to Urban Forestry. *Arboric. Urban For.* **2013**, *39*, 149–156. [CrossRef]
82. Mullaney, J.; Lucke, T.; Trueman, S.J. A review of benefits and challenges in growing street trees in paved urban environments. *Landsc. Urban Plan.* **2015**, *134*, 157–166. [CrossRef]
83. Tan, X.; Hirabayashi, S.; Shibata, S. Estimation of Ecosystem Services Provided by Street Trees in Kyoto, Japan. *Forests* **2021**, *12*, 311. [CrossRef]
84. Gullick, D.; Blackburn, G.; Whyatt, J.; Vopenka, P.; Murray, J.; Abbatt, J. Tree risk evaluation environment for failure and limb loss (TREEFALL): An integrated model for quantifying the risk of tree failure from local to regional scales. *Comput. Environ. Urban Syst.* **2019**, *75*, 217–228. [CrossRef]
85. Martinez-Trinidad, T.; Watson, W.T.; Arnold, M.A.; Lombardini, L.; Appel, D.N. Comparing various techniques to measure tree vitality of live oaks. *Urban For. Urban Green.* **2010**, *9*, 199–203. [CrossRef]
86. Ciftci, C.; Brena, S.F.; Kane, B.; Arwade, S.R. The effect of crown architecture on dynamic amplification factor of an open-grown sugar maple (*Acer saccharum* L.). *Trees* **2013**, *27*, 1175–1189. [CrossRef]
87. James, K.R.; Haritos, N.; Ades, P.K. Mechanical stability of trees under dynamic loads. *Am. J. Bot.* **2006**, *93*, 1522–1530. [CrossRef] [PubMed]
88. Moore, J.R.; Maguire, D.A. Natural sway frequencies and damping ratios of trees: Influence of crown structure. *Trees* **2005**, *19*, 363–373. [CrossRef]
89. Kane, B. Tree failure following a windstorm in Brewster, Massachusetts, USA. *Urban For. Urban Green.* **2008**, *7*, 15–23. [CrossRef]
90. Matheny, N.; Clark, J. Tree risk assessment. *Arborist News* **2009**, *19*, 28–33.
91. Mattheck, C.; Breloer, H. *The Body language of Trees*; HMSO, Department of the Environment: London, UK, 1994.
92. Smiley, E.T.; Matheny, N.; Lilly, S. *Best Management Practices Tree Risk Assessment*; ISA: Champaign, IL, USA, 2011.
93. Ellison, M. Quantified Tree Risk Assessment used in the Management of Amenity Trees. *Arboric. Urban For.* **2005**, *31*, 57–65. [CrossRef]
94. Wessolly, L. Verfahren zur Bestimmung der Stand- und Bruchsicherheit von Bäumen [Methods for determining the safety against uprooting and stem fracture]. *Holz Als Roh-Und Werkst.* **1991**, *49*, 99–104. [CrossRef]
95. Gilman, E.; Masters, F.; Grabosky, J. Pruning Affects Tree Movement in Hurricane Force Wind. *Arboric. Urban For.* **2008**, *34*, 20–28. Available online: https://hort.ifas.ufl.edu/woody/documents/articles/EFG0702.pdf (accessed on 22 October 2022). [CrossRef]
96. Lin, C.J.; Kao, Y.C.; Lin, T.T.; Tsai, M.J.; Wang, S.Y.; Lin, L.D.; Chan, M.H. Application of an ultrasonic tomographic technique for detecting defects in standing trees. *Int. Biodeterior. Biodegrad.* **2008**, *62*, 434–441. [CrossRef]
97. Sheng, R.; Perret, L.; Calmet, I.; Demouge, F.; Guilhot, J. Wind tunnel study of wind effects on a high-rise building at a scale of 1:300. *J. Wind Eng. Ind. Aerodyn.* **2018**, *174*, 391–403. [CrossRef]
98. Li, H.; Zhang, X.; Li, Z.; Wen, J.; Tan, X. A Review of Research on Tree Risk Assessment Methods. *Forests* **2022**, *13*, 1556. [CrossRef]
99. Johnstone, D.; Moore, G.; Tausz, M.; Nicolas, M. The measurement of plant vitality in landscape trees. *Arboric. J.* **2013**, *35*, 18–27. [CrossRef]
100. Callow, D.; May, P.; Johnstone, D.M. Tree Vitality Assessment in Urban Landscapes. *Forests* **2018**, *9*, 279. [CrossRef]
101. Sinn, G.; Wessoly, L. A contribution to the proper assessment of the strength and stability of trees. *Arbor. J.* **1989**, *13*, 45–64. [CrossRef]
102. Rinn, F. Shell-wall thickness and breaking safety of mature trees. *West. Arborist* **2013**, *39*, 40–44. Available online: https://rinntech.info/wp-content/uploads/2019/08/RINN-F.-2013.-Mature-Shell-Walls.-Western-Arborist.pdf (accessed on 29 May 2022).
103. Schubert, S.; Gsell, D.; Dual, J.; Motavalli, M.; Niemz, P. Acoustic wood tomography on trees and the challenge of wood heterogeneity. *Holzforschung* **2009**, *63*, 107–112. [CrossRef]
104. Ostrovský, R.; Kobza, M.; Gažo, J. Extensively damaged trees tested with acoustic tomography considering tree stability in urban greenery. *Trees* **2017**, *31*, 1015–1023. [CrossRef]
105. Arciniegas, A.; Brancheriau, L.; Lasaygues, P. Tomography in standing trees: Revisiting the determination of acoustic wave velocity. *Ann. For. Sci.* **2014**, *72*, 685–691. [CrossRef]
106. Fikos, I.; Vargemezis, G.; Zlotnicki, J.; Puertollano, J.R.; Alanis, P.B.; Pigtain, R.C.; Villacorte, E.; Malipot, G.A.; Sasai, Y. Electrical resistivity tomography study of Taal volcano hydrothermal system, Philippines. *Bull. Volcanol.* **2012**, *74*, 1821–1831. [CrossRef]
107. Sudakova, M.; Terentyeva, E.; Kalashnikov, A. Assessment of health status of tree trunks using ground penetrating radar tomography. *AIMS Math.* **2021**, *7*, 162–179. [CrossRef]

108. Sambuelli, L.; Socco, L.V.; Godio, A. Ultrasonic, electric and radar measurements for living trees assessment. *Boll. Geofis. Teor. Appl.* **2003**, *44*, 253–279. Available online: https://www.researchgate.net/publication/228698615_Ultrasonic_electric_and_radar_measurements_for_living_trees_assessment (accessed on 29 May 2022).
109. Shigo, A.L.; Shortle, W.C. *Spruce Budworms Handbook: Shigometry—A Reference Guide*; Agriculture Handbook 1985 No 646; USA Department of Agriculture, Forest Service, Cooperative State Research Service: DC Washington, DC, USA, 1985; 48p.
110. Seaby, D.A. *Recent Advances in Detection of Wood Decay. Research for Practical Arboriculture*; Hodge, S.J., Ed.; University of York, HMSO: Heslington, UK, 1994; Volume 3, pp. 168–176.
111. Nicolotti, G.; Miglietta, P. Using High-Technology Instruments to Assess Defects in Trees. *Arboric. Urban For.* **1998**, *24*, 297–302. [CrossRef]
112. Shigo, A.L. *Modern Arboriculture*; Shigo & Trees: Barrington, NH, USA, 1991; 421p.
113. Nicolotti, G.; Socco, L.V.; Martinis, R.; Godio, A.; Sambuelli, L. Application and comparison of three tomographic techniques for detection of decay in trees. *J. Arboric.* **2003**, *29*, 66–77. [CrossRef]
114. Al Hagrey, S.A. Geophysical imaging of root-zone, trunk, and moisture heterogeneity. *J. Exp. Bot.* **2007**, *58*, 839–854. [CrossRef]
115. Vössing, K.J.; Niederleithinger, E. Nondestructive assessment and imaging methods for internal inspection of timber. A review. *Holzforschung* **2018**, *72*, 467–476. [CrossRef]
116. Soge, A.O.; Popoola, O.I.; Adetoyinbo, A.A. Detection of wood decay and cavities in living trees: A review. *Can. J. For. Res.* **2021**, *51*, 937–947. [CrossRef]
117. EU Strategy for the Protection of Biological Diversity until 2030 Restoring Nature in Life. Available online: https://eur-lex.europa.eu/legal-content/EN/TXT/?qid=1590574123338&uri=CELEX:52020DC0380 (accessed on 26 August 2020).
118. Projekt Komisji Dotyczący Ogólnego Unijnego Programu Działań w Zakresie Środowiska do 2030 r. Available online: https://data.consilium.europa.eu/doc/document/ST-11987-2020-INIT/en/pdf (accessed on 29 May 2022).
119. Peper, P.J. *New York City, New York—Municipal Forest Resource Analysis*; Center for Urban Forest Research, USDA Forest Service: Davis, CA, USA, 2007. Available online: https://www.fs.usda.gov/research/treesearch/45969 (accessed on 29 May 2022).
120. Szczepanowska, H.B. Trees in the city-green capital of values and ecosystem services. *Man Environ.* **2015**, *39*, 5–28.

land

MDPI

Article

Research on Digital Experience and Satisfaction Preference of Plant Community Design in Urban Green Space

Xinyi Chen, Yuyang Wang, Tao Huang and Zhengsong Lin *

Virtual Landscape Design Laboratory, School of Art & Design, Wuhan Institute of Technology, Wuhan 430205, China
* Correspondence: lzs2020@wit.edu.cn

Abstract: In the context of carbon neutrality, it is increasingly important to reduce carbon and increase sinks, and urban green spaces play an important role in carbon sinks. In this paper, we used virtual reality (VR) and photoplethysmographic (PPG) technology to evaluate subject satisfaction regarding urban green space plant community landscape scenes using physiological eye movement and heart rate variability (HRV) data and psychological data obtained according to positive and negative emotional adjectives (PANA). The results of the study showed the following. (1) The physiological data showed the highest visual interest in single-layer grassland. The compound layer of tree-shrub-grass composite woodland communities resulted in the strongest comfort level. (2) The psychological subjective satisfaction evaluation scores were, in descending order: tree-shrub-grass composite woodland (T-S-G) > single-layer grassland (G) > tree-grass composite woodland (T-G) > single-layer woodland (T). (3) The correlation between interest, comfort, and subjective satisfaction was significant, which verified the feasibility of the model of "interest + comfort + subjective evaluation = comprehensive satisfaction". The results of the study provide theoretical guidance for landscape design based on human perception preferences in the context of carbon neutrality as well as for the implementation of sustainable landscapes to achieve a win–win situation in which carbon sequestration and oxygen release benefits and aesthetics can coexist. The combined physiological and psychological evaluation model can also be applied to other landscapes.

Keywords: carbon fixation and oxygen release; plant communities; virtual reality; heart rate variability; landscape perception

Citation: Chen, X.; Wang, Y.; Huang, T.; Lin, Z. Research on Digital Experience and Satisfaction Preference of Plant Community Design in Urban Green Space. *Land* **2022**, *11*, 1411. https://doi.org/10.3390/land11091411

Academic Editors: Alessio Russo and Giuseppe T. Cirella

Received: 8 August 2022
Accepted: 24 August 2022
Published: 27 August 2022

1. Introduction

The 2015 Paris Agreement marked a new phase in global climate governance, proposing a concerted response to climate change in the form of autonomous national contributions [1]. The agreement states that the ecological risks posed by climate change to the earth and the resulting existential crisis for human beings can only be reduced if the world achieves peak greenhouse gas emissions as soon as possible and net zero greenhouse gas emissions in the second half of this century [2,3]. In September 2020, General Secretary Xi Jinping proposed China's future goals and vision in the general debate of the 75th session of the United Nations General Assembly: "Strive to peak carbon emissions by 2030 and strive to achieve carbon neutrality by 2060" [4].

In the early 1990s, the United States, the United Kingdom, and the Netherlands pioneered research on climate change and carbon sinks [5–7]. The Landscape Biodiversity Planning and Design System (LBPDS) was proposed by AECOM, a leading international design firm, in 2013, to enhance regional biodiversity and habitat benefits through the integration of planning and design, allowing designers to conduct detailed and comprehensive assessments of urban landscapes in a quantitative manner. It guides habitat conservation, restoration, urban form, and landscape improvement in urban environments [8]. According to the habitat theory, scholars such as Jorgensen and Gobster et al. [9,10] have proposed that the aesthetic preferences of humans are derived from survival needs and can drive changes in the landscape, thus affecting the quality of the ecological environment. Therefore, it has been confirmed that analyzing human evaluations of the environment can have positive implications for ecological conservation. In terms of technology and methods, some scholars have applied virtual reality technology to ecological landscapes during the research process. Fabrika et al. [11] carried out case studies and analyses on the ecological visualization of virtual reality technology. Chandler et al. [12] conducted an empirical evaluation of virtual environments and proposed that VR provides a promising approach to facilitate practitioners and policymakers to understand the dynamics of ecological communities. HRV is a reliable reflection of the many physiological factors modulating the normal rhythm of the heart. It provides a powerful means of observing the interplay between the sympathetic and parasympathetic nervous systems [13]. Nkurikiyeyezu and Haynes et al. [14,15] verified that comfort is a personal assessment of a person's satisfaction with the surrounding environment. They proposed the use of human HRV as a proxy for thermal comfort to evaluate human satisfaction with the environment.

Urban green space plant communities mostly comprise artificial plant communities that support important ecological functions in cities [16–18]. Zhang et al. [19] studied the quantitative relationship between carbon sources and carbon sinks between plants during the landscape operation and maintenance phases, which directly reflects the positive ecological benefits of plants. They provided relevant measures to reduce carbon sources and to increase carbon sinks, providing strong evidence supporting the creation of low-carbon landscapes. The landsense ecology theory proposes that the physical perception experience and psychological cognitive experience of human beings be considered and that the theory of urban ecological space planning, construction, and management be explored through the creation of landscape space, which focuses more on the experience feeling from the human perspective and is an effective way to realize the sustainable development of the ecosystem service function [20]. Therefore, on the basis of considering the carbon sequestration benefits of plants, it is also important to consider human perceptions of the environment. Based on this theory, Deng et al. [21] combined plant communities with human visual evaluations and used photos and videos to analyze and obtain human visual preferences for plant landscapes. Technically, the application of virtual reality and visualization technology to landscapes is a need of the present form. To combine VR and landscape, some scholars [22,23] have proposed that design risks could be discovered through human–computer interaction in virtual scenes, reducing unnecessary time and money investments to a certain extent. Duan et al. [24] compared video and photo approaches to assess the effects and differences in human physiological responses while viewing different forms of plant community landscapes. Gao et al. [25] compared three viewing formats: live, photo, and VR, to measure landscape preferences and found that the participants preferred virtual environmental scenes. In terms of landscape evaluation methods, Deng et al. [26] used human physiological and psychological indicators for the visual evaluation of plant landscapes and proved that the method of using human physiological and psychological indicators to rate plant landscapes is reliable. Zhu et al. [27] quantitatively analyzed the visual characteristics of different plant community types in relation to human psychology by monitoring brain wave changes as an intrinsic scientific indicator of human emotional changes using field scene experiments.

A review of the literature reveals that no relevant policies propose specific paths for landscape design planning with the goal of carbon neutrality. There are relatively few studies combining the theory and practice in the areas related to plant community satisfaction. At present, a few scholars have conducted preliminary theoretical studies on landscape design from the perspective of carbon sinks, with existing research areas focusing on the analysis of the carbon sink capacity of different types of plants, the interrelationship between carbon sinks and biodiversity, and the layout of low-carbon landscapes. However, most of the existing studies focus on large-scale planning and strategies, with the planning-level approaches being macroscopic. No specific application methods are proposed for small-scale plant community design based on carbon sequestration and oxygen release. In terms of technology applications, virtual roaming landscapes are still in the exploration stage, although some scholars have carried out virtual roaming designs of the landscape from an ecological perspective, but the scope of such research is still relatively narrow. Such research focuses on analyzing human immersion and less on using it to guide landscape practice. At present, this type of research is related to verifying landscape plant community preferences and is mostly based on traditional photos, videos, and other methods. In terms of research methods, most scholars only use a single visual evaluation method that is psychophysical. These methods are more oriented toward subjective evaluation and lack a certain objective basis. To evaluate future plant landscapes, multiple evaluation methods can be used in combination for analysis from multiple subjective and objective perspectives.

In summary, based on the habitat theory and the landsense ecology theory, it is proposed that human environmental satisfaction can promote the conservation of the ecological environment. Therefore, it is hoped that studying human environmental satisfaction can create a win–win situation in which ecology and aesthetics can coexist. This study suggests that the construction and evaluation of virtual landscape roaming can combine the ecological benefits of plant communities with human preferences to better guide the applications of plant communities in the landscape. The innovative model of "visual interest + physiological comfort + subjective evaluation = comprehensive satisfaction" is proposed to obtain evaluations of environmental satisfaction by combining eye movement data and heart rate data at the physiological level with psychological questionnaires and to build a landscape scenario based on the carbon sequestration and oxygen release ability. Combined with the ErgoLAB platform (Kingfar International Inc, Beijing, China, 2014) to record data and complete the questionnaire at the end of the experiment determining mutual subjective and objective verification, we analyzed how the subjects evaluated their environmental satisfaction with different vegetation cover and different plant community configurations to provide reference ideas for the creation of human-centered ecological landscape design in the future and to create harmony between humans and nature.

2. Materials and Methods

2.1. Study Area

The Guanggu Science Island was selected as the study area (30°29′10.79″ N–114°35′7.23″ E). This island is located at the southernmost point of the East Lake Science City in Wuhan, Hubei Province, and borders Ezhou City along the eastern municipal boundary of Wuhan to the east, the outer ring road of the city to the west, Leopard Creek Lake to the south, and Pear Mountain to the north (Figure 1). It has a subtropical monsoon climate, with abundant rainfall, rain and high temperatures occurring in the same season, and four distinct seasons.

The Guanggu Science Island is a high-tech industrial development zone, and the area is currently under construction, with low vegetation coverage and scattered plant configurations within the site. Creating a good urban green space landscape and creating a beautiful, comfortable, and ecological working and living environment would improve the quality of life of the staff and enable them to relieve stress and enhance their creativity [28]. Meanwhile, the role and status of urban green space with a carbon sink function have received more attention. Bertram et al. [29] also suggested that urban plant communities not only play a role in beautifying the landscape, but they also play a role in cooling and humidifying the environment. Urban green spaces have a large impact on both the ecological environment and human activities, so the landscape design of plant communities functioning as green space in the study area needs to be given attention.

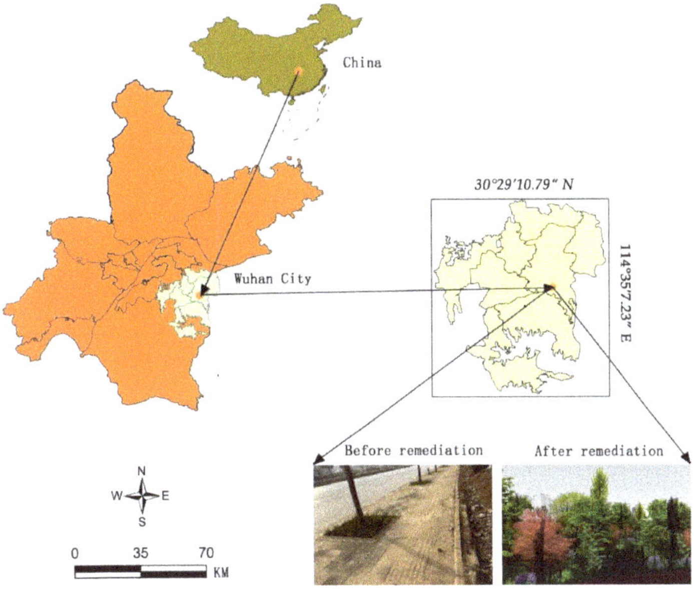

Figure 1. Location overview of study area.

2.2. Data Collection

Based on the field investigation, topographical, vegetation, and architectural data of the study area were compiled and summarized using a DJI Inspire II UAV, rangefinder, and a TITAN360 panoramic camera. A Testo 480 multifunctional meter was fixed at 1.50 m above the ground (this height is the average height of the human head and was chosen to observe the surrounding environment), and the temperature data of the study area and the control area were measured from 7:00 am to 19:00 pm. The average value was obtained and was used as the basis to set the temperature in the subsequent experimental space.

Plant preferences were derived through communication interviews with local residents. The common tree species in and around the study area were identified by combining field surveys and literature research methods, and 15 common plants and their daily carbon sequestration and oxygen release effects within a unit leaf area/single plant were summarized (Table 1) to provide a basis for the configuration of the plant community in the subsequent construction of the virtual scenario.

2.3. Scenario Setting

The configuration of the plant communities on the site was found to be mainly in the form of single-layer woodland (T), single-layer grassland (G), tree-grass composite woodland (T-G), and tree-shrub-grass composite woodland (T-S-G). Combined with the data collected in the previous stage, four plant communities were configured by selecting tree species with a strong carbon sequestration and oxygen release capacity (Table 2). To ensure the consistent ecological benefits of each plant community configuration, the difference in the total daily carbon sequestration and total daily oxygen release between each community was controlled within 500 $g{\cdot}d^{-1}$.

Table 1. Statistical results of the effects of carbon sequestration and oxygen release by plants.

Type	Species	Carbon Sequestration Per Unit Surface Area /g·m⁻²·d⁻¹	Oxygen Release Per Unit Surface Area /g·m⁻²·d⁻¹	Total Daily Carbon Sequestration (g·d⁻¹)	Total Daily Carbon Dioxide Release (g·d⁻¹)	References
Tree	Osmanthus fragrans (Thunb.) Lour.	10.58	7.70	320.04	232.76	[30]
	Cinnamomum camphora (L.) presl	9.19	6.68	445.19	323.99	[30]
	Zelkova serrata (Thunb.) Makino	5.15	3.75	270.92	197.27	[31]
	Koelreuteria paniculata	15.86	11.53	771.92	561.4	[30]
	Magnolia grandiflora L.	7.96	5.79	418.19	304.14	[32]
Shrub	Loropetalum chinense var. rubrum	5.57	5.59	110.57	110.97	[31]
	Acer palmatum Thunb.	3.43	3.00	112.79	98.65	[33]
	Rhododendron simsii Planch.	2.62	2.27	110.88	96.07	[34]
	Ligustrum quihoui Carr.	13.32	9.70	170.06	123.68	[34]
	Nandina domestica.	4.15	3.85	108.84	100.98	[34]
Grass	Cynodon dactylon (L.) Pers.	5.12	3.02	93.79	55.32	[35]
	Radix Ophiopogonis	10.05	7.31	122.01	88.75	[33]
	Axonopus compressus (Sw.) Beauv.	6.81	3.48	117.35	60.48	[35]
	Reineckea carnea (Andrews) Kunth	1.25	0.91	56.14	40.87	[33]
	Coreopsis drummondii Torr. et Gray	7.69	5.59	83.46	68.03	[36]

Table 2. Statistical results of plant communities and plant species.

Plant Community Configuration	Species	Number	Total Carbon Sequestration Per Day (g·d⁻¹)	Total Daily Carbon Sequestration Per Day (g·d⁻¹)
Single-layer woodland (T)	Zelkova serrata (Thunb.) Makino + Koelreuteria paniculata + Magnolia grandiflora L.	5 + 5 + 5	7305.50	5314.05
Single-layer grassland (G)	Radix Ophiopogonis + Axonopus compressus (Sw.) Beauv. + Cynodon dactylon (L.) Pers. + Coreopsis drummondii Torr. et Gray	11 (m^2) +11 (m^2) + 11 (m^2) + 11 (m^2)	7120.50	5352.27
Tree-grass composite woodland (T-G)	Magnolia grandiflora L. + Koelreuteria paniculata + Radix Ophiopogonis	5 + 5 + 5 (m^2)	7312.70	5321.45
Tree-shrub-grass composite woodland (T-S-G)	Cinnamomum camphora (L.) presl + Koelreuteria paniculata + Ligustrum quihoui Carr. + Loropetalum chinense var. rubrum + Radix Ophiopogonis	3 + 3 + 5 + 5 + 9 (m^2)	7506.35	5618.17

According to the designed plant community configuration, 3D Studio Max software was used to build a virtual landscape scene, which was imported into Unity 2019 for VR-plugin program settings, and then, box collider collisions and corresponding scripts were added, so that the eye tracker could detect the object and record eye tracking data. Second, the position of the Camera Rig helmet was adjusted, so that the height of the visual level was consistent with the height of the subject's visual level. Finally, the exe procedure was exported to connect with the ErgoLAB platform. The participants had an immersive experience on the roaming platform.

2.4. Experimental Design

The leader of the experiment asked the subjects to visit the experimental site to familiarize themselves with the experimental environment, procedures, and instructional phases, etc., to ensure the accuracy and reliability of the experiment. The experiment was divided into two parts. For the first group of experiments, subjects were brought to the field in the study area and wore heart rate ear clips attached to their earlobes to monitor the changes in their heart rate variability. The experiment started after ensuring that the subjects' heart rate fluctuations were normal, and during the experiment, the participants were observed for 3 min at rest. After completing the experiment, a subjective questionnaire was filled out to obtain an analysis of each subject's evaluation of their satisfaction with the landscape scene before the renovation.

For the second group of experiments, the temperature was set to the under-tree temperature (Mean (M) = 26.8 °C, standard deviation (SD) = 1.2 °C) measured in areas with high vegetation cover by entering the simulated experimental scene, while the virtual eye movement and PPG experiments were conducted afterward. The experiment was divided into three phases (Figure 2). In the pre-test phase, all of the subjects practiced the instructional phase repeatedly, and it was ensured that the subjects were wearing the helmet correctly and that the heart rate ear clips were linked to the subjects' earlobes and that both were calibrated to the ErgoLAB platform to ensure that the error of the five sight points did not exceed 20 pixel points and that the frequency of heart rate fluctuations was normal; subjects were stationary for 1 min to obtain the LF/HF ratio of their pre-experimental HRV. In the experimental phase, the subject was instructed by the person in charge to perform a virtual immersive roam of the scene according to the instructional phase. The subjects stayed in each of the four different plant community configuration areas for 30 s according to the prescribed route, and there was a 20 s distance between each plant community used to calm the subjects' heart rates. It took 3 min for the subjects to roam around the experimental scene. In the post-test stage, after completing the experiment, a subjective questionnaire was filled out to determine how the subjects evaluated their satisfaction with the landscape scene before the transformation. The SD method was used to quantify the users' feelings about their satisfaction with the plant communities encountered during landscape roaming, and the satisfaction scores of the urban plant green space experience were obtained and evaluated.

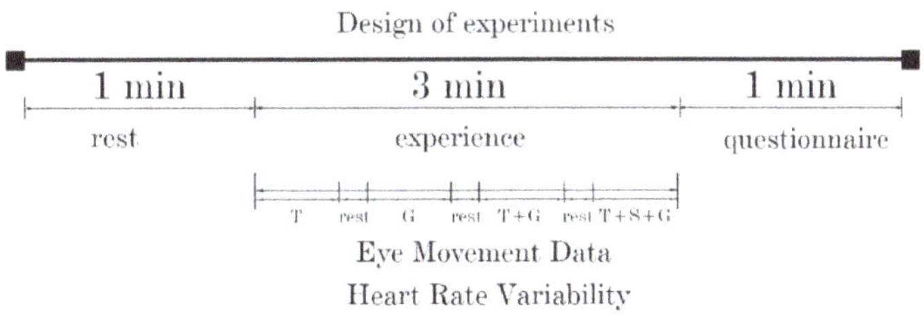

Figure 2. Experiment process.

The above steps were repeated until 88 randomly invited subjects completed the experiment (male: 41, female: 47). The experimental scenario data were monitored using a data monitoring platform (Figure 3).

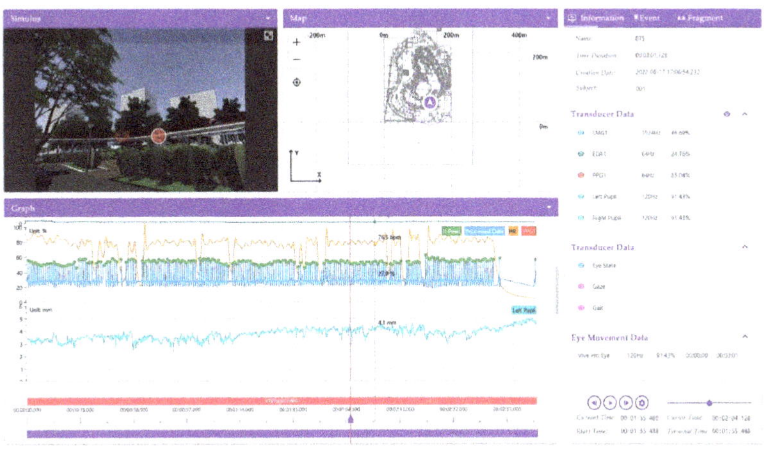

Figure 3. Experimental data monitoring interface.

2.5. *Measurements*

2.5.1. Physiological Measurement

The physiological measurements comprised virtual eye movement data and LF/HF data to monitor the subjects' visual interest and physiological comfort, respectively.

After a 3 min immersion experience in the digital roaming landscape scene, the experimental data collected from the subjects were exported through the ErgoLAB platform and normalized. The degree of interest indicates the weight of subjects' interest in each type of plant community. The higher the weight value, the higher the intensity of subjects' interest in a certain configuration form. This paper calculated the degree of interest according to the fixation time (S), fixation times (S/time), and pupil diameter (mm) of the participants who took part in the immersive experience [37]:

$$\begin{aligned} Y_{ij} &= \frac{\max x_{ij}-x_{ij}}{\max x_{ij}-x_{ij}}; i \in [1,m], j \in [1,n] \text{ and} \\ Y_{ij} &= \frac{x_{ij}-\min x_{ij}}{\max x_{ij}-\min x_{ij}}; i \in [1,m], j \in [1,n] \end{aligned} \tag{1}$$

$$f_{ij} = Y_{ij} / \sum_{j=1}^{n} Y_{ij} \tag{2}$$

$$I_i = \frac{t_i}{\sum\limits_{i=1}^{n} t_{ij}} + \frac{c_i}{\sum\limits_{i=1}^{n} c_{ij}} + \frac{p_i}{\sum\limits_{i=1}^{n} p_{ij}} \tag{3}$$

In the formula, I_i is the ith degree of interest, t_{ij} is the time from the ith fixation point to the jth fixation point, c_{ij} is the number from the ith fixation point to the jth fixation point, p_{ij} is the pupil diameter from the ith fixation point to the jth fixation point, and n is the total number of fixation points.

Qin et al. (2007) [38] pointed out that the physiological parameter heart rate variability was most strongly correlated with environmental satisfaction, and therefore, it was used as a physiological reference index to evaluate the subjects' environmental satisfaction with the selected plants. HRV values change somewhat in different human environments, and in the data analysis, the LF/HF ratio of heart rate is used to reflect the differences in comfort level. When the body is uncomfortable, parasympathetic activity decreases or sympathetic activity increases, and the body sweats or vasoconstriction occurs, causing changes in skin temperature to adapt to the ambient temperature. High-frequency heart rate variability is mainly influenced by parasympathetic activity, while low-frequency heart rate variability is mainly influenced by sympathetic activity, and when combined, the LF/HF ratio can reflect the balance of the sympathetic and parasympathetic nerves [39]. Yao et al. (2007) [40] showed that the LF/HF ratio is close to 1.0 (=0.956) when the human body is more comfortable, and the higher the LF/HF ratio, the lower the satisfaction level with the environment. The lower the heart rate variability ratio, the more comfortable and relaxed a person is in the plant environment. The HRV data obtained at 0–30 s, 50–80 s, 100–130 s, and 150–180 s during the virtual roaming scene were intercepted and compared to the 60 s heart rate variability data obtained before starting the experiment to derive the change values to compare human comfort in the different plant communities.

2.5.2. Psychological Measurement

The semantic differential method (SD) is a psychometric method for the determination of psychological feelings through verbal scales that was proposed by Osgood in 1957 [41]. This paper evaluates user feelings based on a combination of the semantic differential method and the positive affect and negative affect model (PANA model) proposed by Watson et al. [42]. Ten groups of adjectives with opposite meanings were extracted as factors to evaluate people's satisfaction with the environment: settled–worried, comfortable–uncomfortable, pleasant–sentimental, relaxed–tense, liked–disliked, happy–lonely, calm–irritable, satisfied–disgusted, excited–depressed, and eager–disappointed. The neutral factors corresponding to each group of adjectives were stability, comfort, pleasantness, relaxation, fondness, happiness, calmness, contentment, excitement, and longing. The intervals between the pairs of opposite adjectives chosen by the subjects were measured by recording and analyzing the feelings and preferences of the subjects. The questionnaire in this paper uses a five-point Likert scale, which is divided into five levels: not good, not very good, fair, better, and very good. They were assigned values of −2, −1, 0, 1, and 2, respectively [43].

2.5.3. Statistical Analysis

The relationship between the visual interest level, physical comfort level, and subjective satisfaction was further investigated. The interest level data and LF/HF change values obtained from the physiological measurements, as well as evaluation scores obtained from the psychological questionnaire on four different plant communities, were analyzed for correlation. Correlation analysis was performed using spss26.0 and the Pearson formula [44], which is as follows:

$$r = \frac{\sum_{i=1}^{n} (x_i - \overline{x}) - (y_i - \overline{y})}{\sqrt{\sum_{i=1}^{n} (x_i - \overline{x})^2} \sqrt{\sum_{i=1}^{n} (y_i - \overline{y})^2}} \tag{4}$$

In the formula, X_i represents the data of the interest level when viewing the different plant communities, where i = 1, 2, 3,..., n; $\overline{X}$ is the average value of the interest level; Y_i is the value of LF/HF ratio reduction within the different plant communities, where i = 1, 2, 3,..., n; and $\overline{Y}$ is the average value of LF/HF ratio reduction.

2.6. Methods and Logical Structures of Research

In order to scientifically and rationally determine the landscape of the urban green space plant community, the people's satisfaction evaluation must be obtained. This research mainly included four stages: data collection, virtual scene construction, experimental design, and data analysis. The data collected in the early stage were used to design the landscapes of the different plant communities in the study area. VR equipment was used to conduct 3D landscape roaming, and the PPG device was clamped into the earlobe to obtain LF/HF data in HRV. Positive and negative emotion adjective scale was set up to obtain the subjective satisfaction score. Eye movement, heart rate, and questionnaires were used to explore the effects of different plant community landscapes on human physiology and psychology to determine people's satisfaction with the different landscapes. Figure 4 shows the evaluation framework of plant community satisfaction.

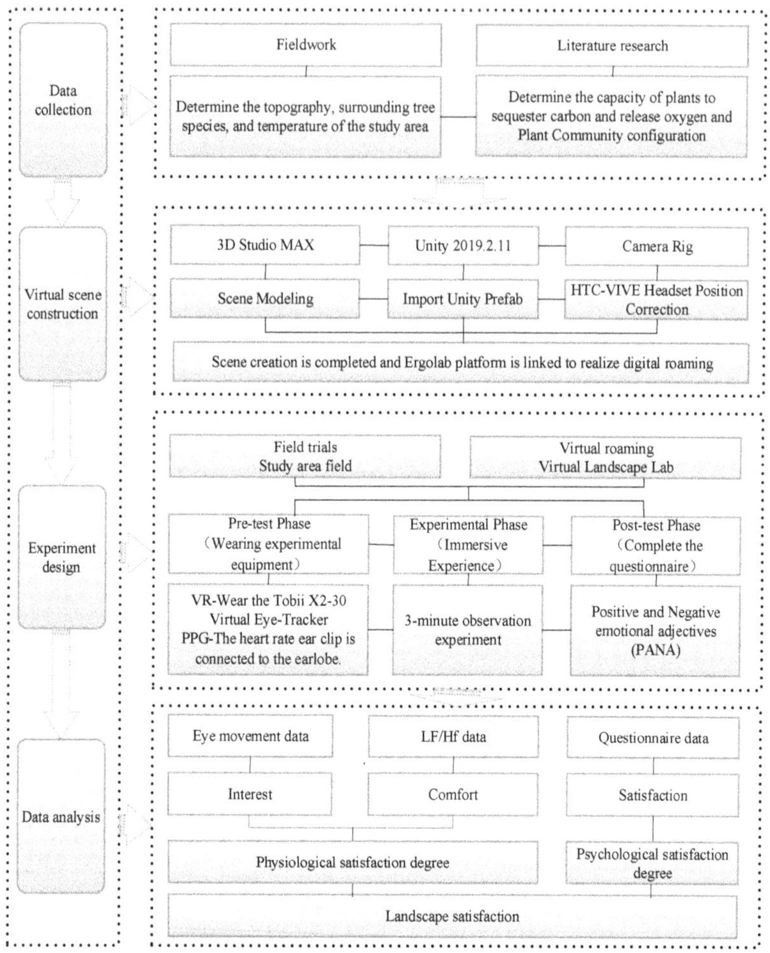

Figure 4. Methodological scheme.

3. Results

3.1. Physiological Results

3.1.1. Effect of Different Plant Communities on the Degree of Interest

Excluding abnormal data, the percentage expressing the trend of the degree of interest of 83 subjects in different plant communities was calculated according to Equations (1)–(3), and the total interest degree was set to 100% (Figure 5).

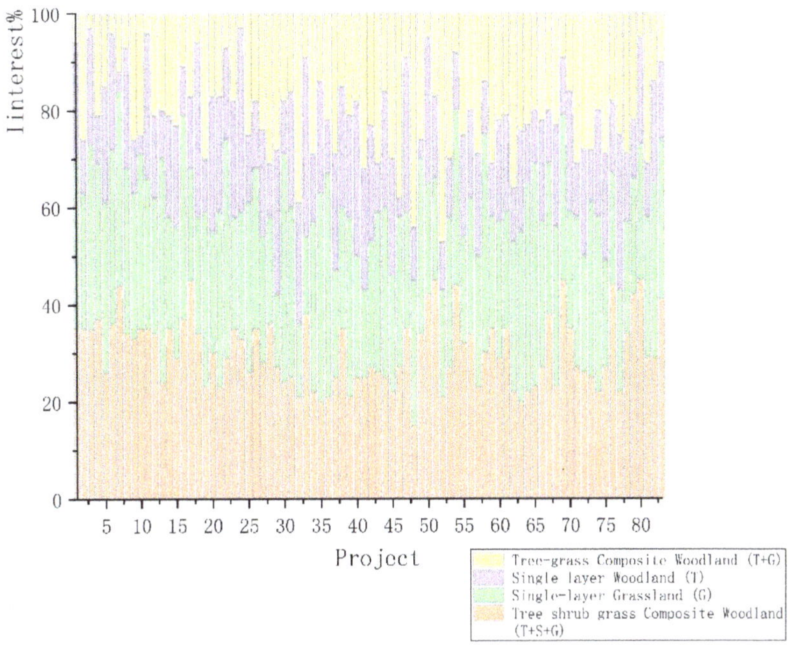

Figure 5. Trends in the degree of interest of different plant communities.

The results of this study found that when the subjects' interest level after the immersion experience was ≥1.130, it indicated that the subjects had a high interest in the plant community configuration shown in the scene [37]. For the single-layer woodland, 60 subjects had an interest level ≥ 1.130, accounting for 72.289% of the total, and 23 subjects had an interest level < 1.130, accounting for 27.711% of the total. For the single-layer grassland, 80 subjects had an interest level ≥ 1.130, accounting for 96.385% of the total, and 3 subjects had an interest level < 1.130, accounting for 3.615% of the total. For the tree-grass composite woodland, 70 subjects had an interest level ≥ 1.130, accounting for 84.337% of the total, and 13 subjects had an interest level < 1.130, accounting for 15.662% of the total; for the tree-shrub-grass composite woodland, 77 subjects had an interest level ≥ 1.130, accounting for 92.771% of the total, and 6 subjects had an interest level < 1.130, accounting for 7.229% of the total.

The data for different plant communities were counted, and the interest levels for the four plant community configurations were as follows: a single-layer woodland community (M = 1.310, SD = 0.232), single-layer grassland community (M = 1.692, SD = 0.311), tree-grass composite woodland community (M = 1.497, SD = 0.357), and tree-shrub-grass composite woodland community (M = 1.572, SD = 0.323) (Figure 6). The level of interest in the plant community was, from high to low: G > T-S-G >T-G> T.

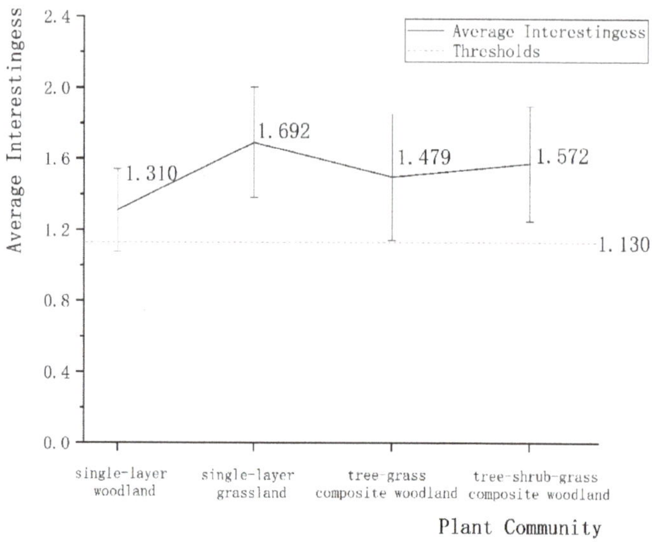

Figure 6. Trends in the mean values of interest for different plant communities.

3.1.2. Effect of Different Types of Vegetation cover and Different Plant Communities on the Comfort Level

In order to analyze the effect of the vegetation landscape on human environmental satisfaction, two scenes with low vegetation cover and high vegetation cover were selected as control groups. Excluding samples with a low data recording rate and data anomalies, a total of 83 samples were counted in the statistical analysis. The mean LF/HF value in the space with low vegetation cover in the field was significantly higher (M = 2.318, SD = 0.958) than that in the high vegetation cover landscape scenario with simulated plant cooling and humidification (M = 1.034, SD = 0.311) (Figure 7). That is to say, in the control group with different vegetation cover, the data showed that the landscape with high vegetation cover could help the autonomic nerves achieve a more balanced state, allowing for a more comfortable physiological response.

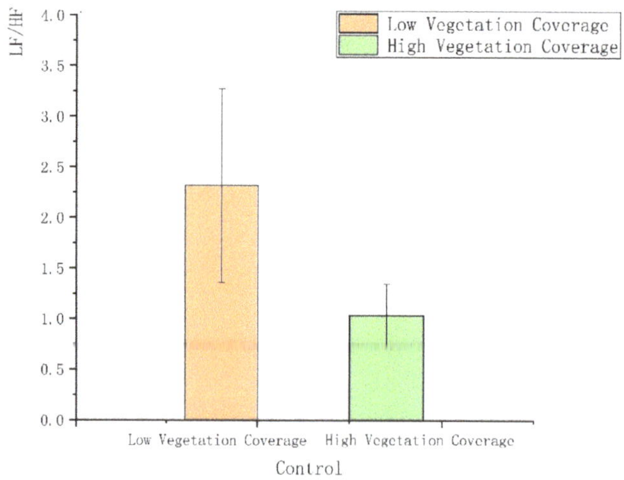

Figure 7. LF/HF values under different vegetation cover scenarios.

The variation in the subjects' LF/HF values after the immersion experience in the four plant community configuration spaces could be organized as follows (Table 3): single-layer woodland community (M = −0.239, SD = 0.739), single-layer grassland community (M = −0.298, SD = 0.431), tree-grass composite woodland community (M = −0.249, SD = 1.045), and tree-shrub-grass composite woodland community (M = −0.301, SD = 0.334).

Table 3. Descriptive statistics of variation in LF/HF values for different plant community configurations.

Plant Community Configuration	M	Min	Max
T	−0.239	−0.871	1.078
G	−0.298	−1.432	1.035
T-G	−0.249	−1.468	1.297
T-S-G	−0.301	−1.029	0.975

3.2. Psychological Results

The 88 questionnaires completed by the subjects after the experiment were collected. There were three questionnaires that were incomplete, and those questionnaires were excluded. Data from the 85 valid questionnaires were counted to determine the mean sum of the scores of the 10 landscape satisfaction evaluation factors for the different plant communities: T = 3.937, G = 11.984, T-G = 5.099, T-S-G = 12.671, and pre-retrofit site = −9.550. The evaluation curves of the landscape elements were combined into a single figure (Figure 8) to visually reflect the differences between different plant cover scenarios and different plant configurations in the evaluation of each landscape element. It was found that the evaluation curve of the scene with low vegetation cover before the renovation was located to the left of the "0" score line and was more inclined to negative adjectives (NA). The evaluation curves of the scenes with high vegetation cover were on the right side of the "0" score line and were more inclined to positive adjectives (PA). The results of the subjective questionnaire were tested for reliability, and it was found that Cronbach's $\alpha = 0.771$, indicating the reliability of the subjective questionnaire.

3.3. Significant Correlation between Interest Level, Comfort Level, and Satisfaction

Some scholars, such as Renshaw et al. (2006) [45] and Lin et al. (2022) [46], have confirmed that the eye movement data determining interest can be used to evaluate landscape satisfaction and that there is a positive correlation. Yao et al. (2008) [40] and Xiong et al. (2018) [47] confirmed that heart rate variability can respond to landscape satisfaction and that there was a negative correlation between the LF/HF ratio and landscape satisfaction. Therefore, the combination of two indicators, interest and comfort, is proposed to derive the people's satisfaction with urban green spaces. Correlation analysis of the experimental data in Section 3.1 revealed a significant correlation between the two (r = −0.977, $p = 0.023$). Correlation analysis of the two types of physiological data, as well as the subjective evaluation scores, revealed a significant positive correlation between interest and the subjective satisfaction scores (r = 0.969, $p = 0.031$); the LF/HF data showed a significantly negative correlation with the subjective satisfaction scores (r = −0.998, $p = 0.001$).

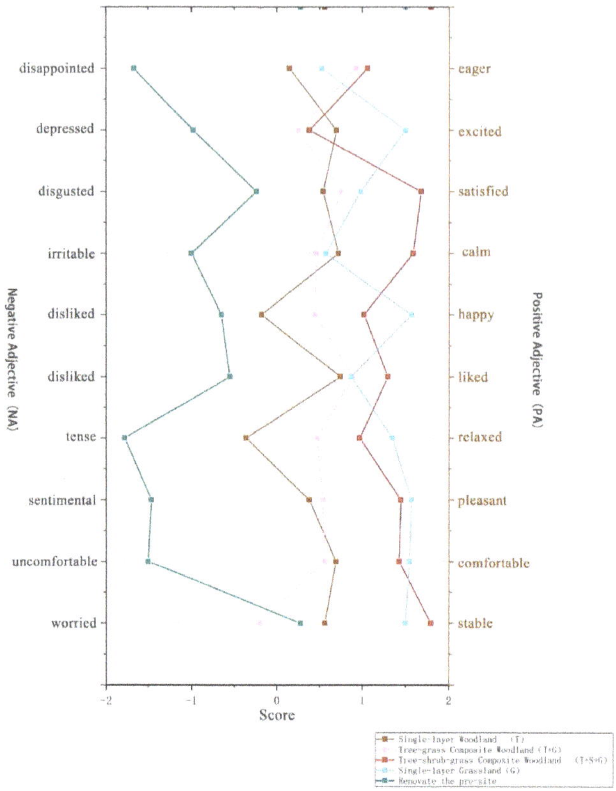

Figure 8. SD evaluation curves of different plant cover scenarios and different plant communities.

4. Discussion

According to the physiological data described in Section 3.1, it was confirmed that the subjects had a high level of physiological satisfaction with the single-story lawn and arboreal plant communities. In the virtual landscape immersion experience, the subjects had a high level of interest in all of the modified plant community designs. The results showed that the subjects' interest levels were, from highest to lowest: G > T-S-G > T-G > T. This is consistent with the findings of Franěk [48]. The change in the LF/HF ratio confirms that people are more comfortable in landscape scenes with high vegetation cover than in those with low vegetation cover, which is consistent with the findings of Qin et al. (2013) [49]. The results showed that in the modified immersion experience, the comfort level was the strongest in the tree-shrub-grass composite woodland, and the comfort levels of the four vegetation communities were, from highest to lowest: T-S-G > G > T-G >T. There are differences between the data on different communities, which indicates that the differences in visual effects and spatial scenes generated by different plant community configurations will affect people's level of satisfaction with urban green spaces. The cleanness and tidiness, openness, and vastness were the favored characteristics of lawns. People's increased interest in lawns may be due to the wide view of single-layer grassland communities, where people can observe diverse landscapes and relax. Moreover, lawns can be used to provide an open space for visitors to play or rest. People like the multilayer tree-shrub-grass configuration probably because of its stable structure, strong spatial hierarchy, and rich diversity of species. Furthermore, the tree-shrub-grass configuration provides a private space for communication and meditation. A combination of plant configuration and functional requirements can help generate higher interest and comfort in the surrounding environment and improve people's satisfaction with the plant communities in urban green space. The plant species configurations assessed in this paper can be applied to the construction of urban green spaces in central China.

According to the results of the psychological evaluation in Section 3.2, it was confirmed that landscapes with high vegetation cover resulted in better environmental satisfaction than landscapes with low vegetation cover. The subjects' satisfaction evaluation scores were, from highest to lowest: T-S-G > G > T-G > T. In the evaluation of the different plant community configurations, the evaluation scores of comfort, relaxation, and pleasantness were higher in single-layer grassland communities; the evaluation scores representing people's comfort, stability, and satisfaction were higher in the tree-shrub-grass composite woodland community. There is a high consistency between the psychological evaluation and physiological data. Through our experiments, it was confirmed that different plant community configurations will have different degrees of influence on people's psychology. The tree-shrub-grass composite woodland has the greatest influence on people, which may be because this kind of landscape is closer to the natural landscape and may act as a stress reliever. The plant community as an integral part of urban green space not only plays an important role physically in adjusting the temperature and humidity, reducing noise, breaking wind, and improving the urban microclimate, but it is also increasingly recognized and verified for its psychological wellbeing benefit. In the post-epidemic era, urban green space is important for providing various environmental services as well as social and psychological benefits. Higher satisfaction with plant communities not only promotes a better ecological conservation of those spaces but also has a healing effect on human psychological health [50].

According to the results of Section 3.3, it was confirmed that there was a significant correlation between interest level and comfort level, thus confirming that, together, the visual interest level and physical comfort constitute satisfaction indicators at the physiological level (i.e., interest + comfort = physiological satisfaction). Moreover, there was also a correlation between each of the data, thus confirming that combining physiological interest and comfort data with subjective psychological evaluation scores is meaningful and constitutes an evaluation of landscape satisfaction (i.e., physiological satisfaction + psychological satisfaction = combined satisfaction). Therefore, the combination of physiological satisfaction and psychological satisfaction constitutes a landscape satisfaction evaluation, and the combination of the two can make the obtained evaluation more comprehensive. From the results of the evaluation of plant community satisfaction, it was confirmed that people's satisfaction evaluations of plant communities can be reflected by eye movement data, heart rate data, and positive and negative emotion scales. This further confirms the "impact of natural scenery or plant landscape on human physiology and psychology" [51,52]. In the process of plant landscape evaluation in the future, by combining the single subjective evaluation and single objective evaluation methods that have previously been used by scholars, a subjective and objective evaluation model can be formed at the visual, physical, and psychological levels. This method of plant landscape evaluation can scientifically and reasonably configure plant communities' landscape, improve the role of small-scale plant community in urban green spaces, and maximize the ecological benefits and aesthetic functions of the plant communities in order to provide a basis for building a livable environment.

5. Conclusions

In this paper, VR technology, PPG technology, and psychological questionnaires were used to evaluate satisfaction with urban green spaces with different vegetation cover and different plant configurations. The application of physiological feedback equipment and eye movement technology meant that participants' satisfaction could be recorded more objectively during the short-term environmental experiences. The current findings confirmed that these new technologies can be used to analyze people's landscape satisfaction of urban green spaces. Secondly, the study revealed that different plant community structures have different impacts on people's satisfaction with urban green spaces. Finally, the results found that the study of people's satisfaction with urban green spaces is conducive to the sustainable development of their design and can stimulate people's awareness of environmental protection. Urban green spaces with high vegetation cover created higher environmental satisfaction, with single-layer grassland and tree-shrub-grass composite woodland community spaces resulting in people experiencing better physiological and psychological feelings, so in the subsequent landscape design, we should pay attention to improving vegetation cover in urban green spaces and creating more urban green space. The combination of plant community types can be flexibly matched according to the actual environment and scene. In different scenes or spaces, combinations and matches are determined according to the needs of designers or users, leading to the creation of a landscape environment, which is suitable for human activities. The results of this study can provide an objective basis for plant community configuration in future urban green spaces and can guide the design of future landscapes. On the basis of ensuring ecological benefits, the way people perceive different landscapes was considered in the present study, thus realizing the need for sustainable landscape design in the age of carbon neutrality and giving full play to the ecological benefits of the carbon sequestration and oxygen release abilities of plant communities while creating a satisfying living environment for people, with the natural environment also playing a healing role. The comprehensive satisfaction evaluation model, which combines interest, comfort, and subjective satisfaction, can also be used to evaluate other aspects of landscape satisfaction.

This paper also has some limitations. Only 88 randomly selected subjects were invited to participate in the experiment, and the effects of age differences and the differences in the professional level of the sample on the data were not considered. Therefore, future studies must be targeted to study the different responses of different nature and age variables. The experiment was conducted in early summer, and although a date with a suitable temperature was chosen for the experiment, the subjects' preferences may still be affected by weather changes, especially if considered through the lens of the field experiment method. Further research and exploration of the effects of different seasonal plant community landscapes on human landscape satisfaction should be conducted. Temperature and humidity can only be controlled in an indoor simulation scenario, which lacks the ability to simulate natural wind speed, solar radiation, and other factors, and the technology for simulating field scenarios needs to be strengthened. Only common tree species with a high carbon sink capacity were selected for the plant species configuration, and the influence of different plant species, forms, and colors on the subjects' experience was not fully considered. In the future, more specific aspects should be studied in detail.

Author Contributions: Conceptualization, X.C. and T.H.; methodology, Z.L.; software, X.C.; validation, Z.L.; formal analysis, X.C. and Y.W.; writing—original draft preparation, X.C., T.H., and Y.W.; writing—review and editing, Z.L.; All authors have read and agreed to the published version of the manuscript.

Funding: This research was supported by the Major Project of the National Social Science Fund (NSSF) of China "Theoretical and Practical Innovations for Evaluation Systems of Art Theory"(No. 21ZD11), and the Culture & Tourism IP Design based on AR/VR technology (No. 202002064005).

Institutional Review Board Statement: Not applicable.

Informed Consent Statement: All participants in the study gave informed consent.

Data Availability Statement: Not applicable.

Conflicts of Interest: The authors declare no conflict of interest.

References

1. Agreement, P. Paris agreement. In Proceedings of the 21st Session of the Conference of the Parties to the United Nations Framework Convention on Climate Change, Paris, France, 30 November–11 December 2015.
2. Zhao, X.; Ma, X.; Chen, B.; Shang, Y.; Song, M. Challenges toward carbon neutrality in China: Strategies and countermeasures. *Resour. Conserv. Recycl.* **2022**, *176*, 105–959. [CrossRef]
3. Savaresi, A. The Paris agreement: A new beginning? *J. Energy Nat. Resour. Law* **2016**, *13*, 3–29.
4. Xi, J.P. Statement by H.E. Xi Jinping President of the People's Republic of China at the General Debate of the 75th Session of the United Nations General Assembly. *Peace* **2020**, *3*, 5–7.
5. Schroeder, P. Can intensive management increase carbon storage in forests? *Environ. Manag.* **1991**, *15*, 475–481. [CrossRef]
6. Heathwaite, A.L. Disappearing peat-regenerating peat? The impact of climate change on British peatlands. *Geogr. J.* **1993**, *159*, 203–208. [CrossRef]
7. Nabuurs, G.; Mohren, G. Carbon in Dutch forest ecosystems. *Neth. J. Agric. Sci.* **1993**, *41*, 309–326. [CrossRef]
8. Felson, A.J. The design process as a framework for collaboration between ecologists and designers. In *Resilience in Ecology and Urban Design: Linking Theory and Practice for Sustainable Cities*; Pickett, S.T.A., Cadenasso, M.L., McGrath, B., Eds.; Springer: New York, NY, USA, 2013; pp. 365–382.
9. Jorgensen, A. Beyond the view: Future directions in landscape aesthetics research. *Landsc. Urban Plan.* **2011**, *100*, 353–355. [CrossRef]
10. Gobster, P.H.; Nassauer, J.I.; Daniel, T.C.; Fry, G. The shared landscape: What does aesthetics have to do with ecology? *Landsc. Ecol.* **2007**, *22*, 959–972. [CrossRef]
11. Fabrika, M.; Valent, P.; Scheer, L'. Thinning trainer based on forest-growth model, virtual reality and computer-aided virtual environment. Environ. *Model. Softw.* **2018**, *100*, 11–23. [CrossRef]
12. Chandler, T.; Richards, A.E.; Jenny, B.; Dickson, F.; Huang, J.; Klippel, A.; Neylan, M.; Wang, F.; Prober, S.M. Immersive landscapes: Modelling ecosystem reference conditions in virtual reality. *Landsc Ecol.* **2022**, *37*, 1293–1309. [CrossRef]
13. Rajendra Acharya, U.; Paul Joseph, K.; Kannathal, N.; Lim, C.M.; Suri, J.S. Heart rate variability: A review. *Med. Biol. Eng. Comput.* **2004**, *44*, 1031–1051. [CrossRef]
14. Nkurikiyeyezu, K.N.; Suzuki, Y.; Tobe, Y.; Lopez, G.F.; Itao, K. Heart rate variability as an indicator of thermal comfort state. In Proceedings of the 2017 56th Annual Conference of the Society of Instrument and Control Engineers of Japan (SICE), Kanazawa, Japan, 19–22 September 2017; pp. 1510–1512.
15. Haynes, B. The impact of office comfort on productivity. *J. Facil. Manag.* **2008**, *6*, 37–51. [CrossRef]
16. Fang, J.; Guo, Z.; Piao, S.; Chen, A. Terrestrial vegetation carbon sinks in China, 1981–2000. *Sci. China Earth Sci* **2007**, *50*, 1341–1350. [CrossRef]
17. Aronson, M.F.J.; Lepczyk, C.A.; Evans, K.L.; Goddard, M.A.; Lerman, S.B.; MacIvor, J.S.; Nilon, C.H.; Vargo, T. Biodiversity in the city: Key challenges for urban green space management. *Front. Ecol. Environ.* **2017**, *15*, 189–196. [CrossRef]
18. Strohbach, M.W.; Arnold, E.; Haase, D. The carbon footprint of urban green space—A life cycle approach. *Landsc. Urban Plan.* **2012**, *104*, 220–229. [CrossRef]
19. Zhang, Y.; Meng, W.; Yun, H.; Xu, W.; Hu, B.; He, M.; Zhang, L. Is urban green space a carbon sink or source?A case study of China based on LCA method. *Environ. Impact Assess. Rev.* **2022**, *94*, 106766. [CrossRef]
20. Zhao, J.; Liu, X.; Dong, R.; Shao, G. Landsenses ecology and ecological planning toward sustainable development. *Int. J. Sustain. Dev. World Ecol.* **2015**, *23*, 293–297. [CrossRef]
21. Deng, L.; Luo, H.; Ma, J.; Huang, Z.; Sun, L.-X.; Jiang, M.-Y.; Zhu, C.-Y.; Li, X. Effects of integration between visual stimuli and auditory stimuli on restorative potential and aesthetic preference in urban green spaces. *Urban For. Urban Green.* **2020**, *53*, 126–702. [CrossRef]
22. Song, J.; Huang, S. Virtual Reality (VR) technology and landscape architecture. In Proceedings of the MATEC Web of Conferences, Padang, Indonesia, 2–4 May 2018.
23. Pei, L. Green urban garden landscape design and user experience based on virtual reality technology and embedded network. *Environ. Technol. Innov.* **2021**, *24*, 101–738. [CrossRef]
24. Duan, Y.; Li, S. Effects of Plant Communities on Human Physiological Recovery and Emotional Reactions: A Comparative Onsite Survey and Photo Elicitation Study. *Int. J. Environ. Res. Public Health* **2022**, *19*, 721. [CrossRef]
25. Gao, T.; Liang, H.; Chen, Y.; Qiu, L. Comparisons of Landscape Preferences through Three Different Perceptual Approaches. *Int. J. Environ. Res. Public Health* **2019**, *16*, 4754. [CrossRef] [PubMed]
26. Deng, L.; Li, X.; Luo, H.; Fu, E.K.; Ma, J.; Sun, L.-X.; Huang, Z.; Cai, S.Z.; Jia, Y. Empirical study of landscape types, landscape elements and landscape components of the urban park promoting physiological and psychological restoration. *Urban For. Urban Green.* **2020**, *48*, 126–488. [CrossRef]
27. Zhu, H.Z.; Yang, F.; Bao, Z.Y.; Nan, X.G. A study on the impact of Visible Green Index and vegetation structures on brain wave change in residential landscape. *Urban For. Urban Green.* **2021**, *64*, 22. [CrossRef]
28. Choumert, J.; Salani, J. Provision of urban green spaces: Some insights from economics. *Landscape Res.* **2008**, *33*, 331–345. [CrossRef]
29. Bertram, C.; Rehdanz, K. The Role of Urban Green Space for Human Well-Being. *Ecol. Econ.* **2015**, *120*, 139–152. [CrossRef]

30. Gong, Y.; Luo, X. Experimental Study on the Carbon Sequestration Benefit in Urban Residential Green Space Based on Urban Ecological Carrying Capacity. *Sustainability* **2022**, *14*, 7780. [CrossRef]
31. Zhang, J.; Sui, Y.H. Selection and design of ornamental plants for low-carbon urban landscape: A review. *Appl. Mech. Mater.* **2012**, *174*, 2314–2317. [CrossRef]
32. Yang, Q.; Yao, D.X.; Li, S.Q.; Zhang, Z.G.; Cheng, Y.C.; Zhao, K. The research progress on carbon fixation and oxygen release of phytoremediation. *J. Coal. Sci. Eng.* **2012**, *18*, 196–200. [CrossRef]
33. Chen, J.; Wang, P.; Gao, M.; Hou, W.; Liao, H. Carbon sequestration capacity of terrestrial vegetation in China based on satellite data. *J. Chin. Econ. Bus. Stud.* **2022**, *20*, 109–124. [CrossRef]
34. Dan, W. Study on Plant Carbon Sequestration Capacity and Ecological Landscape Configuration. In Proceedings of the WCT, Beijing, China, 18–21 June 2018; pp. 969–974.
35. Xue, X.; Zhang, J.; Sun, Y.; Zhuang, J.; Wang, Y. Study of carbon sequestration & oxygen release and cooling & humidifying effect of main greening tree species in Shanghai. *J. Nanjing For. Univ. Nat. Sci. Ed.* **2016**, *40*, 81–86.
36. Zhang, H.; Wang, L. Species Diversity and Carbon Sequestration Oxygen Release Capacity of Dominant Communities in the Hancang River Basin, China. *Sustainability* **2022**, *14*, 5405. [CrossRef]
37. Wang, X.H.; Yuan, Y.C.; Jin, Y.C. Research on Readers' Interests in Patent Literature and Its Topology Representation. *J. Libr. Inf. Serv.* **2016**, *60*, 11.
38. Qin, J.; Zhou, X.; Sun, C.; Leng, H.; Lian, Z. Influence of green spaces on environmental satisfaction and physiological status of urban residents. *Urban. For. Urban. Green.* **2013**, *12*, 490–497. [CrossRef]
39. Malliani, A.; Pagani, M.; Lombardi, F.; Cerutti, S. Cardiovascular neural regulation explored in the frequency domain. *Circulation* **1991**, *84*, 482–492. [CrossRef]
40. Yao, Y.; Lian, Z.; Liu, W.; Shen, Q. Experimental study on physiological responses and thermal comfort under various ambient temperatures. *Physiol. Behav.* **2008**, *93*, 310–321. [CrossRef]
41. Osgood, C.E. Semantic differential technique in the comparative study of cultures. *Am. Anthropol.* **1964**, *66*, 171–200. [CrossRef]
42. Watson, D.; Tellegen, A. Toward a consensual structure of mood. *Psychol. Bull.* **1985**, *98*, 219. [CrossRef]
43. Joshi, A.; Kale, S.; Chandel, S.; Pal, D. Likert scale: Explored and explained. *Br. J. Appl. Sci. Technol.* **2015**, *7*, 396–403. [CrossRef]
44. Benesty, J.; Chen, J.; Huang, Y.; Cohen, I. Pearson Correlation Coefficient. In *Noise Reduction in Speech Processing*; Springer: Berlin/Heidelberg, Germany, 2009; Volume 2, pp. 1–4.
45. Renshaw, J.A.; Finlay, J.; Webb, N. Getting a measure of satisfaction from eyetracking in practice. In *CHI'06 Extended Abstracts on Human Factors in Computing Systems*; ACM: New York, NY, USA, April 2006.
46. Lin, Z.; Wang, Y.; Ye, X.; Wan, Y.; Lu, T.; Han, Y. Effects of Low-Carbon Visualizations in Landscape Design Based on Virtual Eye-Movement Behavior Preference. *Land* **2022**, *11*, 782. [CrossRef]
47. Xiong, J.; Lian, Z.; Zhang, H. Physiological response to typical temperature step-changes in winter of China. *Energy Build.* **2017**, *138*, 687–694. [CrossRef]
48. Franěk, M.; Petruálek, J.; Efara, D. Eye movements in viewing urban images and natural images in diverse vegetation periods. *Urban For. Urban Green.* **2019**, *46*, 126–477. [CrossRef]
49. Qin, J.; Sun, C.; Zhou, X.; Leng, H.; Lian, Z. The effect of indoor plants on human comfort. *Indoor Build. Environ.* **2014**, *23*, 709–723. [CrossRef]
50. Moreira, T.C.L.; Polize, J.L.; Brito, M.; da Silva Filho, D.F.; Chiavegato Filho, A.D.P.; Viana, M.C.; Andrade, L.H.; Mauad, T. Assessing the impact of urban environment and green infrastructure on mental health: Results from the São Paulo Megacity Mental Health Survey. *J. Expo. Sci. Environ. Epidemiol.* **2022**, *32*, 205–212. [CrossRef] [PubMed]
51. Ulrich, R.S.; Simons, R.F.; Losito, B.D.; Fiorito, E.; Miles, M.A.; Zelson, M. Stress recovery during exposure to natural and urban environments. *J. Environ. Psychol.* **1991**, *11*, 201–230. [CrossRef]
52. Chang, C.; Chen, P. Human response to window views and indoor plants in the workplace. *HortScience* **2005**, *40*, 1354–1359. [CrossRef]

land

MDPI

Article

Analysis of Anthropogenic Disturbances of Green Spaces along an Urban–Rural Gradient of the City of Bujumbura (Burundi)

Henri Kabanyegeye [1,2], Yannick Useni Sikuzani [3], Kouagou Raoul Sambieni [4,5], Didier Mbarushimana [6], Tatien Masharabu [1] and Jan Bogaert [2,*]

1 Research Centre for Natural and Environmental Sciences, University of Burundi, Bujumbura P.O. Box 2700, Burundi
2 Gembloux Agro-Bio Tech, University of Liège, Passage des Déportés 2, 5030 Gembloux, Belgium
3 Ecology, Ecological Restoration and Landscape Research Unit, Faculty of Agricultural Sciences, University of Lubumbashi, Lubumbashi P.O. Box 1825, Congo
4 Regional Post-Graduate Training School on Integrated Management of Tropical Forests and Lands (ERAIFT), Kinshasa P.O. Box 15373, Congo
5 Faculty of Architecture, University of Lubumbashi, Lubumbashi P.O. Box 1825, Congo
6 Burundian Office for Environmental Protection, Bujumbura P.O. Box 2757, Burundi
* Correspondence: j.bogaert@uliege.be; Tel.: +32-473-86-32-65

Abstract: Bujumbura city has diversified but unevenly distributed green spaces. The typology and anthropogenic disturbances of these green spaces are still unknown. This study presents a typology of green spaces along the urban–rural gradient through a literature review. It assesses the presence of anthropogenic disturbances through inventories in 100 m × 100 m grids. Data reveal that Bujumbura's green spaces are made up of green squares concentrated exclusively in urban areas, cemeteries present in peri-urban areas and sports green spaces observable all along the urbanization gradient. These green spaces are more exposed to trampling, which is more present in administrative entities with a peri-urban morphological status, as opposed to various constructions in administrative entities with an urban status. Finally, significative pairwise associations of anthropogenic disturbances were observed. The results show the need to protect these green spaces from all kinds of anthropogenic disturbances by raising the eco-responsible awareness of the population and the municipal authorities.

Keywords: green spaces; Bujumbura; peri-urbanization; anthropogenic disturbances; urban ecology

Citation: Kabanyegeye, H.; Sikuzani, Y.U.; Sambieni, K.R.; Mbarushimana, D.; Masharabu, T.; Bogaert, J. Analysis of Anthropogenic Disturbances of Green Spaces along an Urban–Rural Gradient of the City of Bujumbura (Burundi). *Land* **2023**, *12*, 465. https://doi.org/10.3390/land12020465

Academic Editors: Alessio Russo and Giuseppe T. Cirella

Received: 13 December 2022
Revised: 2 February 2023
Accepted: 3 February 2023
Published: 12 February 2023

1. Introduction

Urbanization has become a global phenomenon as a result of the urban transition that started in the second half of the 20th century [1,2]. Indeed, in 1950, only 30% of the world's population lived in urban areas, a proportion that has increased to 57% in 2021 [3]. Urbanized areas, which occupy nearly six million square kilometers, or 1.17% of the Earth's surface, are expected to increase by more than one million square kilometers by 2030 [4].

In developing countries, particularly in Sub-Saharan Africa, urbanization is currently happening at a high speed due to the combined effects of rural–urban migration, natural population growth and civil wars [5]. In addition, this demographic explosion in urbanized areas takes place in a context of absence of planning that has prevailed since the 1950s [6]. The result is an accelerated urban spatial dynamic materialized through a dual process of densification of already built spaces and generally of an anarchic extension of the built environment in an area that conserves its rural characteristics, leading to the creation of urban and peri-urban areas, respectively [7,8]. This situation is not without environmental consequences.

Indeed, through uncontrolled urban spatial expansion, urbanization puts significant pressure on natural ecosystems through their fragmentation, leading to a regression of the green space coverage [9,10] and an alteration of their ecological functionality [8,11–14]. Yet, green spaces, considered as vegetated surfaces in urbanized environments, are essential in urbanized environments to mitigate several scourges, such as temperature increase, biodiversity loss, and degradation of social interactions, health conditions and the wellbeing of inhabitants [15–18]. However, the degree of influence of green spaces on the wellbeing of urban dwellers varies according to several criteria, such as the type of green space and the types and magnitudes of anthropogenic disturbances [19–21].

Several typologies of green spaces have been proposed based on function, ownership status and physical characteristics [22–29]. Other typologies have been based on the degree of naturalness by distinguishing between natural, rural and peri-urban green spaces [30]. However, these typologies focus rather on European and Asian cities, and rarely on African cities where the few existing typologies focus only on the city of Kumasi in Ghana [31] and the cities of Kinshasa [7] and Lubumbashi [12,13] in the Democratic Republic of the Congo. No typology, however, has a universal scope due to the high variability in environmental and urban planning contexts, a situation that makes it difficult to compare studies [32].

In African cities, green spaces in urbanized areas are subject to anthropogenic disturbances [8,13]. These disturbances are consequent to the observed gaps in their land tenure security, the uncooperative attitude of the general public, the low level of coordination among stakeholders involved in their management and the low priority given to them in the municipal budgets [33,34].

Bujumbura city, the political and economic capital of the Republic of Burundi, is not spared by this alarming trend since its spatial growth is marked by the densification of existing central neighborhoods, and by the progressive destruction of green spaces to install new habitats and other infrastructures such as roads and monuments [5,35]. In addition, housing estates are being created on the outskirts of the city without any overall plan or coherence in order to contain the constantly increasing demand for housing [5], amplifying the anthropogenic disturbance of the various residual green spaces in the city, which are unevenly distributed and whose ecological functionality is still poorly known [35]. It has recently been known that the recent green spaces typology of Bujumbura was only based on soil moisture (terra firma and wetland green spaces) and socio-economic factors (domestic garden) [36]. There is, on the one hand, a need for a more detailed typological analysis of the public green spaces of Bujumbura to improve and adapt their management. On the other hand, adapting the green spaces preservation measures is important to characterize the anthropogenic disturbances to which they are subjected, and which are assumed to act in a synergistic manner. However, considering the spatial pattern, the ecological functionalities and management of green spaces change according to the extent of urbanization [13], which is crucial to study green spaces within cities by considering separately urban and peri-urban areas.

Thus, the purpose of our study is based on three hypotheses: (i) the green spaces of Bujumbura city are not only less diversified in types but also more concentrated in the urban area due to the proximity of diverse public services involved in their management; (ii) due to building extension and modernization, the most important anthropogenic disturbances of the green spaces of the city of Bujumbura are constructions, which are present in zones with urban morphological status; (iii) there are associations of anthropogenic disturbances in the studied green spaces in the urban and peri-urban zones.

2. Materials and Methods

2.1. Study Area

The city of Bujumbura was created in 1897 on the borders of Tanganyika lake by the Germans on a site called Kajaga. Since then, it was known as the economic and political capital of the country until February 2019 [37]. Located in the western part of Burundi (Figure 1), between latitudes 3°30′ and 3°51′ S and longitudes 29°31′ and 29°42′ E, it covers an area of 10,462 hectares, whereas it was estimated at 30 hectares in 1907 [38]. It is administratively split into three communes (Muha, Mukaza and Ntahangwa) that cover thirteen administrative entities or urban zones. A municipality, whether urban or rural, is a decentralized territorial authority with legal personality and organic and financial autonomy [39]. A group of rural municipalities form a province; a group of urban municipalities compose a city. A rural municipality is subdivided into zones and census hills while the urban municipality is subdivided into zones and districts. Consequently, a zone is a deconcentrated district of the municipality, an intermediate between the municipality and the census hill or district.

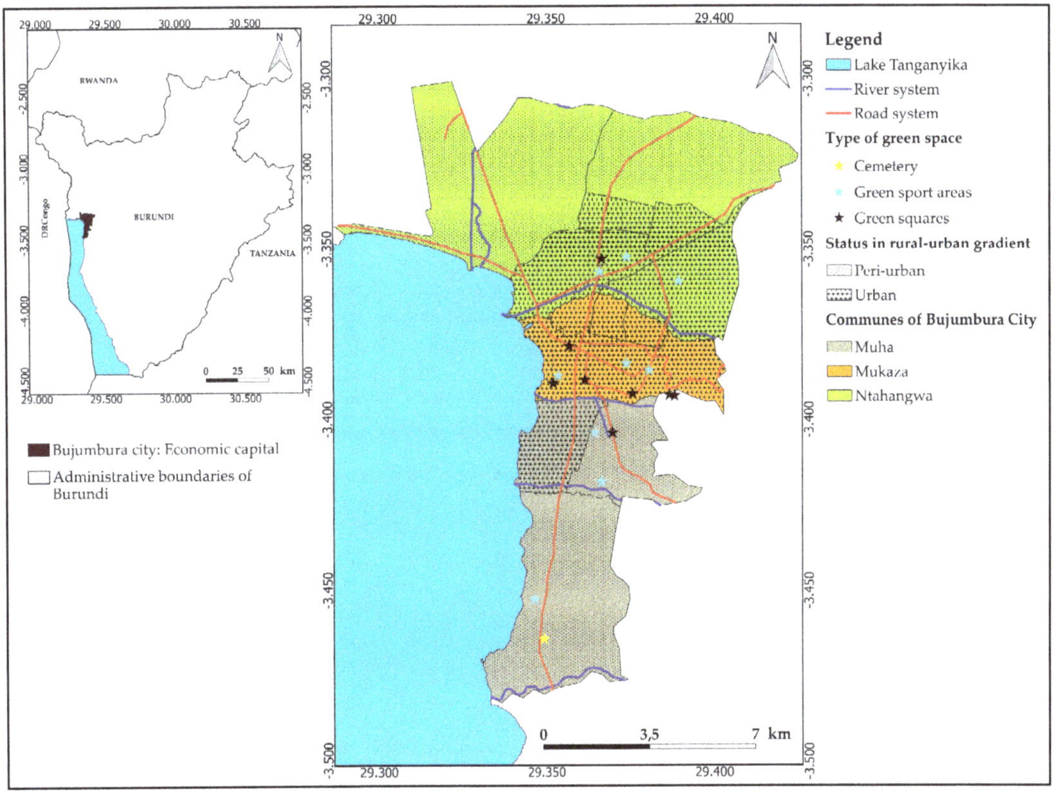

Figure 1. Location of Bujumbura city in the Republic of Burundi, near Tanganyika Lake. The municipalities of the city have both morphological statuses of urban and peri-urban areas. Within these areas, three types of green spaces have been found and located.

The climate of Bujumbura is characterized by a dry season and a rainy season, each with two variations: the long dry season (June to mid-September), the short dry season (mid-December to mid-February), the long rainy season (mid-February to May) and the short rainy season (mid-September to December). The average annual temperature is above 23 °C with a total rainfall of between 800 and 1000 mm [40]. The relief is characterized by a plain that rises from west to east, with an average altitude of 820 m. Ntahangwa is the largest river flowing through the city. If clay soils dominate the north of this river, the south is dominated by sandy soils [35]. Water-logged or flooded during the rainy season, clay soils become very hard during the dry season and show shrinkage cracks. The natural vegetation of the region is savanna, miombo woodland and wetlands [41]. Miombo woodland is mainly dominated by trees of the genus *Brachystegia* and *Uapaca*. The undergrowth is very sparse and is dominated by grass species [42]. According to the phytogeographical classification of [43], Burundi is covered by two regional centers of endemism: the Afromontane region (an archipelago-like regional center of endemism) and the Lake Victoria Regional Mosaic. Bujumbura belongs to the Lake Victoria Regional Mosaic. The natural vegetation has almost disappeared in the Bujumbura region, except in the Rusizi Natural Reserve (5932 ha), located 15 km north of the city, and in Kibira National Park, the largest montane rain forest reserve of the country (40,000 ha), located less than 35 km to the northeast [36]. This is due, among other reasons, to the use of charcoal as the main energy source for the urban population of Bujumbura [44]. It has previously been demonstrated that, when studying the spontaneous vegetation in the city of Bujumbura, ruderal plant assemblages dominate in the most urbanized areas, while in the outskirts of the city, the vegetation shows similarities to natural plant assemblages in the region [41]. Fishing, industry, administration and urban agriculture are the main socio-economic activities carried out by the city's population, estimated at 497,166 inhabitants in 2008 [45] and 792,504 inhabitants in 2022 [46]. In this city, urban agriculture has considerable importance. It is dominant in 50% of the urban territory that is not built. All food plants adapted to the soils and climate of the region are grown in these spaces (e.g., maize, cassava, soybeans, peanuts and vegetables), mainly for consumption at the household level. Cultivation of rice in the swamps northwest of the city (Kiyange and Carama) and cotton is an important source of income for urban farmers. Urban horticulture is also under development within this city, and it currently receives support (since November 2001) from the Burundian Ministry of Agriculture and the Food Agriculture Organization (FAO). It has been reported that the average number of children per household and the literacy rate were about 4.8 and 82.6%, respectively, in 2017 [47]. Traditional practitioners of Burundian medicine collect further plants from these wetlands and from terra firma vegetation to treat various diseases of humans and livestock [48].

2.2. Segmentation of the Urban–Rural Gradient Zones

In order to analyze the morphological status of the different administrative entities composing the city, Bujumbura was gridded by 698 square grids of 25 ha each, enabling the observation of urbanization morphology on Google Earth [49]. A flowchart based on the morphological characteristics of the different zones along the urban–rural gradient was used to qualify each of the sub-zones as urban or peri-urban [50]. Urban zones combine high densities and proportions of built-up zones as opposed to peri-urban zones that have some discontinuity of the built-up area at lower proportions [50]. In the context of Bujumbura (Figure 1), the zones with an urban morphological status area are characterized by popular neighborhoods and medium standing, very dense buildings and paved streets. Unfortunately, with houses that date from 1850 to 1983, they can have a small part with no marked roads. Zones with peri-urban morphological status are under development and generally not characterized by a clear road infrastructure. The older part of the peri-urban area consists of houses that date from 1983.

2.3. Data Collection

2.3.1. Green Space Sampling

To facilitate a parcel analysis [50], we first made a pre-identification on Google Earth of the green spaces that have a minimum surface of one hectare. These were identified in the field in November 2020, in a rather random and simple way by following the boulevards, pavements and avenues of the city [7,34,51,52]. The geographical coordinates of the sampled green spaces were mapped using Arc GIS 10.2 software (Figure 1).

2.3.2. Typological Analysis

The inventory focused on public green spaces without taking into account their management type. The literature review of existing green space inventories as well as the field observations allowed the establishment of an adapted typology.

In the absence of a single typology that covers all these particularities, we combined four existing typologies to characterize the green spaces of Bujumbura: (i) the typology of the Association of Engineers of the Cities of France (1995) described by [53], (ii) the typology of [22] in France, (iii) the typology implemented by [23] in Belgium and (iv) the typology used by [7] in the Democratic Republic of the Congo. Among the existing typologies, those of [7,13,31] were used in this study since they are adapted to the socio-ecological context of the city since they have the particularity of classifying green spaces on the land use.

2.3.3. Definition and Inventory of Anthropogenic Disturbances of Bujumbura's Green Spaces and Their Indicators

Definitions of the types of anthropogenic disturbances and their indicators were developed prior to the fieldwork. They were based on exploratory observations and contacts with the population living close to the green spaces. Indeed, exchanges with local populations, through focus groups of 9 to 12 people, made it possible to identify and validate the types of anthropogenic disturbance. In total, ten focus groups were organized, as more than eight new indicators were no longer cited. Eight disturbance types were selected (Table 1) and their identification consisted of noting the presence or absence of the corresponding indicator(s) in each grid cell in which green spaces were identified. The size of the selected grid cells allowed for a parcel-based and sub-square kilometer analysis [50]. The number of grid cells for each green space depended on its area and shape. In total, 18 green spaces were identified. These green spaces were located in five administrative entities with different morphological statuses. A total of 14 green spaces, with 111 grid cells, were located in entities with urban morphological status. Four green spaces, located in two entities with peri-urban morphological status, were covered by 48 grid cells.

Table 1. List of types and indicators of anthropogenic disturbances of green spaces in the city of Bujumbura, based on exploratory observations made from September to October 2021.

Types of Disturbance	Indicators
Buildings	Presence of houses, monuments, parking lots, tiles, sports surfaces, bleachers, makeshift housing
Crops	Presence of crop plants
Rubble	Presence of construction debris
Flooding	Presence of stagnant water from washing vehicles, carpets or tents
Military installations	Presence of tents, obstacles, shelters, trenches, etc.
Household waste	Unauthorized dumping of household waste
Improvised parking lots	Presence of vehicles, motorcycles and bicycles
Bare floors	Observation of the ground without vegetation cover, mainly due to trampling and livestock grazing

Based on the study of [54], the data analysis was based on the model which describes disturbance processes according to a set of characteristics; notably, type of disturbance (classification of events or observations), frequency (extent of the disturbance), the spatial structure of the disturbance (in this case the urban–rural gradient) and synergy (the interaction with other disturbances). Relative frequencies were calculated for each type of anthropogenic disturbance from the presence-absence data [54]. To test that the most important anthropogenic disturbances of the green spaces of the city of Bujumbura are constructions, mainly present in zones with urban morphological status, the Friedman test, an alternative procedure to the analysis of variance method for repeated measures [55,56], was performed under SPSS to compare the different types of disturbance in terms of their frequencies [53,57]. It allowed the comparison of more than two matched samples by determining whether the values of the samples were different from those of the other samples [55,58]. When the null hypothesis of the Friedman test, assuming there is no difference between the frequencies of occurrence of the eight disturbance types along the urbanization gradient, was rejected, post hoc tests were applied [55,56] to compare pairwise the different types of disturbance in order to identify those disturbances that would be different from each other in terms of frequency of occurrence. To verify the existence of associations of anthropogenic disturbances in the studied green spaces in the urban and peri-urban zones, a Chi-square test of independence was performed to assess the presence of positive or negative associations between disturbance types [49].

3. Results

3.1. Typology of Green Spaces along the Urban–Rural Gradient of the City of Bujumbura

The literature review and the field observations revealed that the green spaces in Bujumbura are mainly made up of green sports areas (stadiums and sports centers), green spaces accompanying roads, green spaces accompanying public buildings (whose role is to enhance the building), green spaces accompanying industrial and commercial establishments, green spaces of social and educational establishments (gardens of nurseries, primary and secondary schools and universities), trees of alignment, green squares, cemeteries, traffic circles and nurseries. These green spaces can be grouped into three distinct types: green squares, green sports areas and cemeteries, and are rather small in size, only up to 20 hectares for the largest green space (Table 2).

Table 2. Spatial data of the 18 green spaces present along the urban–rural gradient of the Bujumbura city. GS: green space, EA: administrative entity, GC: geographic coordinates, SL: southern latitude, EL: eastern longitude, MS: morphological status, NC: number of grid cells, P: perimeter, TG: type of green space, GS: green squares, GA: green sports area, CE: cemetery.

GS	AE	GC		MS	NC	Area (m^2)	P (m)	TG
		SL	EL					
1	Rohero	3°22′42.43″	29°21′24.24″	U	35	207,800.00	2011.89	GS
2	Kanyosha	3°27′09.54″	29°20′47.20″	PU	18	97,441.96	1274.12	GA
3	Musaga	3°24′13.74″	29°22′10.29″	PU	10	56,600.50	1034.88	GS
4	Kanyosha	3°27′50.94″	29°20′57.39″	PU	10	47,051.50	965.45	CE
5	Rohero	3°23′32.67″	29°22′31.64″	U	11	62,860.08	932.60	GS
6	Gihosha	3°21′33.89″	29°23′21.23″	U	9	43,474.50	859.68	GA
7	Rohero	3°23′00.59″	29°22′24.81″	U	9	49,677.99	846.90	GA
8	Rohero	3°23′13.48″	29°21′12.48″	U	6	39,878.50	842.18	GA
9	Rohero	3°23′21.44″	29°21′07.40″	U	6	39,236.00	833.60	GS
10	Rohero	3°23′34.76″	29°23′17.02″	U	8	35,719.96	795.65	GS
11	Ngagara	3°21′10.58″	29°21′58.83″	U	4	37,514.10	787.11	GS
12	Ngagara	3°21′23.97″	29°21′56.81″	U	6	30,754.55	757.94	GA
13	Rohero	3°23′33.83″	29°23′11.44″	U	7	31,144.50	742.86	GS
14	Musaga	3°24′13.73″	29°21′51.95″	PU	6	22,272.00	653.29	GA
15	Rohero	3°23′07.56″	29°22′49.81″	U	4	14,071.00	477.12	GA
16	Ngagara	3°21′08.67″	29°22′25.13″	U	4	11,440.00	432.09	GA
17	Musaga	3°25′05.41″	29°21′58.85″	PU	4	10,054.93	421.35	GA
18	Rohero	3°23′17.93″	29°21′41.29″	U	2	10,680.51	405.18	GS

In relation to the number of green spaces found in urban and peri-urban zones, cemeteries are absent in urban zones but present in peri-urban zones in small proportions (20.0%). Green sports areas are present in urban (46.2%) and peri-urban (60.0%) zones. Green squares are also present in urban (53.8%) and peri-urban zones (20.0%). Considering the importance of each type of green space as a proportion of the total area, cemeteries represent 20.2% in the zones with peri-urban morphological status, against 55.6% and 24.3% for green sports areas and green squares, respectively. In the zones with urban morphology, green squares dominate green sports areas (69.15% vs. 30.82%) while cemeteries are absent (Figure 2).

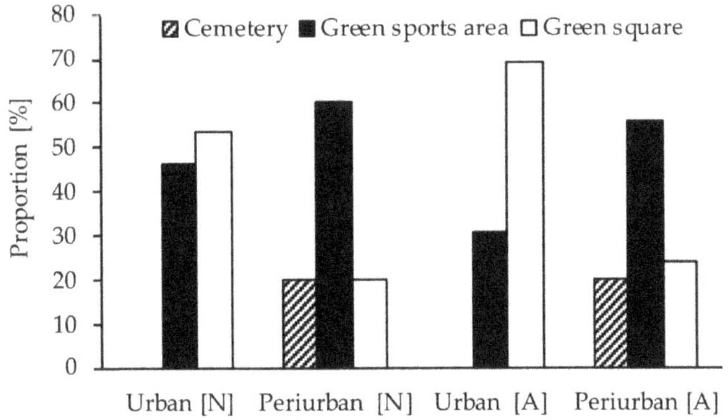

Figure 2. Presence and abundance of green spaces in the urban and peri-urban parts of Bujumbura expressed as a proportion of the number of green spaces (N) or as a function of the total area (A) of the green spaces.

3.2. *Types and Indicators of Disturbance of Green Spaces in Bujumbura*

3.2.1. Distribution of the Types of Anthropogenic Disturbances of the Green Spaces of Bujumbura Observed in the Field

A study of the distribution of the eight types of anthropogenic disturbance on the identified green spaces indicated different frequencies of occurrence according to the urbanization gradient considered. In the administrative entities with an urban morphological status, bare soil and constructions were the most frequent types of disturbance, with proportions of about 79.3% and 47.7%, respectively, of the total number of grid cells inventoried. The other types of disturbance showed low or zero frequencies, as is the case of rubble. In the administrative entities with a peri-urban morphological status, bare soil, constructions and crops were the dominant types of disturbance with proportions of about 85.4%, 35.4% and 39.6% of the total inventoried grid cells, respectively. Rubble and military installations showed low frequencies while flooding, garbage and improvised parking lots showed zero frequencies (Figure 3).

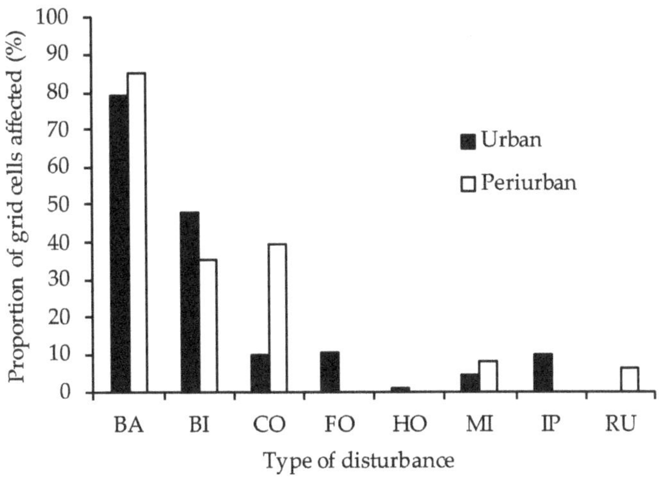

Figure 3. Proportion of grid cells affected by type of disturbance in sampled green spaces in urban and peri-urban zones of the city of Bujumbura. BA: bare ground, BI: building, CO: crops, FO: flooding, HO: household waste, MI: military installations, IP: improvised parking lot, RU: rubble.

In addition, the results of the Friedman test with $\chi^2 = 368.86$ ($p < 0.001$) in urban and $\chi^2 = 146.43$ ($p < 0.001$) in peri-urban entities showed a highly significant difference between the frequencies of the eight disturbance types along the urbanization gradient. The results of the post hoc test indicated that the disturbance types "buildings" and "bare soil" showed significant differences ($p < 0.05$) from the other disturbance types in terms of frequencies of occurrence. For the other frequencies regarding crops, rubble, flooding, military installations, garbage and improvised parking lots, significant differences were not observed.

3.2.2. Association between the Types of Anthropogenic Disturbance of the Green Spaces of Bujumbura

In the parts of the city with a peri-urban morphological status, one single association was observed, that of which between buildings and bare soil. In the parts of the city with an urban morphological status, buildings were linked to military installations and improvised parking. An association between crops and military installations was also observed; the military installations interact with the improvised parking lots (Table 3).

Table 3. Results of χ^2 independence tests for association of anthropogenic disturbances in Bujumbura green spaces (* = $p < 0.05$). The baseline data are from gridded inventories of these disturbance types conducted on 18 green spaces in the city of Bujumbura of at least 1 ha in area in November 2020. BA: bare ground, BI: building, CO: crops, FO: flooding, HO: household waste, MI: military installations, IP: improvised parking lot, RU: rubble.

	Peri-urban zones					
	BA	RU	CO	MI		
BI	4.49 *	1.76	2.84	2.39		
BA		0.90	3.48	0.75		
RU			2.10	2.62		
CO				0.39		
			Urban zones			
	BA	FO	HO	CO	MI	IP
BI	3.49	1.12	0.92	0.63	0.52 *	9.12 *
BA		0.14	0.26	0.05	1.37	1.01
FO			0.12	1.48	0.64	3.43
HO				0.11	0.05	0.11
CO					14.71 *	1.34
MI						5.31 *

4. Discussion

4.1. Typology of Green Spaces in Bujumbura

Several typologies of green spaces have been proposed and no inventory is able to cover all the particularities that exist in every city around the world due to the differences in natural conditions (geomorphological, climatic, biological), history and social composition [32]. In a European context, the authors of [59] clearly showed differences between green spaces as parks and gardens; natural and semi-natural spaces; green corridors; allotments, community gardens and urban farms; outdoor sports facilities; amenity green spaces; provisions for children and young people; cemeteries, churchyards and other burial grounds; and other public spaces. The authors of [60] produced an inventory of 25 urban green space types, falling into 10 subgroups and 4 main groups (amenity green spaces, functional green spaces, semi-natural habitats and linear green spaces). Other inventories are based on usage [61], based mainly on scale [62] or cover informal urban green spaces [28]. Some typologies combine urban green spaces with other open spaces such as squares, pedestrian areas, or cycling areas [63]. In China, the authors of [24] reclassified green spaces into nine types (public parks, plaza-green spaces, nurseries, green buffer zones, attached green spaces, residential green spaces, roadside green spaces, riparian green spaces and scenery forests). This classification is based on urban green space functions, land use and property information. In Africa, the authors of [31] revealed that there are different forms of urban green spaces and grouped them into seven categories (semi-private spaces; parks, street trees and roadside plantations; public green areas; public and private tree plantations on vacant lots; green belts; woodlands and peri-urban farming areas; rangelands and forests close to urban areas; natural forests under urban influence; and trees planted for environmental protection and aesthetic purposes).

Furthermore, social initiatives, technological advances, environmental awareness and the creativity of urban planners and city dwellers are perpetually leading to new types of urban green spaces [32]. In this study, we combined four existing typologies, which have the advantage of associating each type of green space with the corresponding land use, as well as the status of the owner [7]. Consequently, these typologies allowed us to classify the green spaces of the city of Bujumbura into green squares, green sports spaces and cemeteries on the basis of their land use. These green spaces are present with varying frequencies along the urbanization gradient. In fact, green squares are present exclusively in urban entities, while sport green areas are present in urban and peri-urban zones and cemeteries are present in peri-urban zones only, hence confirming our first hypothesis. Indeed, the distribution of green spaces in Bujumbura city is based on the age of the different neighborhoods of the city and their socio-economic diversity [35,64]. Thus, in the Rohero zone, the oldest, inherited from the colonization period and characterized by a high standard of living, there are many green squares and sports green areas. In contrast, in the Kanyosha zone, the most recent city part and characterized by a precarious standard of living, only a few sports green areas and a cemetery are present.

4.2. *Anthropogenic Disturbances of Green Spaces in Bujumbura*

Urban green spaces are among the most degraded ecosystems in African cities [31], in particular, due to the rapid and unplanned expansion of built-up areas [8,11]. In this context, the desire to transform the architectural image of Bujumbura city, the objective of the public authorities to provide sports infrastructure and the phenomenon of gentrification [64] were at the origin of the construction of new infrastructures such as hotel residences, offices, commercial buildings and banks. In addition, the concern to commemorate the history of Burundi following the socio-political events that have characterized it since its independence, together with the objective to protect military strategic places in the city, are at the origin of the fragmentation and the shrinking of some emblematic green spaces in favor of monuments and military positions. In the African context, this situation is quite similar to the one observed in the municipalities of Treichville in Abidjan (Ivory Coast), in Lubumbashi and in Kolwezi (both situated in the southeastern part of the Democratic Republic of the Congo) where green spaces are threatened by the construction of shops, informal businesses, the installation of car garages, residential landmarks, etc., [65–67]. However, the installation of military positions in green spaces is a particularity of Bujumbura city, in order to protect strategic positions in the city. The authors of [68] specify that peri-urbanization and associated actions on green spaces will continue until 2090 without any planning.

In addition, with the repeated passages of residents on green spaces, bare soil areas appear rapidly, leading to a modification of soil properties by sealing them [69,70]. However, our results underline that the green spaces of Bujumbura are not only threatened by bare soil resulting from trampling following improvised passages, but also from various constructions along the urbanization gradient at different frequencies. Indeed, residents of green spaces consider these areas as shortcuts, thus testifying to the presence of trampling [70]. These observations confirm partly our second hypothesis according to which various constructions as well as bare soil, especially due to trampling, is visible along the urbanization gradient, while military positions are only present in administrative entities with an urban morphological status. These perturbations are largely due to the incivility and absence of eco-responsible behavior of local populations who consider these spaces free of any restriction [65].

In Sub-Saharan Africa, agriculture is a growing activity for survival in peri-urban areas, but its development threatens (semi-)natural vegetation elements and formations, which are considered crucial for sustainable development. This corroborates our findings that crop disturbances are significantly more present in peri-urban green spaces in Bujumbura and confirm the second part of our second hypothesis. The authors of [54,55,71] argue that populations living in peri-urban areas seek to ensure their survival, by any means, in a context of economic degradation and with little concern for the preservation of vegetation. This is typically the case of the military forces who install small fields next to their positions, a practice also observed concerning the guards of sports infrastructures.

4.3. Association between Types of Anthropogenic Disturbances of Green Spaces in Bujumbura

The authors of [72] state that human disturbances often interact synergistically, hence confirming our results. Indeed, within the green spaces of Bujumbura, each time constructions such as monuments or sports facilities were erected, access paths were laid out and parking lots were created. In addition, the Republic of Burundi has experienced a decade of socio-political instability leading to a certain degree of insecurity in the cities, including Bujumbura [73,74]. To improve this situation, military positions have been set up in green spaces, which required the construction of housing or shelters, in addition to parking lots for their vehicles on which water could stagnate after cleaning them. Similarly, according to the UNDP, 64.9% of the Burundi population lives on less than USD 0.0196 (BIF 41.054) per adult per month in the context of acute food insecurity [75]. As a result, there is an economy of scavenging characterized by, among other things, the development of vegetable gardens by the military, near their positions in green spaces, a situation also observed in Lubumbashi in the southeastern part of the Democratic Republic of the Congo [13]. The tendency for vegetable gardening development is also another phenomenon observed within sports infrastructures. Indeed, the guards arrange vegetable gardens for the production of food for their respective families, due to the scarcity of the land [76], thus supporting our third hypothesis related to the association of anthropogenic disturbances. It corroborates the findings of [54] that certain anthropogenic disturbances are often related to other types of degradation. Urban developments coupled with anthropogenic actions have been reported by [77], who state that they are the main constraints to the sustainability of urban vegetation.

5. Conclusions

This study presented a typology of the green spaces of Bujumbura city. It also identified the anthropogenic disturbances to which these spaces are subjected. A literature review revealed that green squares, green sports areas and cemeteries are among the types of green spaces encountered in Bujumbura. On the basis of the inventories, it appears that green spaces are exposed to anthropogenic disturbances, notably agriculture in peri-urban areas, as opposed to the development of military positions and the construction of monuments, buildings and sports facilities, which are more pronounced in urban areas. Trampling as a type of anthropogenic disturbance is present on green spaces in urban and peri-urban areas. The synergies between the different types of anthropogenic disturbances amplify their negative impact.

This study is a continuation of our research on green spaces in the city of Bujumbura and on the analysis of the anthropogenic pressures to which they are exposed. It was written to raise awareness. All stakeholders (municipal technical services, local administrations, non-profit associations and sports club associations) in charge of managing green spaces should invest more in protecting their green spaces against all kinds of disturbances. They should also promote environmental education in order to safeguard them and the ecosystem goods and services that they provide.

Author Contributions: Conceptualization, H.K., Y.U.S., T.M. and J.B.; methodology, H.K., Y.U.S., K.R.S., T.M. and J.B.; software, H.K.; validation, H.K., Y.U.S., K.R.S., T.M. and J.B.; formal analysis, H.K., Y.U.S., K.R.S., T.M. and J.B.; investigation, H.K. and D.M.; resources, H.K., Y.U.S., T.M. and J.B.; data curation, H.K.; writing—original draft preparation, H.K.; writing—review and editing, H.K., Y.U.S., K.R.S., D.M., T.M. and J.B.; visualization, H.K., Y.U.S., K.R.S., D.M., T.M. and J.B.; supervision, H.K., Y.U.S., K.R.S., D.M., T.M. and J.B.; project administration, H.K., T.M. and J.B.; funding acquisition, J.B. All authors have read and agreed to the published version of the manuscript.

Funding: This research was funded by the Académie de Recherche et d'Enseignement Supérieur-Commission de la Coopération au Développement (ARES-CCD, PSRCI-UB programme and PRD CHARLU).

Data Availability Statement: Not applicable.

Conflicts of Interest: The authors declare no conflict of interest.

References

1. Turan, S.Ö.; Kadioğullari, A.I.; Günlü, A. Spatial and temporal dynamics of land use pattern response to urbanization in Kastamonu. *Afr. J. Biotechnol.* **2010**, *9*, 640–647.
2. Halleux, J.M. Les territoires périurbains et leur développement dans le monde: Un monde en voie d'urbanisation et de périurbanisation. In *Territoires Périurbains: Développement, Enjeux et Perspectives dans les Pays du Sud*; Bogaert, J., Halleux, J.M., Eds.; Presses Agronomiques de Gembloux: Gembloux, Belgium, 2015; pp. 43–61.
3. United Nations. *Departement of Economic and Social Affairs, Population Division (UNDSA), World Urbanization Prospects: The 2018 Revision*; United Nations: New York, NY, USA, 2019.
4. Tian, G.; Liu, J.; Xie, Y.; Yang, Z.; Zhuang, D.; Niu, Z. Analysis of spatio-temporal dynamic pattern and driving forces of urban land in China in 1990s using TM images and GIS. *Cities* **2005**, *22*, 400–410. [CrossRef]
5. Kabanyegeye, H.; Useni, S.Y.; Sambieni, K.R.; Masharabu, T.; Havyarimana, F.; Bogaert, J. Trente-trois ans de dynamique spatiale de l'occupation du sol de la ville de Bujumbura, République du Burundi. *Afrique Sci.* **2021**, *18*, 203–2015.
6. Vermeiren, K.; Van Rompaey, A.; Loopmans, M.; Serwajja, E.; Mukwaya, P. Urban growth of Kampala, Uganda: Pattern analysis and scenario development. *Landsc. Urban Plan.* **2012**, *106*, 199–206. [CrossRef]
7. Sambieni, K.R.; Useni, S.Y.; Kaleba, S.C.; Moyene, A.B.; Kankumbi, F.M.; Nzuzi, F.L.; Occhiuto, R.; Bogaert, J. Les espaces verts en zone urbaine et périurbaine de Kinshasa en République Démocratique du Congo. *Tropicultura* **2018**, *36*, 478–491. [CrossRef]
8. Useni, S.Y.; André, M.; Mahy, G.; Kaleba, S.C.; Kankumbi, F.M.; Bogaert, J. Interprétation paysagère du processus d'urbanisation à Lubumbashi: Dynamique de la structure spatiale et suivi des indicateurs écologiques entre 2002 et 2008. In *Anthropisation des Paysages Katangais*; Bogaert, J., Colinet, G., Mahy, G., Eds.; Presses Universitaires de Liège-Agronomie-Gembloux: Gembloux, Belgium, 2018; pp. 281–296.
9. Sikuzani, Y.U.; Kaleba, S.C.; Halleux, J.M.; Bogaert, J.; Kankumbi, F.M. Caractérisation de la croissance spatiale urbaine de la ville de Lubumbashi (Haut-Katanga, RD Congo) entre 1989 et 2014. *Tropicultura* **2018**, *36*, 99–108.
10. Sikuzani, Y.U.; Kaleba, S.C.; Khonde, C.N.; Mwana, Y.A.; Malaisse, F.; Bogaert, J.; Kankumbi, F.M. Vingt-cinq ans de monitoring de la dynamique spatiale des espaces verts en réponse à l'urbanisation dans les communes de la ville de Lubumbashi (Haut-Katanga, RD Congo). *Tropicultura* **2017**, *35*, 300–311. [CrossRef]
11. Alberti, M. The effects of urban patterns on ecosystem function. *Int. Reg. Sci. Rev.* **2005**, *28*, 168–192. [CrossRef]
12. Useni Sikuzani, Y.; Malaisse, F.; Cabala Kaleba, S.; Kalumba Mwanke, A.; Yamba, A.M.; Nkuku Khonde, C.; Bogaert, J.; Munyemba Kankumbi, F. Tree diversity and structure on green space of urban and peri-urban zones: The case of Lubumbashi City in the Democratic Republic of Congo. *Urban For. Urban Green.* **2019**, *41*, 67–74. [CrossRef]
13. Useni Sikuzani, Y.; Sambiéni Kouagou, R.; Maréchal, J.; Ilunga wa Ilunga, E.; Malaisse, F.; Bogaert, J.; Munyemba Kankumbi, F. Changes in the spatial pattern and ecological functionalities of green spaces in Lubumbashi (the Democratic Republic of Congo) in relation with the degree of urbanization. *Trop. Conserv. Sci.* **2018**, *11*, 1–17. [CrossRef]
14. Maréchal, J.; Useni, S.Y.; Bogaert, J.; Munyemba, K.F.; Mahy, G. La perception par des experts locaux des espaces verts et de leurs services écosystémiques dans une ville tropicale en expansion: Le cas de Lubumbashi. In *Anthropisation des Paysages Katangais*; Bogaert, J., Colinet, G., Mahy, G., Eds.; Les Presses Universitaires de Liège: Liège, Belgium, 2018; pp. 59–69.
15. Rasidi, M.H.; Jamirsah, N.; Said, I. Urban green space design affects urban residents'social interaction. *Procedia-Soc. Behav. Sci.* **2012**, *68*, 464–480. [CrossRef]
16. Buyadi, S.N.A.; Mohd, W.M.N.W.; Misni, A. Impact of land use changes on the surface temperature distribution of area surrounding the National Botanic Garden, Shah Alam. *Procedia-Soc. Behav. Sci.* **2013**, *101*, 516–525. [CrossRef]
17. Robert, A.; Yengué, J.L. When Allotment gardens become urban green spaces like others, providing cultural ecosystem services. *Environ. Ecol. Res.* **2017**, *5*, 453–460. [CrossRef]
18. Wood, E.; Harsant, A.; Dallimer, M.; Cronin de Chavez, A.; McEachan, R.R.; Hassall, C. Not all green space is created equal: Biodiversity predicts psychological restorative benefits from urban green space. *Front. Psychol.* **2018**, *9*, 1–13. [CrossRef]

19. Gonzales, L.P.; Magnaye, D.C. Measuring the urban biodiversity of green spaces in a highly urbanizing environment and its implications for human settlement resiliency planning: The case of Manila city, Philippines. *Procedia Environ. Sci.* **2017**, *37*, 83–100. [CrossRef]
20. Xiao, X.D.; Dong, L.; Yan, H.; Yang, N.; Xiong, Y. The influence of the spatial characteristics of urban green space on the urban heat island effect in Suzhou Industrial Park. *Sustain. Cities Soc.* **2018**, *40*, 428–439. [CrossRef]
21. Aram, F.; Higueras García, E.; Solgi, E.; Mansournia, S. Urban Green Space Cooling Effect in Cities. *Heliyon* **2019**, *5*, e01339. [CrossRef]
22. Jancel, R. Typologie Des Espaces Verts. In *Les Plantes dans la Ville: Angers (France)*; Les Colloques, N., Ed.; COLLOQUES-INRA: Paris, France, 1997; pp. 69–80.
23. Hermy, M.; Cornelis, J.; Hermy, M.; Cornelis, J. Towards a monitoring method and a number of multifaceted and hierarchical biodiversity indicators for urban and suburban parks. *Landsc. Urban Plan.* **2000**, *49*, 149–162. [CrossRef]
24. Kong, F.; Nakagoshi, N. Changes of urban green spaces and their driving forces: A case study of Jinan City, China. *J. Int. Dev. Coop.* **2005**, *11*, 97–109. [CrossRef]
25. Guerin, M. *Etablissement d'une Typologie pour la Sélection des Espaces verts de la Région Centre du Programme SERVEUR*; Université d'Orléans. M. QUODVERTE Philippe: Orléans, France, 2013.
26. Neuenschwander, N.; Hayek, U.W.; Grêt-Regamey, A. Integrating an urban green space typology into procedural 3D visualization for collaborative planning. *Comput. Environ. Urban Syst.* **2014**, *48*, 99–110. [CrossRef]
27. Hansen, R.; Rall, E.L.; Chapman, E.; Rolf, W.; Pauleit, S. *Urban Green Infrastructure Planning: A Guide for Practitioners*; Green Surge; University of Copenhagen: Copenhagen, Denmark, 2017.
28. Rupprecht, C.D.; Byrne, J.A. Informal urban greenspace: A typology and trilingual systematic review of its role for urban residents and trends in the literature. *Urban For. Urban Green.* **2014**, *13*, 597–611. [CrossRef]
29. Amontcha, A.A.M.; Djego, J.G.; Lougbegnon, T.O.; Sinsin, B.A. Typologie et répartition des espaces verts publics dans le grand Nokoué (Sud Bénin). *Eur. Sci. J. ESJ* **2017**, *13*, 79–97. [CrossRef]
30. Mili, M. *Module: Espaces Verts*; Université Mohamed Boudiaf de M'sila, Institut de Gestion des Techniques Urbaines: M'Sila, Algeria, 2018.
31. Mensah, C.A. Urban green spaces in Africa: Nature and challenges. *Int. J. Ecosyst.* **2014**, *4*, 1–11. [CrossRef]
32. Rall, L.; Niemela, J.; Pauleit, S.; Pintar, M.; Lafortezza, R.; Santos, A.; Železnikar, Š. A Typology of Urban Green Spaces, Eco-System Services Provisioning Services and Demands. Report D3 1, UE. 2015. Available online: https://assets.centralparknyc.org/pdfs/institute/p2p-upelp/1.004_Greensurge_A+Typology+of+Urban+Green+Spaces.pdf (accessed on 27 October 2022).
33. Mensah, C.A. Is Kumasi still a garden city? Land use analysis between 1980–2010. *J. Environ. Ecol.* **2014**, *5*, 2. [CrossRef]
34. Osseni, A.A.; Mouhamadou, T.; Tohoain, B.A.C.; Sinsin, B. SIG et gestion des espaces verts dans la ville de Porto Novo au Bénin. *Tropicultura* **2015**, *33*, 146–156.
35. Kabanyegeye, H.; Masharabu, T.; Sikuzani, Y.U.; Bogaert, J. Perception sur les espaces verts et leurs services écosystémiques par les acteurs locaux de la ville de Bujumbura (République du Burundi). *Tropicultura* **2020**, *38*, 1–17. [CrossRef]
36. Bigirimana, J. Plant Diversity Patterns, Processes and Conservation Value in Bujumbura (Burundi), an Urban Ecology Study in Sub-Saharan Africa. Ph.D. Thesis, Université libre de Bruxelles, Brussels, Belgium, 2012.
37. République du Burundi, Cabinet du Président. *Loi n°1/04 du 04 Février 2019 Portant Fixation de la Capitale Politique et de la Économique du Burundi*; République du Burundi, Cabinet du Président: Bujumbura, Burundi, 2019.
38. République du Burundi, Ministère des Finances. Analyse des facteurs de risques, évaluation des dommages et propositions pour un relèvement et une reconstruction durables. In *Evaluation Rapide Conjointe Suite à la Catasrtophe des 9-10 Février 1014 aux Alentours de Bujumbura*; République du Burundi, Ministère des Finances: Bujumbura, Burundi, 2014.
39. République du Burundi, Présidence de la République. *Loi n°1/33 du 28 Novembre 2014 Portant Révision de la loi n°1/20 du 25 Janvier 2010 Portant Organisation de l'administration Communale*; République du Burundi, Présidence de la République: Bujumbura, Burundi, 2014.
40. République du Burundi, Ministère de l'Aménagement du Territoire, du Territoire et de l'Environnement. *Plan d'Action National d'Adaptation Aux Changements Climatiques (PANA)*; République du Burundi, Ministère de l'Aménagement du Territoire, du Territoire et de l'Environnement: Bujumbura, Burundi, 2007.
41. Bigirimana, J.; Bogaert, J.; De Canniere, C.; Lejoly, J.; Parmentier, I. Alien plant species dominate the vegetation in a city of Sub-Saharan Africa. *Lands. Urb. Plan.* **2011**, *100*, 251–267. [CrossRef]
42. Hakizimana, P.; Bangirinama, F.; Masharabu, T.; Habonimana, B.; De Cannière, C.; Bogaert, J. Caractérisation de la végétation de la forêt dense de Kigwena et de la forêt claire de Rumonge au Burundi. *Bois For. Trop.* **2012**, *312*, 43–52. [CrossRef]
43. White, F. *The Vegetation of Africa: A Descriptive Memoir to Accompany the UNESCO/AETFAT/UNSO Vegetation Map of Africa*; UNESCO: Paris, France, 1983.
44. Bangirinama, F.B.; Nzitwanayo, B.; Hakizimana, P. Utilisation du charbon de bois comme principale source d'énergie de la population urbaine: Un sérieux problème pour la conservation du couvert forestier au Burundi. *Bois For. Trop.* **2016**, *328*, 45–53. [CrossRef]
45. République du Burundi. Ministère de l'intérieur, Bureau Central du Recensement. *Recensement Général de la Population et de l'habitat du Burundi 2008, Synthèse des Résultats Définitifs*; République du Burundi. Ministère de l'intérieur, Bureau Central du Recensement: Bujumbura, Burundi, 2011.

46. République du Burundi. *Projections Démographiques 2010–2050 Niveau National et Provincial, ProService Des Etudes et Statistiques Démographiques*; République du Burundi: Bujumbura, Burundi, 2017.
47. République du Burundi. Ministère de la Santé Publique et de la Lutte Contre le Sida, Institut de Statistiques et d'Études Économiques du Burundi. *Enquête Démographique et de Santé Au Burundi 2016–2017: Rapport de Synthèse*; République du Burundi. Ministère de la Santé Publique et de la Lutte Contre le Sida, Institut de Statistiques et d'Études Économiques du Burundi: Bujumbura, Burundi, 2018.
48. Byavu, N.; Henrard, C.; Dubois, M.; Malaisse, F. Phytothérapie traditionnelle des bovins dans les élevages de la plaine de la Ruzizi. *Biotechnol. Agron. Soc. Environ.* **2000**, *4*, 135–156.
49. Sambieni, K.R.; Messina Ndzomo, J.P.; Biloso Moyene, A.; Halleux, J.M.; Occhiuto, R.; Bogaert, J. Les statuts morphologiques d'urbanisation des communes de Kinshasa en République Démocratique du Congo. *Tropicultura* **2018**, *3*, 520–530.
50. André, M.; Mahy, G.; Lejeune, P.; Bogaert, J. Vers une synthèse de la conception et une définition des zones dans le gradient urbain-rural. *Biotechnol. Agron. Soc. Environ.* **2014**, *18*, 61–74.
51. Polorigni, B.; Radji, R.; Kokou, K. Perceptions, tendances et préférences en foresterie urbaine: Cas de la ville de Lomé au Togo. *Eur. Sci. J.* **2014**, *10*, 261–277.
52. Vroh, B.T.A.; Tiebre, M.S.; N'Guessan, K.E. Diversité végétale urbaine et estimation du stock de carbone: Cas de la commune du Plateau Abidjan, Côte d'Ivoire. *Afrique Sci. Rev. Int. des Sci. Technol.* **2014**, *10*, 329–340.
53. Bougé, F. *Caractérisation des Espaces verts Publics en Fonction de leur Place dans le Gradient Urbain Rural. Cas d'étude : La Trame verte de l'Agglomération Tourangelle*; Projet de Fin d'Etudes-Université François: Tours, France, 2009.
54. Rakotondrasoa, O.L.; Ayral, A.; Stein, J.; Rajoelison, G.L.; Ponette, Q.; Malaisse, F.; Ramamonjisoa, B.S.; Raminosoa, N.; Verheggen, F.J.; Poncelet, M.; et al. Analyse des facteurs anthropiques de dégradation des bois de tapia (Uapaca bojeri) d'Arivonimamo. In *Les vers à Soie Malgaches. Enjeux Écologiques et Socio-Économiques*; Verheggen, F., Bogaert, J., Haubruge, E., Eds.; Les Presses Agronomiques de Gembloux: Gembloux, Belgium, 2013; pp. 151–162.
55. Demšar, J. Statistical comparisons of classifiers over multiple data sets. *J. Mach. Learn. Res.* **2006**, *7*, 1–30.
56. Pereira, D.G.; Afonso, A.; Medeiros, F.M. Overview of Friedman's test and post-hoc analysis. *Commun. Stat. Simul. Comput.* **2015**, *44*, 2636–2653. [CrossRef]
57. Rakotondrasoa, O.L.; Malaisse, F.; Rajoelison, G.L.; Rabearisoa, M.R.; Ramamonjisoa, B.S.; Raminosoa, N.; Verheggen, F.J.; Poncelet, M.; Haubruge, E.; Bogaert, J. Identification des indicateurs de dégradation de la forêt de tapia (Uapaca bojeri) par une analyse sylvicole. *Tropicultura* **2013**, *31*, 10–19.
58. Calavas, D.; Sulpice, P.; Lepetitcolin, E.; Bugnard, F. Appréciation de la fidélité de la pratique d'une méthode de notation de l'état corporel des brebis dans un cadre professionnel. *Vet. Res.* **1998**, *29*, 129–138.
59. Bell, S.; Montarzino, A.; Travlou, P. Mapping research priorities for green and public urban space in the UK. *Urban For. Urban Green* **2007**, *6*, 103–115. [CrossRef]
60. Swanwick, C.; Dunnett, N.; Woolley, H. Nature, role and value of green space in towns and cities: An overview. *Built Environ.* **2003**, *29*, 94–106. [CrossRef]
61. Hofmann, M.; Gerstenberg, T. A user-generated typology of urban green spaces. In Proceedings of the 17th International Conference of the European Forum on Urban Forestry (EFUF), Lausanne, Switzerland, 3–7 June 2014.
62. Byrne, J.; Sipe, N. Green and Open Space Planning for Urban Consolidation—A Review of the Literature and Best Practice. *Urban Reseach Program, Issues Paper 11*; Griffith University, 2010. Available online: https://core.ac.uk/download/pdf/143882947.pdf (accessed on 27 October 2022).
63. DTLR–Department for Transport, Local Government and the Regions. *Green Spaces, Better Places. Final Report of the Urban Green Spaces Taskforce*; DTLR–Department for Transport, Local Government and the Regions: London, UK, 2002.
64. Toyi, O. Politiques Publiques Urbaines de l'habitat dans la ville de Bujumbura de 1962 a 2009. Ph.D. Thesis, Université de Pau et des Pays de l'Adour, Pau, France, 2012.
65. Gauthier, K.B.A.A.; Célestin, A.Y. Problématique de la gestion durable des espaces verts publics urbains à Abidjan: Cas de Treichville (côte d'ivoire). *Revue Scientifique du Tchad* **2017**, *1*, 63–72.
66. Useni, Y.S.; Malaisse, F.; Yona, J.M.; Mwamba, T.M.; Bogaert, J. Diversity, use and management of household-located fruit trees in two rapidly developing towns in Southeastern D.R. Congo. *Urban For. Urban Green* **2021**, *63*, 127220. [CrossRef]
67. Useni Sikuzani, Y.; Mpibwe Kalenga, A.; Yona Mleci, J.; N'Tambwe Nghonda, D.; Malaisse, F.; Bogaert, J. Assessment of street tree diversity, structure and protection in planned and unplanned neighborhoods of Lubumbashi City (DR Congo). *Sustainability* **2022**, *14*, 3830. [CrossRef]
68. Muteya, H.K.; Nghonda, D.D.N.; Malaisse, F.; Waselin, S.; Sambiéni, K.R.; Kaleba, S.C.; Kankumbi, F.M.; Bastin, J.F.; Bogaert, J.; Sikuzani, Y.U. Quantification and Simulation of Landscape Anthropization around the Mining Agglomerations of Southeastern Katanga (DR Congo) between 1979 and 2090. *Land* **2022**, *11*, 850. [CrossRef]
69. Lorenz, K.; Lal, R. Biogeochemical C and N cycles in urban soils. *Environ. Internat.* **2009**, *35*, 1–8. [CrossRef] [PubMed]
70. Boukelouha, R.; Labii, B. Re-Définir la Marchabilité Urbaine Une revue de littérature. *Les Ann. l'université d'Alger 1* **2019**, *33*, 776–800.
71. Trefon, T.; Cogels, S. La gestion des ressources naturelles dans les zones périurbaines d'Afrique centrale: Une approche privilégiant les parties prenantes. *Cad. Estud. Africanos* **2007**, *13–14*, 101–126. [CrossRef]

72. Peres, C.A.; Barlow, J.; Laurance, W.F. Detecting anthropogenic disturbance in tropical forests. *Trends Ecol. Evol.* **2006**, *21*, 227–229. [CrossRef] [PubMed]
73. Havyarimana, F.; Masharabu, T.; Kouao, J.K.; Bamba, I.; Nduwarugira, D.; Bigendako, M.-J.; Hakizimana, P.; Mama, A.; Bangirinama, F.; Banyankimbona, G.; et al. La Dynamique Spatiale de La Foret Situee Dans La Reserve Naturelle Forestiere de Bururi Au Burundi. *Tropicultura* **2017**, *35*, 158–172.
74. Havyarimana, F.; Bamba, I.; Barima, Y.S.S.; Masharabu, T.; Nduwarugira, D.; Bigendako, M.J.; Bogaert, J. La contribution des camps de déplacés à la dynamique paysagère au sud et au sud-est du Burundi. *Tropcultura* **2018**, *36*, 243–257.
75. République du Burundi. *Rapport National sur le Développement Humain*; Programmes; République du Burundi: Bujumbura, Burundi, 2019.
76. Bogaert, J.; Biloso, A.; Vranken, I.; André, M. Peri-urban dynamics: Landscape ecology perspectives. In *Territoires Périurbains: Développement, Enjeux et Perspectives dans les Pays du Sud*; Bogaert, J., Halleux, J.M., Eds.; Presses agronomiques de Gembloux: Gembloux, Belgium, 2015; pp. 63–73.
77. Aziz, O.A.; Brice, S.; Ismaïla, T.I. Analyse des contraintes de viabilite de la vegetation urbaine: Cas des arbres d'alignement dans la ville de porto-novo au Bénin. *Eur. Sci. J.* **2014**, *10*, 1–15.

land

MDPI

Article

Barriers to Native Plantings in Private Residential Yards

Amélie Davis [1,2,*] and Jessica Stoyko [2]

1 Department of Economics and Geosciences, United States Air Force Academy, Colorado Springs, CO 80921, USA
2 Department of Geography, Miami University, Oxford, OH 45056, USA
* Correspondence: amelie.davis@afacademy.af.edu or davis.amelie@miamioh.edu

Abstract: In urban areas, private yards can make up large portions of the available "green space" which can be used to provide resources for many species, including birds, and pollinators. If residents are persuaded or willing to plant certain native plants, the aggregate effect of these plantings could be hugely beneficial for key pollinator species. The objectives of this study are to uncover impediments to adding different types of pollinator-beneficial plants to private yards, as well as ascertain which incentives to plant these native plants might be most persuasive, and finally determine if there are procedural knowledge gaps in how to plant, care for, or where to purchase three pollinator-beneficial plants. In this study, we randomly selected properties in two counties in southwestern Ohio along two gradients: parcel size and parcel valuation (as a proxy for income). Two hundred surveys were distributed and 113 were returned (57% response rate). We find that, in aggregate, respondents do not have a strong intent to plant these native plants, especially *Asclepias syriaca* (a milkweed that serves as host plant to the iconic monarch butterfly; *Danaus plexippus*) and, surprisingly, the intent to plant these does not differ statistically even when help with costs, labor, or the provision of online resources are offered. We also find that the reported knowledge of where to purchase wildflowers is significantly higher than how to care for them and how to plant them. Lastly, respondents are much more confident in how to take care of trees compared to the three pollinator-beneficial plants shown in the survey. We discuss the implications of these findings for outreach and extension purposes.

Keywords: human-environment interactions; ecosystem services; education; habitat; bees; urban; peri-urban; garden; intent; pro-environmental behavior

Citation: Davis, A.; Stoyko, J. Barriers to Native Plantings in Private Residential Yards. *Land* **2023**, *12*, 114. https://doi.org/10.3390/land12010114

Academic Editors: Alessio Russo and Giuseppe T. Cirella

Received: 15 December 2022
Revised: 28 December 2022
Accepted: 29 December 2022
Published: 30 December 2022

1. Introduction

As humans continue to appropriate Earth's resources [1], finding novel ways to provide habitat and resources for species other than humans is needed. One of many potential approaches is to add native plants to private yards or gardens in urban, suburban and exurban (i.e., peri-urban) areas. Indeed, yards can positively contribute to species richness and abundance of some species [2,3], especially if they contain native plants [4]. In particular, private yards can provide enough floral resources and diverse plants necessary to maintain bee communities [5]. They have the potential to provide many more needed resources for bees, but also iconic species such as the monarch butterfly, i.e., *Danaus plexippus* [6–8]. However, this opportunity depends on what is planted (or not) in these private gardens as not all green spaces are created equal.

Given that residential gardens operate as a complex adaptive socio-ecological system, the plants that are found in yards are a result of a multitude of interacting and multiscalar human and ecological drivers [9,10]. The decision to add native plants in particular (or keep the native species that were planted by a previous owner, or the landscape architect/contractor) can be viewed through the lens of pro-environmental behavior model [11] and will depend on many factors including attitudinal and structural factors (e.g., plant aesthetics, time, values), the influence of formal and informal institutions (e.g., homeowners association rules; [9]), impact, procedural, and normative knowledge [12], knowledge of the plants' ecological value, social norms [13], income [14], and habits [11] to name a few.

While the pro-environmental behavior model (or variations of it) is generally accepted, the weight of influence of the variables that comprise it of varies depending on which pro-environmental behavior is considered. Furthermore, Kollmuss and Agyeman [11] specify how there can exist a gap between a) environmental knowledge, values, attitudes, and awareness, and b) pro-environment choices and behaviors. This gap can be explained by the presence of barriers to executing the "desired" behavior including external incentives (socioeconomic and cultural factors for example), strength of internal incentives (e.g., value system), negative or insufficient feedback about the particular pro-environmental behavior, old behavior patterns (i.e., habits), lack of time, weak environmental consciousness, and knowledge gaps [11]. In order to design effective programs to elicit pro-environmental behaviors, it is important to know the strength of the aforementioned drivers as well as the perceived or real benefits and barriers to desired behaviors and the perceived or real benefits and barriers to the competing (undesired) behaviors [15].

Known barriers to adding native plants to landscapes include their lack of availability in wholesale and retail nurseries [16], preferences for some homeowners for large flowers and wide green leaves [17], the thought that yards that include these plants can seem "messy" [18,19], pressure from social norms [13,20], the "belief that native plants belong in the city [14], influence of socioeconomic factors [14,21], maintenance needs [22] and people's familiarity with the plants [23], but also characteristics of yards themselves including yard size [24]. While much has been done in this regard few studies compare people's desire to plant different native plants, as well as procedural knowledge of planting native plants and how that interacts with intent to plant these plants.

Consequently, ascertaining people's intent to plant various native plants within the context of the timing of when to plant them, or whether some barriers such as help with purchasing the seeds, or the labor that needs to go into planting or knowledge of how to plant, purchase and care for these plants is needed. Additionally, understanding if certain native plants are more of less desired could be useful as well. Finally, trying to uncover knowledge gaps surrounding key aspects of adding these plants to yards can also help in designing programs to help homeowners add natives to their gardens.

The objectives of this study are to uncover impediments to adding different types of pollinator-beneficial plants to private yards, as well as ascertain which incentives to plant these native plants might be most persuasive, and finally determine if there are procedural knowledge gaps in how to plant, care for, or where to purchase these pollinator beneficial plants. We focus on the following pollinator beneficial plant types: purple coneflower (*Echinacea purpurea*), common milkweed (*Asclepias syriaca*), and wildflowers (multiple *spp.*). We ask the following questions:

1. Is there a difference in mean stated willingness to add the pollinator beneficial plants shown in the survey based on offers of different types of assistance (e.g., help with cost or labor) or the timing of when to plant pollinator beneficial plants (this season or next time changes will be made to respondents' landscaping)?
2. Is the self-disclosed knowledge of how to plant, care for, and purchase *E. purpurea* statistically different, i.e., do the respondents report they are more or less knowledgeable about planting, caring, or purchasing *E. purpurea*? Similarly, is the self-disclosed knowledge of how to plant, care for, and purchase *A. syriaca* or wildflowers statistically different? Additionally, is the self-disclosed knowledge of how to plant and care for trees significantly different?
3. Does self-disclosed knowledge of how to plant, care for, and purchase pollinator beneficial plants differ between the different plant types shown in the survey, e.g., do the respondents feel more knowledgeable about planting *E. purpurea* compared to wildflowers or *A. syriaca* and so forth? For this question set, we also compare some of the answers to the self-disclosed knowledge of planting and caring for trees as a comparison group.

4. What are the most frequently cited barriers to adding the plants shown in the survey to respondents' yards?

The results of this research could be used to develop and tailor outreach and extension materials and programs designed to increase pollinator beneficial plantings in the Midwestern United States.

2. Methods

2.1. Study Area and Sampling Approach

The study area consists of two counties in the midwestern United States of America (U.S.A.): Darke and Miami counties. Both are located in Southwest Ohio and sit within the Dayton-Springfield-Sidney Combined Statistical Area delineated by the U.S. Census Bureau. Combined Statistical Areas are defined by the Census Bureau as "the county or counties (or equivalent entities) associated with at least one core (urbanized area or urban cluster) of at least 10,000 population, plus adjacent counties having a high degree of social and economic integration with the core as measured through [strong] commuting ties". Both Darke and Miami counties contain medium-sized cities (Greenville, Troy, and Piqua) and exhibit suburban to rural urbanization patterns, especially moving north, i.e., away from Dayton, Ohio (Figure 1). Except for the urbanized areas with strong suburban and exurban (peri-urban) development, the region is mostly agricultural, dominated by corn and soybeans, with some forest remnants.

Using parcel data obtained from Darke and Miami Counties' assessors in 2019 we selected parcels for which the land-use code was single-family residential. From the 81,859 parcels located in both counties, we were left with 16,390 candidate properties. From those, we randomly selected 200 properties along two strata, i.e., property valuation and parcel size, so that we could ensure a range of respondents from various socio-economic backgrounds and property types. The selected households received a survey that shows pollinator beneficial plants native to the region but of differing appearance, management needs, and pollinator service. These were purple coneflower (*Echinacea purpurea*), common milkweed (*Asclepias syriaca*), and wildflowers (multiple *spp.*).

We ascertained respondents' knowledge of each plant and willingness to add each plant to their yards by asking respondents to rate on a 5-point Likert scale of "Strongly Disagree," "Disagree," "Neither Agree Nor Disagree," "Agree," and "Strongly Agree" whether they: (1) know how to plant, (2) how to care for, (3) where to purchase seed, (4) whether they would add (or add more) of the plants shown in the survey to their yard next year, (5) whether they would add (or add more) of these plants to their yard next time they made changes to their landscape, (6) would likely add these plants if help with seed cost was provided, (7) would likely add these if online resources regarding the purchase, planting and care of these plants was received, and (8) would likely add these plants if help with labor is received. For comparison purposes, we also asked some of the aforementioned questions for trees in general. The full survey instrument can be found in Appendix A.

Surveys were dropped off or left with respondents in Darke and Miami counties, Ohio. Selected residents were contacted up to four times total. All initial surveys were distributed from mid-July to mid-August 2019. The first contact consisted of dropping off the survey at people's homes without prior contact or notification. If a resident did not answer, the clear bag containing the survey and a $2 bill (as incentive) was hung on the doorknob for the respondent to find later. The second contact consisted of a postcard, sent to all respondents, thanking those who took the survey and gently nudging those who did not, to take the survey. The third contact consisted of another copy of the survey mailed to all the non-respondents two weeks after the second contact. The fourth and final contact was another reminder and thank you postcard. The final postcard emphasized the importance of the residents' individual answers and explained this was the last time we were going to be contacting them. The verbiage for the contact letters and postcards followed examples found in Dillman et al. (2014).

We obtained approval prior to conducting the research through Miami University's Institutional Review Board (IRB). Our project reference number is 03270e.

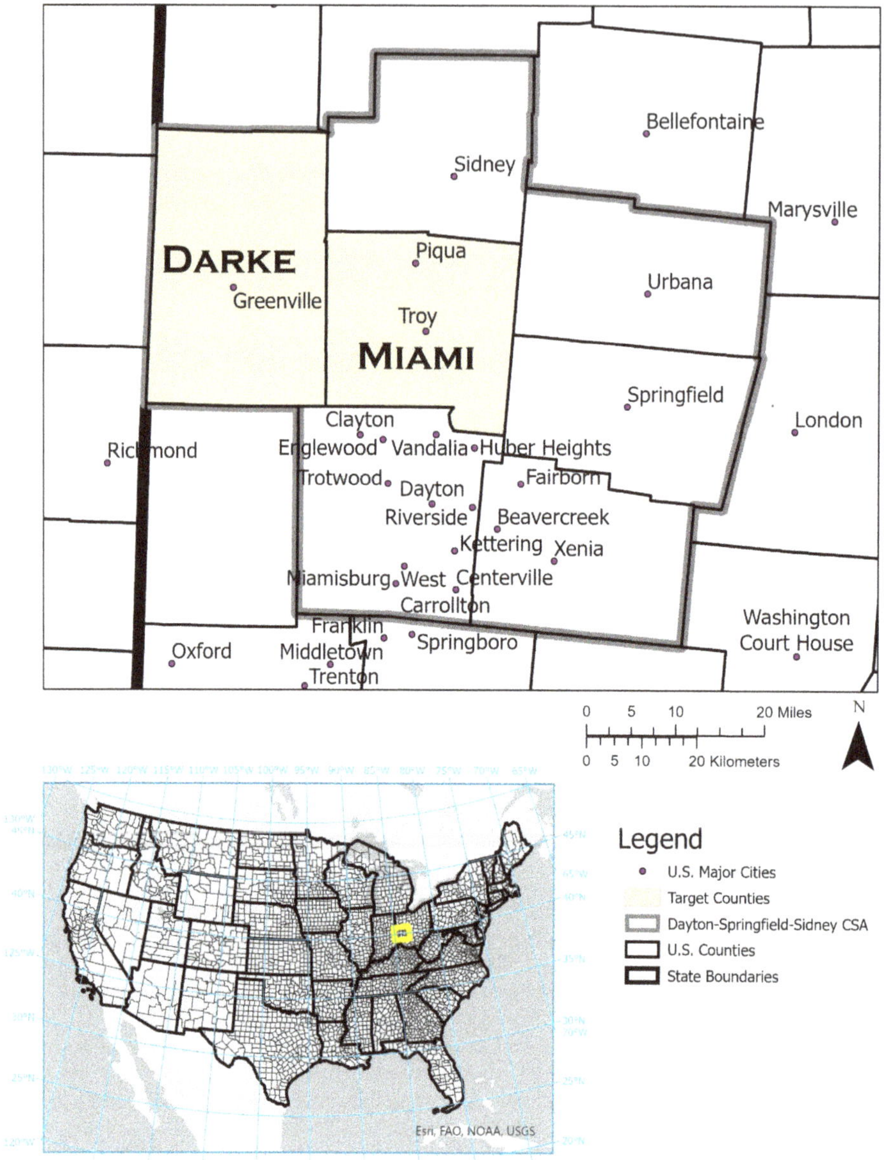

Figure 1. Study area. Study area map displaying the two counties (Darke and Miami counties in Ohio, U.S.A) from which single-family homes were randomly selected. Additionally, shown are states for the conterminous U.S.A. and the Dayton-Springfield-Sidney Combined Statistical Area (CSA) as designated by the United States Census Bureau, as well as cities in the U.S.A. with more than 10,000 inhabitants. The yellow box in the lower map shows the location of the study region with respect to the United States of America (and within the State of Ohio).

2.2. Statistical Analysis

We examined whether the response variables were statistically different for the different groups using one-way Analyses of Variances (ANOVAs) with repeated measures, performed using the *lme* package in R [25]. The stated willingness to add the proposed plants in the following year varied from 1 to 5 with 1 standing for "Strongly Disagree" and 5 "Strongly Agree", therefore a mean of three stands for neutral responses. For the ANOVAs where the group means were significantly different, we performed post hoc multi-comparison tests with a Bonferroni adjustment for multiple comparisons to determine which group means differed statistically. Prior to performing the ANOVAs, we visually assessed that the data in each group were normal (or for the one-way ANOVAs with repeated measures we examined whether the difference between groups was normal). We used Bartlett tests to check for heteroscedasticity.

Lastly, to categorize the reasons respondents wrote as reasons why they might not want to add the various plant choices shown in the survey, we read their answers twice and grouped the answers into similar concerns and counted the occurrences. Note that not all respondents chose to answer these questions, and that some respondents wrote in multiple reasons.

Statistical analyses were performed in R version 3.6.1 (R Core Team, 2019). Aside from those mentioned above, we used the following R packages in our analyses: *foreign, ggplot2, psych, dplyr, Rmisc, multcomp*.

3. Results

Out of 200 surveys distributed, we received 113 responses (57% response rate). Below, we summarize the research questions and results, including summaries of the statistical analyses if applicable.

3.1. Is There a Difference in Mean Reported Willingness to Add the Pollinator Beneficial Plants Shown in the Survey Based on Offers of Different Types of Assistance (e.g., Help with Cost or labour) or the Timing of When to Plant Pollinator Beneficial Plants (This Season or Next Time Changes Will Be Made to Respondents' Landscaping)?

According to a one-way analysis of variance (ANOVA) with repeated measures, the mean reported willingness to plant *E. purpurea* does not differ between various offers of assistance or timing ($F(4,425) = 0.93$, $p = 0.45$; Table 1). Similarly, two separate one-way ANOVAs with repeated measures indicate that the timing or offer of various types of assistance does not affect the reported willingness to plant *A. syriaca* ($F(4,423) = 0.96$, $p = 0.43$) or wildflowers ($F(4,427) = 1.12$, $p = 0.35$; Table 1).

3.2. Is the Self-Disclosed Knowledge of How to Plant, Care for, and Purchase E. purpurea Statistically Different, i.e., Do the Respondents Report They Are More or Less Knowledgeable about Planting, Caring, or Purchasing E. purpurea? Similarly, Is the Self-Disclosed Knowledge of How to Plant, Care for, and Purchase A. syriaca or Wildflowers Statistically Different? And Is the Self-Disclosed Knowledge of How to Plant and Care for Trees Significantly Different?

We find no difference in the self-disclosed knowledge of how to plant, care for, and purchase both *E. purpurea* and *A. syriaca* ($F(2,214) = 1.25$, $p = 0.29$, and $F(2,210) = 1.40$, $p = 0.25$, respectively). This is not the case for wildflowers ($F(2,216) = 7.72$, $p = 0.0006$, Table 2) or for trees ($W = 55$, $p = 0.003$, Table 2). Note that a Wilcoxon signed rank test had to be performed to ascertain whether planting and caring for trees was different because the difference between the two was severely non-normal. The mean reported knowledge of where to purchase wildflowers ($M = 3.5$, $SD = 1.1$) is significantly higher than how to care for them ($M = 3.3$, $SD = 1.2$) and how to plant them ($M = 3.2$, $SD = 1.2$; Table 2). Similarly, respondents' reported knowledge of how to plant trees ($M = 4.1$, $SD = 0.9$) is significantly higher than how to care for trees ($M = 4.0$, $SD = 0.9$; Table 2).

Table 1. Summary statistics for reported intent to plant the native plants shown in the survey when various types of assistance or options for different timing of plantings are given. Mean +/− 1 standard deviation (SD) and median for stated willingness (i.e., reported intent) to plant purple coneflower, common milkweed and wildflowers if assistance is provided with costs, labor, or provision of online resources; mean +/− 1 standard deviation (SD) and median for stated willingness to add the particular plant within the next year, or the next time landscaping changes are made to the yard. Higher values indicate that the respondent is more willing to plant the plants shown in the survey. The scale varies from 1 to 5, with 5 standing for "Strongly agree," and 1 "Strongly disagree" so that a value of three is neutral ("Neither agree nor disagree"). All means are below three, but especially for common milkweed. The number of respondents who answered each question is provided in the column labelled n.

Plant Type	Variable	Mean +/− 1SD	Median	n	Min–Max
Purple coneflower (*Echinacea purpurea*)	Help with cost	2.8 +/− 1.2	3	108	1–5
	Help with labor	2.7 +/− 1.2	3	107	1–5
	Plant when change yard next	2.9 +/− 1	3	107	1–5
	Plant next year	2.7 +/− 1	3	107	1–5
	Provide online resources	2.8 +/− 1.1	3	108	1–5
Common milkweed (*Asclepias syriaca*)	Help with cost	2.4 +/− 1.1	2	107	1–5
	Help with labor	2.4 +/− 1.2	2	107	1–5
	Plant when change yard next	2.3 +/− 1	2	107	1–5
	Plant next year	2.3 +/− 1	2	106	1–5
	Provide online resources	2.5 +/− 1.2	3	107	1–5
Wildflowers (multiple *spp.*)	Help with cost	2.9 +/− 1.3	3	109	1–5
	Help with labor	2.7 +/− 1.2	3	107	1–5
	Plant when change yard next	2.9 +/− 1.1	3	108	1–5
	Plant next year	2.8 +/− 1	3	108	1–5
	Provide online resources	2.8 +/− 1.2	3	108	1–5

3.3. Does Self-Disclosed Knowledge of How to Plant, Care for, and Purchase Pollinator Beneficial Plants Differ between the Different Plant Types Shown in the Survey, e.g., Do the Respondents Feel More Knowledgeable about Planting E. purpurea Compared to Wildflowers or A. syriaca, and so Forth

Knowledge of where to purchase the different plants in the survey does differ significantly (F(2,208) = 44.63, $p < 2 \times 10^{-16}$, Table 3.) Respondents are significantly less likely to know where to purchase *A. syriaca* (M = 2.6, SD = 1.2) compared to *E. purpurea* (M = 3.5, SD = 1.1) as well as wildflowers (M = 3.5, SD = 1.1). Knowledge of how to care for the plants shown in the survey as well as trees also differs statistically between groups (F(3,312) = 53.94, $p < 2 \times 10^{-16}$, Table 3). Respondents are significantly less sure how to care for *A. syriaca* (M = 2.7, SD = 1.2) compared to *E. purpurea* (M = 3.4, SD = 1.2), and are much more confident in how to take care of trees (M = 4.1, SD = 0.9; Table 3) compared to the other three plants shown in the survey. The same patterns can be seen in the responses for how to plant these plants. The mean reported knowledge of how to plant the various pollinator beneficial plant choices shown in the survey differs statistically (F(3,314) = 53.23, $p < 2 \times 10^{-16}$) with respondents stating they are much more knowledgeable about how to plant trees (M = 4.0, SD = 0.9), than *E. purpurea* (M = 3.4, SD = 1.2) and *A. syriaca* (M = 2.6, SD = 1.1). Overall, respondents say are much less knowledgeable about planting *A. syriaca* compared to the other plants shown in the survey (Table 3).

Table 2. Summary statistics for reported knowledge of planting, caring for, and purchasing the three pollinator-beneficial plants shown in the survey (as well as for trees as a comparison group). Mean +/− 1 standard deviation (SD) and median for reported knowledge of where to purchase, how to plant, and how to care for purple coneflower, common milkweed and wildflowers. Note that the "where to purchase trees" question was not included in the survey instrument. We conducted three one-way repeated-measures ANOVAs to test whether the mean reported knowledge about how to plant, care for, or purchase each of the plants shown in the survey was different. The groups are the reported knowledge categories. The plant types were *Echinacea purpurea, Asclepias syriaca*, and wildflower prairie (multiple species of native plants). Higher values indicate that the respondent believes they are more knowledgeable. The scale varies from 1 to 5, with 5 indicating that the respondent strongly agrees with the statement (e.g., "I know how to care for wildflowers"). A value of one indicates the respondent strongly disagrees with the statement so that a value of three is neutral ("Neither agree nor disagree"). There is no significant difference between the knowledge questions for purple coneflower or common milkweed so not further tests were conducted. For the ANOVA for the wildflowers, the p values reported for the post hoc test of multiple comparisons is Bonferroni adjusted. ^ To compare responses for caring and planting for trees since the difference between reported knowledge of planting and caring for trees is severely non-normal and there were only two groups a Wilcoxon signed rank test is used. The number of respondents who answered each question is provided in the column labelled n. Asterisks *, **, and *** denote 5%, 1%, and 0.1% levels of significance, respectively when testing the null hypothesis that the coefficients equal zero.

Plant Type	Variable	Mean +/− 1SD	Median	n	Min-Max	Tukey's HSD Comparison Plant	Care
Purple coneflower (*Echinacea purpurea*)	Know how to plant	3.4 +/− 1.2	4	108	1–5		
	Know how to care for	3.4 +/− 1.2	4	108	1–5		
	Know where to purchase	3.5 +/− 1.1	4	108	1–5		
Common milkweed (*Asclepias syriaca*)	Know how to plant	2.6 +/− 1.1	2	107	1–5		
	Know how to care for	2.7 +/− 1.2	3	108	1–5		
	Know where to purchase	2.6 +/− 1.2	2	105	1–5		
Wildflowers (multiple *spp.*)	Know how to plant	3.2 +/− 1.2	3	109	1–5		
	Know how to care for	3.3 +/− 1.2	4	109	1–5	0.2593	
	Know where to purchase	3.5 +/− 1.1	4	109	1–5	0.0488 *	<0.001 ***
Trees	Know how to plant	4.03 +/− 0.9	4	104	1–5		
	Know how to care for	4.12 +/− 0.9	4	105	1–5	0.003 ** ^	

3.4. What Are the Most Frequently Cited Barriers to Adding the Plants Shown in the Survey to Respondents' Yards?

The most oft-stated reasons respondents mention why not to add the plants shown in the survey are a general dislike for the appearance of the plants (especially for *A. syriaca*, and *E. purpurea*), that the plants require or are perceived to require more maintenance than the respondent is willing or has the ability to expend, that the plant(s) spread too much on their own, and for *A. syriaca* that it is considered a weed or invasive plant (Figure 2). Please note that these are reasons specified by the respondents in an open-ended portion of the survey, i.e., there were no predetermined answer choices for this question on the survey. Overall, the reasons respondents list as to why they might not want to add *E. purpurea*, *A. syriaca*, and wildflowers to their landscaping overwhelmingly have to do with visual appeal and appearance, maintenance considerations or lack of knowledge of how to care for these plants, and considerations about the plants spreading beyond the planted area, as well as for *A. syriaca* concerns about it being a weed, or an invasive, or toxic to pets and livestock.

Table 3. Statistical comparisons for the reported knowledge of how to plant, care for, and purchase for the different native plants shown in the survey as well as for trees. Mean +/− 1 standard deviation (SD) and median for reported knowledge of where to purchase, how to plant, and how to care for purple coneflower, common milkweed and wildflowers (*Echicanea purpurea*, *Asclepias syriaca*, and multiple *spp.* of wildflowers, respectively). Note that the "where to purchase trees" question was not included in the survey instrument. We conducted a one-way repeated-measures ANOVA to examine whether the mean reported knowledge about how to plant, *E. purpurea*, *A. syriaca*, and wildflowers were different. The same analyses were conducted for the other knowledge statements, i.e., "know how to care for" and "know where to purchase". Higher values indicate that the respondent believes they are more knowledgeable. The scale varies from 1 to 5, with 5 indicating that the respondent strongly agrees with the statement (e.g., "I know how to care for wildflowers"). A value of one indicates the respondent strongly disagrees with the statement so that a value of three is neutral ("Neither agree nor disagree"). The *p* values reported for the post hoc test of multiple comparisons are Bonferroni adjusted. The number of respondents who answered each question is provided in the column labelled n. Asterisks *** denote 0.1% levels of significance.

Question	Plant Type	Mean +/− 1SD	Median	n	Min–Max	Tukey's HSD Comparison		
						E. purpurea	*A. syriaca*	Tree
Know how to plant	*E. purpurea*	3.4 +/− 1.2	4	108	1–5			
	A. syriaca	2.6 +/− 1.1	2	107	1–5	1.65×10^{-10} ***		
	Tree	4 +/− 0.9	4	104	1–5	1.58×10^{-8} ***	$<2 \times 10^{-16}$ ***	
	Wildflowers	3.2 +/− 1.2	3	109	1–5	1	1.51×10^{-8} ***	1.62×10^{-10} ***
Know how to care for	*E. purpurea*	3.4 +/− 1.2	4	108	1–5			
	A. syriaca	2.7 +/− 1.2	3	108	1–5	4.39×10^{-12} ***		
	Tree	4.1 +/− 0.9	4	105	1–5	7.70×10^{-8} ***	$<2 \times 10^{-16}$ ***	
	Wildflowers	3.3 +/− 1.2	4	109	1–5	1	3.48×10^{-8} ***	1.14×10^{-11} ***
Know where to purchase	*E. purpurea*	3.5 +/− 1.1	4	108	1–5			
	A. syriaca	2.6 +/− 1.2	2	105	1–5	6.66×10^{-15} ***		
	Wildflowers	3.5 +/− 1.1	4	109	1–5	1	$<2 \times 10^{-16}$ ***	

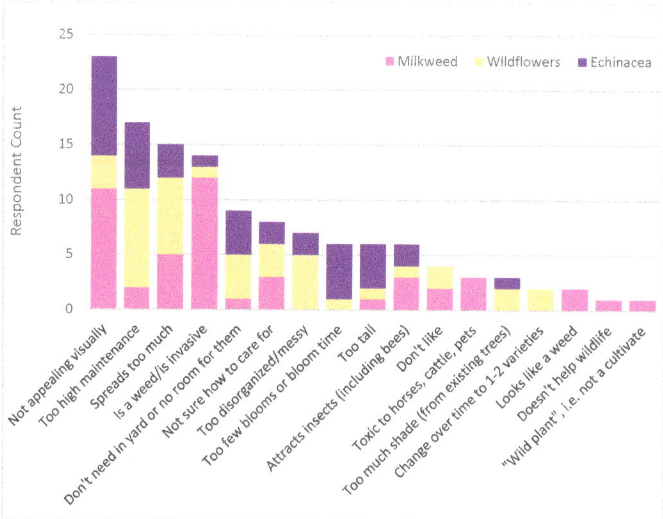

Figure 2. Key barriers to adding the pollinator beneficial plants shown in the survey. Number of times (counts) a respondent stated a particular reason as to why they might not like to add *Asclepias syriaca* (Common Milkweed; pink bars), *Echninacea purpurea* (purple coneflower; purple bars), or wildflowers (multiple *spp.*; orange bars) to their yard. This was an open-ended question on the survey instrument. Out of the 113 respondents to the survey, 48 chose to write at least one reason down in the answer box provided for that purpose for *E. purpurea*, 57 for *A. syriaca*, and 54 for wildflowers.

4. Discussion

The fact that many respondents mentioned they did not care for the plants' visual appearance is expected as these plants do not conform to the traditional aesthetics of yards with neat (weed-free) lawns and ornamental showy flowers, yet the desire for those types of gardens is prevalent across the U.S.A. [26]. Others have found that in the Southwestern United States native landscapes were perceived as unattractive and high maintenance, but "86% [of respondents said they] would use native plants if native plants were attractive" [27]. These results were influenced by education and time spent on horticultural activities. Similarly, in a study of ecological and socioeconomic factors that affect plant richness composition in residential gardens in Minneapolis MN, the most important characteristics for planting decisions were "aesthetics, wildlife, neatness, and food provision" [16].

Unfortunately, most of the undesired characteristics of the plants shown in the survey (and other pollinator beneficial plants) seem difficult to counter although some organizations (e.g., the Cincinnati Zoo and Botanical Garden®) do produce plant guides that promote adding plants that would bloom throughout the Spring, Fall, and Summer (https://cincinnatizoo.org/horticulture/plant-for-pollinators/resources/). Ostensibly, this is to provide forage for many insect pollinators during their active season, but can also help with a flower garden's appearance over the growing season. Some of the guides also specify plant height, spread, size of flowers typically and thus can help alleviate some residents' concerns as well. Finding ways to keep maintenance needs low for large plantings of these plants (especially weed-free and if possible lower to the ground) seems desired [22]. Perhaps a combination of these plants and groundcover could help counter some of these impediments and while research along those lines has been conducted in orchards, i.e., agricultural settings [28,29], to the best of our knowledge this is an active research need in urban settings (although see [30]).

Correcting the misconceptions surrounding *A. syriaca* with the general public is greatly needed especially since another study found similar concerns in large property owners and farmers of SW Ohio [6]. Indeed, *A. syriaca* is neither a weed nor an invasive plant in this region [31]. It is a native plant that, for full disclosure, can propagate by "sending up new shoots from roots that spread outward from the parent plant. This clonal reproduction allows their populations to expand over time, and plants may spread out of their original area. [...] Despite the vegetative growth, many of these species are unlikely to create an ongoing and unmanageable weed problem for roadside managers (or for other land managers, homeowners, and others)" [32]. As to the fears of it being toxic to livestock and pets, this needs to be reframed. Indeed, even the ASPCA's website states that milkweed can be toxic to dogs, cats, and horses and can cause their death [33] with no allusion to the fact that it is highly unpalatable to those species, that it would have to be ingested in large quantities to have those effects, that livestock only eat it if there is nothing else for them to eat, and that toxicity varies by milkweed species [34,35]. With that said, perhaps the survey respondents in this study just did not want to take the chance of accidental large quantity ingestion by pets which of course is understandable, but, for the purposes of this study, rather unsurmountable.

SW Ohioans were in general slightly unwilling to plant the native plants shown in the survey. Surprisingly, there is no difference in the mean stated willingness to plant these pollinator beneficial plants regardless of whether the timing of planting was within the following year or the next time landscaping changes were made, or whether assistance with cost, labor, or the provision of online resources were offered. This is especially true for *A. syriaca* as, collectively, respondents moderately disagreed that they were willing to plant it. Overall, these findings are rather problematic as they may indicate that the aesthetic and other considerations previously mentioned are paramount in these residents' minds and since milkweeds *spp.* are the host plants for the monarch butterfly no substitutes can be used in this case. It also may mean that for the organizations that currently provide such incentives to interested homeowners, they are reaching a population that was already going to make the changes to their landscape regardless of the assistance. If this is the case, in the long term, it may be more difficult to find willing homeowners. We do not imply that initiatives that provide these types of incentives should be halted especially since they are one known effective strategy to help change behavior [36]; however, it seems wise to find other mechanisms by which to entice residents to put these plants and other pollinator beneficial plants in their yards. One example of an approach that worked for native plants in Dunedin (New Zealand), was to use biodiversity assessments and dialog with homeowners as it led to an increase in knowledge and positive shift in attitude towards native plant species [37]. After participating in the program, one quarter of participants incorporated native plants in their yard and 13% stated it was to improve biodiversity [37]. Another mechanism would include passing policies aimed at improving conservation efforts for pollinators, which is greatly needed [38,39], including in urban areas [40]. It is critical that these policies be enforced [41] and integrate the "perspective of farmers and local communities on the need to conserve pollinators, alongside scientific understanding" [42].

Other than for *A. syriaca*, reported knowledge of how to plant, care for or purchase the plants shown in the survey is strong although, for wildflowers and trees, survey respondents were significantly less certain of their abilities to plant these species compared to caring for them or purchasing them. Educational materials, including demonstrations, videos, and hands-on workshops might be valuable in this regard. Reported knowledge of purchasing, planting, and caring for *A. syriaca* is significantly lower than for the other plants shown in the survey and should be addressed if the vision of 1.3 billion new stems is to be achieved [8]. Whether or not these plants or seeds are widely available at nurseries and other retailers is a separate but important barrier to overcome [16,43–45].

Reported knowledge of how to plant and care for trees is high. Given the mistakes homeowners are known to make when it comes to tree planting and care [46,47] this is a bit alarming. It might be beneficial to educate homeowners about the needs of trees and it also may be an opportunity. Some trees provide abundant resources for pollinators [48–50] and given their general acceptability to residents, the fact they are easier to maintain, that their aesthetic value is generally high, that the area underneath can be mowed around and not take up large portions of a yard's surface area, and that they provide many other ecosystem services, promoting pollinator beneficial tree plantings could be a straightforward way to provide floral resources for many pollinators and nesting resource for some. Of course, this proposed solution would not serve the monarch butterfly populations (*Danaus plexippus*).

Limitations and Future Research Needs

It is probable that the individuals who took the time to respond to this survey may have more affinity to gardening than non-respondents. Despite all the measures we took to increase response rate, 43% of selected households did not return a completed survey. Given that individuals who like gardening and view it as a hobby are associated with more wildlife supporting yards [22] the mean responses to the questions about intentions to plant the native plant species shown in the survey and the plant and gardening knowledge responses may be higher than they would be for the general population and this should be considered while interpreting the results presented herein.

We solely surveyed residents of single-family homes in suburban and peri-urban southwestern Ohio (USA). More research is needed to uncover whether these findings are universal across geographies, vary by yard size, structure, and classification (e.g., urban, versus suburban, versus exurban). An interesting area of research would be to examine whether different typologies of residents exist with regard to adding native plants to private yards as that could help tailor education and outreach programs for different constituents. Additionally, we examine people's intention to add native plantings to their yards but there exists a gap between intention and behavior [51]. Conducting experiments to ascertain the width of that gap is greatly needed, although see Champine et al. [52] for such a study in Fort Collins, Colorado, USA and Knapp et al. [53] for empirical data on this gap in the

United Kingdom. Lastly, if efforts to add pollinator-beneficial plants in yards are wildly successful, it will be important to know whether (and in what proportion) residents might take measures to control stinging insects in their yard. While insect pollinator knowledge and appreciation has increased over time [54], a study of residents in a county in SW Ohio found that approximately 5% of residents surveyed specified they would not like to see these plants in their yards because the plants attracted bees [55]. To the best of our knowledge, whether this sentiment translates to actively trying to eradicate these insects from one's yard is unknown but should be examined closely.

5. Conclusions

The objectives of this study were to uncover impediments to adding different types of pollinator-beneficial plants to private yards, as well as ascertain which incentives to plant these native plants might be most persuasive, and finally determine if there were procedural knowledge gaps in how to plant, care for, or where to purchase these pollinator beneficial plants. Unfortunately, getting residents of single-family homes to voluntarily add *A. syriaca* (common milkweed) to private yards in quantities that are needed to support a thriving monarch butterfly population may prove to be difficult given SW Ohio residents seem to dislike the appearance of the plant, its mode of propagation, and have some preconceived notions as well as somewhat erroneous concerns about the plant itself (e.g., invasiveness). Focusing instead on planting large areas of milkweed (*Asclepias* spp.) along roadsides, on agricultural and government land, and exploring payments for ecosystem services options might be necessary.

Overall, SW Ohio residents' intent to add the pollinator-beneficial plants shown in the survey to the yards of single-family homes is not high and, surprisingly, does not differ when incentives are offered; even when help with costs, labor, or the provision of online resources are proposed. Finding other ways to entice homeowners to plant native plants is necessary, especially given the gap between strong behavioral intentions to plant native plants and their actual addition to yards [52]. Experimenting with different approaches to elicit this pro-environmental behavior is needed, such as the one described in van Heezik et al. [37].

Procedural knowledge, i.e., knowledge of how to plant and care for *A. syriaca*, *E. purpurea*, and wildflowers is significantly lower than for trees. More education and outreach in this regard may be warranted. However, education and outreach materials and programs should consider ways to increase the visual appeal of the plantings (by for example providing "cues to care" [19]) and decrease maintenance needs (especially compared to the mindlessness and ease of taking care of turfgrass). Convenience is important when attempting to convince someone to adopt a pro-environmental behavior [36], especially planting wildlife friendly yards [22].

The results of this research could be used to design more effective education and outreach programs for Midwesterners, keeping in mind that the incentives proposed in this study are unlikely to substantially change the intent to plant these pollinator-beneficial plants. Other mechanisms (e.g., through policy and/or payment for ecosystem services) will most likely be needed to markedly increase the area of private yards devoted to native plants in general but milkweeds in particular in suburban/rural Southwest Ohio (USA).

Author Contributions: Conceptualization, A.D.; methodology, A.D and J.S.; software, A.D.; validation, A.D.; formal analysis, A.D.; investigation, A.D. and J.S.; resources, A.D.; data curation, J.S.; writing—original draft preparation, A.D. and J.S.; writing—review and editing, A.D. and J.S.; visualization, A.D.; supervision, A.D.; project administration, A.D.; funding acquisition, A.D. All authors have read and agreed to the published version of the manuscript.

Funding: This research received no external funding.

Institutional Review Board Statement: This research was approved through Miami University's Institutional Review Board (IRB). Our project reference number is 03270e.

Informed Consent Statement: Informed consent was obtained from all subjects involved in the study.

Data Availability Statement: Anonymized survey responses can be requested by emailing the corresponding author.

Acknowledgments: The views expressed in this article are those of the author and do not necessarily reflect the official policy or position of the United States Air Force Academy, the Air Force, the Department of Defense, or the U.S. Government (PA#: USAFA-DF-2022-911). We are extremely grateful to the survey participants for taking the time to complete the survey. We are also thankful for funding from Miami University to complete this study. Without these elements this research would not be possible. Lastly, we want to extend a huge thank you to Debbi White, the departmental assistant for the Geography Department at Miami University for overseeing the stuffing of the survey packets and helping us navigate reimbursements for the field work. Her competence and can-do attitude as well as incessant hard work is immensely appreciated.

Conflicts of Interest: The authors declare no conflict of interest.

Appendix A

Your Landscaping Preferences!

Completing the survey is easy:

1. **Keep the $2 bill and marker** as a small token of appreciation.
2. Fill out the survey. **This survey should be filled out by an adult (18 years or older) who makes most of the landscaping choices at this address.**
3. Fold it in half, place it in the prepaid envelope and **mail it back to us!** Please remember to **include the map** (but keep the marker).

This survey should be filled out by the **person in this household who makes most of the landscaping and gardening decisions** and is over 18 years old.

For the purposes of this research the **front yard** is defined as the section of your yard between the main road and the house, while the **backyard** is the section from the house to the property line furthest from the main road. The **main road** is the road that meets the driveway.

Please note that the plant selections shown in this survey will all grow in our region given our climatic conditions.

First Question

A. Of the three options shown below, <u>which one (if any) do you think is the MOST beautiful</u>? Please check ONE.

- ☐ 1. Purple coneflower
- ☐ 2. Wildflower prairie
- ☐ 3. Common Milkweed
- ☐ I do not think any of these landscapes are beautiful

B. These are purple coneflowers (*Echinacea purpurea*)

Bloom: June to August
Height: 2 to 5 feet; Spread: 2 to 5 feet

Do you already have this plant in your yard?
Please check ONE.
□ Yes, it was here when I moved in
□ Yes, I planted it here
□ No
□ I don't know

Please write possible reasons you might like to see purple coneflowers in your yard:
____
____

Please write possible reasons you might not like to see purple coneflowers in your yard:
____
____

Please Check the most appropriate box:	Strongly Disagree	Disagree	Neither Agree Nor Disagree	Agree	Strongly Agree
I know how to plant coneflowers	□	□	□	□	□
I know how to care for coneflowers	□	□	□	□	□
I know where to purchase coneflowers or its seeds	□	□	□	□	□
I will add (or add more) coneflowers to my yard in the next year	□	□	□	□	□
I will add (or add more) coneflowers to my yard the next time I make changes to my landscape	□	□	□	□	□
I would likely add coneflowers if I received help with the cost of seed for this plant	□	□	□	□	□
I would likely add coneflowers if I received online resources regarding plant purchase, planting, and care	□	□	□	□	□
I would likely add coneflowers if I received help with the labor (i.e. tilling and planting)	□	□	□	□	□

C. These are wildflowers (multiple species of plants)

Bloom time: May to September
Height: varies but usually between 2-4 feet
Multiple plants that bloom at different times so that there is continuous and varied color.

Do you already have wildflowers in your yard?
□ Yes, it was here when I moved in
□ Yes, I planted it here
□ No
□ I don't know

Please write possible reasons you might like to see wildflowers in your yard:
____
____

Please write possible reasons you might not like to see wildflowers in your yard:
____
____

Please Check the most appropriate box:	Strongly Disagree	Disagree	Neither Agree Nor Disagree	Agree	Strongly Agree
I know how to plant wildflowers	□	□	□	□	□
I know how to care for wildflowers	□	□	□	□	□
I know where to purchase wildflowers or their seeds	□	□	□	□	□
I will add (or add more) wildflowers to my yard in the next year	□	□	□	□	□
I will add (or add more) wildflowers to my yard the next time I make changes to my landscape	□	□	□	□	□
I would likely add wildflowers if I received help with the cost of seed for these plants	□	□	□	□	□
I would likely add wildflowers if I received online resources regarding plant purchase, planting, and care	□	□	□	□	□
I would likely add wildflowers if I received help with the labor (i.e. tilling and planting)	□	□	□	□	□

D. This is common milkweed (*Asclepias syriaca)*

Bloom Time: June to August

Height: 2.00 to 3.00 feet; Spread: 0.75 to 1.00 feet

Do you already have this plant in your yard?
Please check ONE.

□ Yes, it was here when I moved in

□ Yes, I planted it here

□ No

□ I don't know

Please write possible reasons you might like to see common milkweed (*Asclepias syriaca*) in your yard:

____

____

Please write possible reasons you might not like to see common milkweed (*Asclepias syriaca*) in your yard:

____

____

Please Check the most appropriate box:	Strongly Disagree	Disaree	Neither Agree Nor Disagree	Agree	Strongly Agree
I know how to plant milkweed	□	□	□	□	□
I know how to care for milkweed	□	□	□	□	□
I know where to purchase milkweed or its seeds	□	□	□	□	□
I will add (or add more) milkweed to my yard in the next year	□	□	□	□	□
I will add (or add more) milkweed to my yard the next time I make changes to my landscape	□	□	□	□	□
I would likely add milkweed if I received help with the cost of seed for this plant	□	□	□	□	□
I would likely add milkweed if I received online resources regarding plant purchase, planting, and care	□	□	□	□	□
I would likely add milkweed if I received help with the labor (i.e. tilling and planting)	□	□	□	□	□

E. Map Instructions
Please use the map of this property which was provided in your packet

If you WOULD NOT consider planting any of the plants shown previously in the survey within the next year, please check this box □ and SKIP to section F

If you WOULD consider planting the plants shown previously WITHIN THE NEXT YEAR, please complete the following 4 steps:

1) **Draw the area or areas where you might plant one of the three plant choices shown in the survey** with the pen we provided on the map we provided.
2) **Label each area** with:
   - □ **"PC"** if you intend to plant **purple coneflowers**
   - □ **"CW"** if you intend to plant **common milkweed**
   - □ **"WP"** if you intend to plant **wildflower prairie**.

If the label will not fit in the area you drew please write the letters next to the area with an arrow pointing to it.

3) **Mark with an X any/all locations where you will plant trees WITHIN THE NEXT YEAR**
4) Please answer the following questions:
   - **Were there any areas that you would consider planting but that were too small to represent on the map.** □ Yes □ No
   - **The area(s) I drew on the map is/are currently covered in (please check all that apply):**
     □ Pasture □ Lawn □ Trees □ Honeysuckle □ Weeds □ Crops
     □ Flower beds □ Hay □ Other (please specify):______________________

F. Your Current Landscaping Practices

Which of the following are present on this property?
□ Lawn □ Pasture/Hay □ Trees □ Shrubs □ Vegetable plot
□ Hard surface (e.g. driveway, patio, decking, gravel) □ Flower beds
□ Row crops □ Other (please state): ______________________

How many trees have you planted since you've lived on this property: _______trees planted

How many trees have you removed since you've lived on this property: ______trees removed

Do you keep bees or let someone keep bees on this property? □ Yes □ No

In the summer months, how much time do you spend mowing on average each week?
________________hours per week on average in summer spent mowing

Excluding mowing, in the summer months, how much time do you spend gardening, weeding, and/or doing other yard work each week on average?
__________________hours per week on average in summer

Who takes care of your mowing?
□ Self □ Other person living in household □ A person (or company) you hire
□ I/We do not have any lawn □ Other __________________

Who takes care of your landscaping?
□ Self □ Person living in household □ A person (or company) you hire
□ I/We do not have any landscaping needs □ Other __________________

Is your lawn mostly under tree cover, i.e. full shade? □ Yes □ No

Please check the most appropriate box	Strongly Disagree	Disagree	Neither Agree nor Disagree	Agree	Strongly Agree
I would be willing to mow my lawn every two weeks regardless of rainfall	☐	☐	☐	☐	☐
I would be willing to allow white clover to grow in my front yard	☐	☐	☐	☐	☐
I would be willing to allow white clover to grow in my backyard	☐	☐	☐	☐	☐
I would be willing to allow dandelions to grow in my front yard	☐	☐	☐	☐	☐
I would be willing to allow dandelions to grow in my backyard	☐	☐	☐	☐	☐
I want my front yard to look neat and tidy	☐	☐	☐	☐	☐
I want my backyard to look neat and tidy	☐	☐	☐	☐	☐
I know how to plant trees	☐	☐	☐	☐	☐
I know how to take care of trees	☐	☐	☐	☐	☐
I think my yard is a reflection of me	☐	☐	☐	☐	☐
I enjoy mowing	☐	☐	☐	☐	☐
I enjoy gardening	☐	☐	☐	☐	☐
I enjoy planting trees	☐	☐	☐	☐	☐
I enjoy caring for trees	☐	☐	☐	☐	☐
I enjoy being outside	☐	☐	☐	☐	☐

G. Parcel Characteristics

Do you own a parcel that shares a property line with this property? □ Yes □ No

Do you own a zero turn radius mower? □ Yes □ No

Do you have a pond on the property? □ Yes □ No

Which of the following would you like to see more of on your property (check all that apply)?

□ Bats □ Bees □ Beetles □ Birds □ Butterflies □ Deer

□ Foxes □ Moths □ Spiders □ Squirrels □ Other: ______________

Is anyone in your household allergic to bee stings? □ Yes □ No □ I Don't know

Do you have any pets that spend time in your yard? □ Yes □ No □ I Don't know

Do you have white clover in your front yard? □ Yes □ No □ I Don't know

Do you have white clover in your backyard? □ Yes □ No □ I Don't know

Do you have dandelions in your front yard? □ Yes □ No □ I Don't know

Do you have dandelions in your backyard? □ Yes □ No □ I Don't know

In the last year, have you or someone else applied chemicals to control weeds or unwanted plants in the outdoor areas of your home?
□ Yes □ No □ I don't know

If yes, to which part(s) of your yard have you or someone else applied chemicals to in order to control weeds or unwanted plants (check all that apply):
□ Lawn □ Flower Beds □ Vegetable Garden □ Pasture
□ Other (please specify): ____

In the last year, have you or someone else applied chemicals to control insects or other pests in the outdoor areas of your home?
□ Yes □ No □ I don't know

If yes, to what part(s) of your yard have you or someone else applied chemicals in order to control insects or other pests (check all that apply):
□ Lawn □ Flower Beds □ Vegetable Garden □ Pasture
□ Other (please specify): ____

In the last year, have you or someone else applied fertilizer on this property?
□ Yes □ No □ I don't know

If yes, what part(s) of your yard have you or someone else applied fertilizer to (check all that apply):
□ Lawn □ Flower Beds □ Vegetable Garden □ Potted Plants
□ Other (please specify): ____

Is this property a working farm? □ Yes □ No □ I Don't know

Do you keep livestock on this property? □ Yes □ No □ I Don't know

Do you keep horses on this property? □ Yes □ No □ I Don't know

Do you own a tractor? □ Yes □ No □ I Don't know

Do you grow hay on this property? □ Yes □ No □ I Don't know

Last page next!

BACKGROUND INFORMATION.

This section is for statistical purposes only.
As a reminder, all survey responses will be strictly confidential.

How many years have you lived in your current home? ______ years

Is some or all the land on this property under a conservation easement? □ Yes □ No

What year was your house built (approximately)?

□ <1900 □ 1901-1940 □ 1941-1960 □ 1961-1980 □ 1981-2000 □ 2001-2010 □ 2010-2019

Does your neighborhood have a Homeowners Association? □ Yes □ No □ I don't know

Do you rent or own your home? □ Rent □ Own

How many people live in your home (including you) that are age 18 or older? _______people

How many people live in your home that are younger than age 18? ___________ people

What is the employment status of the main earner in your household?

□ Full-time □ Part-time □ Not working □ Retired □ Student

What is the highest level of education you have completed?

□ Less than High School □ High School/GED □ Certificate Program

□ Associates Degree □ Some College □ 4-year College Degree

□ Graduate Degree □ Prefer not to answer

In what year were you born? _________

Are you married or in a domestic partnership? □ Yes □ No □ Prefer not to answer

What is your gender? □ Male □ Female □ Non Binary □ Prefer not to answer

How do you identify your race/ethnicity?

□ White □ Hispanic □ African American □ American Indian

□ Asian/Pacific Islander □ Other □ Prefer not to answer

What was your total household income last year, before taxes? Please check ONE answer

□ Less than $25,000 □ $50,001 up to $100,000 □ More than $150,001

□ $25,001 up to $50,000 □ $100,001 up to $150,000 □ Prefer not to answer

Do you have any final comments you would like to share with us?

____

____

____

If resources mentioned throughout this survey become available, and you would like us to share them with you, please provide your email address: ____________________

Thank you greatly for completing our survey! Please remember to include the map along with the survey in the prepaid return envelope.

If you misplaced the return envelope we provided to send this survey back to us, please mail it to Dr. Amelie Davis, Dept. of Geography, Miami University, 250 S. Patterson Ave, Oxford OH, 45056.

References

1. Haberl, H.; Erb, K.-H.; Krausmann, F. Human appropriation of net primary production: Patterns, trends, and planetary boundaries. *Annu. Rev. Environ. Resour.* **2014**, *39*, 363–391. [CrossRef]
2. Aronson, M.F.J.; La Sorte, F.A.; Nilon, C.H.; Katti, M.; Goddard, M.A.; Lepczyk, C.A.; Warren, P.S.; Williams, N.S.G.; Clilliers, S.; Clarkson, B. A global analysis of the impacts of urbanization on bird and plant diversity reveals key anthropogenic drivers. *Proc. R. Soc. B Biol. Sci.* **2014**, *281*, 20133330. [CrossRef] [PubMed]
3. Lerman, S.B.; Warren, P.S. The conservation value of residential yards: Linking birds and people. *Ecol. Appl.* **2011**, *21*, 1327–1339. [CrossRef] [PubMed]

4. Burghardt, K.T.; Tallamy, D.W.; Shriver, W.G. Impact of Native Plants on Bird and Butterfly Biodiversity in Suburban Landscapes. *Conserv. Biol.* **2009**, *23*, 219–224. [CrossRef] [PubMed]
5. Lowenstein, D.M.; Matteson, K.C.; Xiao, I.; Silva, A.M.; Minor, E.S. Humans, bees, and pollination services in the city: The case of Chicago, IL (USA). *Biodivers. Conserv.* **2014**, *23*, 2857–2874. [CrossRef]
6. Davis, A.Y.; Lonsdorf, E.V.; Shierk, C.R.; Matteson, K.C.; Taylor, J.R.; Lovell, S.T.; Minor, E.S. Enhancing pollination supply in an urban ecosystem through landscape modifications. *Landsc. Urban Plan.* **2017**, *162*, 157–166. [CrossRef]
7. Derby Lewis, A.; Bouman, M.J.; Winter, A.M.; Hasle, E.A.; Stotz, D.F.; Johnston, M.K.; Klinger, K.R.; Rosenthal, A.; Czarnecki, C.A. Does Nature Need Cities? Pollinators Reveal a Role for Cities in Wildlife Conservation. *Front. Ecol. Evol.* **2019**, *7*, 220. [CrossRef]
8. Thogmartin, W.E.; López-Hoffman, L.; Rohweder, J.; Diffendorfer, J.; Drum, R.; Semmens, D.; Black, S.; Caldwell, I.; Cotter, D.; Drobney, P.; et al. Restoring monarch butterfly habitat in the Midwestern US:'all hands on deck'. *Environ. Res. Lett.* **2017**, *12*, 074005. [CrossRef]
9. Cook, E.M.; Hall, S.; Larson, K.L. Residential landscapes as social-ecological systems: A synthesis of multi-scalar interactions between people and their home environment. *Urban Ecosyst.* **2012**, *15*, 19–52. [CrossRef]
10. Chowdhury, R.R.; Larson, K.; Grove, M.; Polsky, C.; Cook, E.; Onsted, J.; Ogden, L. A Multi-Scalar Approach to Theorizing Socio-Ecological Dynamics of Urban Residential Landscapes. *Cities Environ.* **2011**, *4*, 6. [CrossRef]
11. Kollmuss, A.; Agyeman, J. Mind the Gap: Why do people act environmentally and what are the barriers to pro-environmental behavior? *Environ. Educ. Res.* **2002**, *8*, 239–260. [CrossRef]
12. Schultz, P.W. Knowledge, information, and household recycling: Examining the knowledge-deficit model of behavior change. In *New Tools for Environmental Protection: Education, Information, and Voluntary Measures*; National Academy Press: Washington, DC, USA, 2002.
13. Nassauer, J.I.; Wang, Z.; Dayrell, E. What will the neighbors think? Cultural norms and ecological design. *Landsc. Urban Plan.* **2009**, *92*, 282–292. [CrossRef]
14. Wheeler, M.M.; Larson, K.L.; Bergman, D.; Hall, S.J. Environmental attitudes predict native plant abundance in residential yards. *Landsc. Urban Plan.* **2022**, *224*, 104443. [CrossRef]
15. McKenzie-Mohr, D. *Fostering Sustainable Behavior: An Introduction to Community-Based Social Marketing*; New Society Publishers: Gabriola Island, BC, Canada, 2011.
16. Cavender-Bares, J.; Cubino, J.P.; Pearse, W.D.; Hobbie, S.E.; Lange, A.J.; Knapp, S.; Nelson, K.C. Horticultural availability and homeowner preferences drive plant diversity and composition in urban yards. *Ecol. Appl.* **2020**, *30*, e02082. [CrossRef] [PubMed]
17. Kendal, D.; Williams, K.J.; Williams, N.S. Plant traits link people's plant preferences to the composition of their gardens. *Landsc. Urban Plan.* **2012**, *105*, 34–42. [CrossRef]
18. Li, J.; Nassauer, J.I. Cues to care: A systematic analytical review. *Landsc. Urban Plan.* **2020**, *201*, 103821. [CrossRef]
19. Nassauer, J.I. Messy ecosystems, orderly frames. *Landsc. J.* **1995**, *14*, 161–170. [CrossRef]
20. Uren, H.V.; Dzidic, P.L.; Bishop, B.J. Exploring social and cultural norms to promote ecologically sensitive residential garden design. *Landsc. Urban Plan.* **2015**, *137*, 76–84. [CrossRef]
21. Luck, G.W.; Smallbone, L.T.; O'Brien, R. Socio-Economics and Vegetation Change in Urban Ecosystems: Patterns in Space and Time. *Ecosystems* **2009**, *12*, 604–620. [CrossRef]
22. Larson, K.L.; Lerman, S.B.; Nelson, K.C.; Narango, D.L.; Wheeler, M.M.; Groffman, P.M.; Hall, S.J.; Grove, J.M. Examining the potential to expand wildlife-supporting residential yards and gardens. *Landsc. Urban Plan.* **2022**, *222*, 104396. [CrossRef]
23. Herzog, T.R.; Herbert, E.J.; Kaplan, R.; Crooks, C. Cultural and developmental comparisons of landscape perceptions and preferences. *Environ. Behav.* **2000**, *32*, 323–346. [CrossRef]
24. Marco, A.; Barthelemy, C.; Dutoit, T.; Bertaudière-Montes, V. Bridging Human and Natural Sciences for a Better Understanding of Urban Floral Patterns: The Role of Planting Practices in Mediterranean Gardens. *Ecol. Soc.* **2010**, *15*. [CrossRef]
25. Bates, D.; Mächler, M.; Bolker, B.; Walker, S. Fitting Linear Mixed-Effects Models Using lme4. *J. Stat. Softw.* **2015**, *67*, 48. [CrossRef]
26. Larson, K.L.; Nelson, K.C.; Samples, S.R.; Hall, S.J.; Bettez, N.; Cavender-Bares, J.; Groffman, P.M.; Grove, M.; Heffernan, J.B.; Hobbie, S.E.; et al. Ecosystem services in managing residential landscapes: Priorities, value dimensions, and cross-regional patterns. *Urban Ecosyst.* **2016**, *19*, 95–113. [CrossRef]
27. Lockett, L.; Montague, T.; McKenney, C.; Auld, D. Assessing Public Opinion on Water Conservation and Water Conserving Landscapes in the Semiarid Southwestern United States. *HortTechnology* **2002**, *12*, 392–396. [CrossRef]
28. Karamaouna, F.; Kati, V.; Volakakis, N.; Varikou, K.; Garantonakis, N.; Economou, L.; Birouraki, A.; Markellou, E.; Liberopoulou, S.; Edwards, M. Ground cover management with mixtures of flowering plants to enhance insect pollinators and natural enemies of pests in olive groves. *Agric. Ecosyst. Environ.* **2019**, *274*, 76–89. [CrossRef]
29. Saunders, M.E.; Luck, G.W.; Mayfield, M.M. Almond orchards with living ground cover host more wild insect pollinators. *J. Insect Conserv.* **2013**, *17*, 1011–1025. [CrossRef]
30. Masierowska, M.; Stawiarz, E.; Rozwałka, R. Perennial ground cover plants as floral resources for urban pollinators: A case of Geranium species. *Urban For. Urban Green.* **2018**, *32*, 185–194. [CrossRef]
31. Spreading Milkweed, not Myths | U.S. Fish & Wildlife Service. *FWS.gov*. 23 June 2021. Available online: https://www.fws.gov/story/spreading-milkweed-not-myths (accessed on 12 December 2022).

32. Hopwood, J. "Roadsides as Habitat for Pollinators: Are Milkweeds Really Weeds? Xerces Society for Intertebrates. 2018. Available online: https://xerces.org/sites/default/files/2018-05/15-057_01_XercesSoc_Pollinators%2BRoadsides_Are-Milkweeds-Really-Weeds_web.pdf (accessed on 12 December 2022).
33. *Milkweed.* ASPCA. Available online: https://www.aspca.org/pet-care/animal-poison-control/toxic-and-non-toxic-plants/milkweed (accessed on 12 December 2022).
34. Milkweed (Asclepias spp.): USDA ARS. Available online: https://www.ars.usda.gov/pacific-west-area/logan-ut/poisonous-plant-research/docs/milkweed-asclepias-spp/ (accessed on 12 December 2022).
35. Nelson, M.; Alfuth, D. Ornamental Plants Toxic to Animals: Milkweed. University of Wisconsin-Madison Extension. 2021. Available online: https://hort.extension.wisc.edu/files/2021/03/Milkweed.pdf (accessed on 12 December 2021).
36. Schultz, P.W. Strategies for Promoting Proenvironmental Behavior. *Eur. Psychol.* **2014**, *19*, 107–117. [CrossRef]
37. Van Heezik, Y.M.; Dickinson, K.J.M.; Freeman, C. Closing the Gap: Communicating to Change Gardening Practices in Support of Native Biodiversity in Urban Private Gardens. *Ecol. Soc.* **2012**, *17*. [CrossRef]
38. Hall, D.M.; Steiner, R. Insect pollinator conservation policy innovations at subnational levels: Lessons for lawmakers. *Environ. Sci. Policy* **2019**, *93*, 118–128. [CrossRef]
39. Dicks, L.V.; Viana, B.; Bommarco, R.; Brosi, B.; Arizmendi, M.D.C.; Cunningham, S.A.; Galetto, L.; Hill, R.; Lopes, A.V.; Pires, C.; et al. Ten policies for pollinators. *Science* **2016**, *354*, 975–976. [CrossRef] [PubMed]
40. Hipólito, J.; Coutinho, J.; Mahlmann, T.; Santana, T.B.R.; Magnusson, W.E. Legislation and pollination: Recommendations for policymakers and scientists. *Perspect. Ecol. Conserv.* **2021**, *19*, 1–9. [CrossRef]
41. Senapathi, D.; Goddard, M.A.; Kunin, W.E.; Baldock, K.C.R. Landscape impacts on pollinator communities in temperate systems: Evidence and knowledge gaps. *Funct. Ecol.* **2017**, *31*, 26–37. [CrossRef]
42. Gemmill-Herren, B.; A Garibaldi, L.; Kremen, C.; Ngo, H.T. Building effective policies to conserve pollinators: Translating knowledge into policy. *Curr. Opin. Insect Sci.* **2021**, *46*, 64–71. [CrossRef]
43. Coombs, G.; Gilchrist, D.; Watson, P. An assessment of the native and invasive horticultural plants sold in the mid-Atlantic region. *Nativ. Plants J.* **2020**, *21*, 74–82. [CrossRef]
44. Hooper, V.H.; Endter-Wada, J.; Johnson, C.W. Theory and Practice Related to Native Plants: A Case Study of Utah Landscape Professionals. *Landsc. J.* **2008**, *27*, 127–141. [CrossRef]
45. White, A.; Fant, J.B.; Havens, K.; Skinner, M.; Kramer, A.T. Restoring species diversity: Assessing capacity in the U.S. native plant industry. *Restor. Ecol.* **2018**, *26*, 605–611. [CrossRef]
46. Conway, T.M. Tending their urban forest: Residents' motivations for tree planting and removal. *Urban For. Urban Green.* **2016**, *17*, 23–32. [CrossRef]
47. Kuhns, M.R.; Reiter, D.K. Tree care and topping beliefs, knowledge, and practices in Six Western US Cities. *J. Arboric.* **2009**, *35*, 122.
48. Hausmann, S.L.; Petermann, J.S.; Rolff, J. Wild bees as pollinators of city trees. *Insect Conserv. Divers.* **2016**, *9*, 97–107. [CrossRef]
49. Mach, B.M.; Potter, D.A. Quantifying bee assemblages and attractiveness of flowering woody landscape plants for urban pollinator conservation. *PLoS ONE* **2018**, *13*, e0208428. [CrossRef]
50. Somme, L.; Moquet, L.; Quinet, M.; Vanderplanck, M.; Michez, D.; Lognay, G.; Jacquemart, A.-L. Food in a row: Urban trees offer valuable floral resources to pollinating insects. *Urban Ecosyst.* **2016**, *19*, 1149–1161. [CrossRef]
51. Ajzen, I. The theory of planned behavior: Frequently asked questions. *Hum. Behav. Emerg. Technol.* **2020**, *2*, 314–324. [CrossRef]
52. Champine, V.M.; Jones, M.S.; Lischka, S.; Vaske, J.J.; Niemiec, R.M. Understanding individual and diffusion behaviors related to native plant gardening. *J. Environ. Psychol.* **2022**, *81*, 101798. [CrossRef]
53. Knapp, J.L.; Phillips, B.B.; Clements, J.; Shaw, R.F.; Osborne, J.L. Socio-psychological factors, beyond knowledge, predict people's engagement in pollinator conservation. *People Nat.* **2021**, *3*, 204–220. [CrossRef]
54. Hall, D.M.; Martins, D.J. Human dimensions of insect pollinator conservation. *Curr. Opin. Insect Sci.* **2020**, *38*, 107–114. [CrossRef]
55. Davis, A.; Herron, O.; Dumyahn, S. Uncovering the potential for exurban properties and small working farms in the Midwestern United States to provide food and refuge for pollinators. *Urban Ecosyst.* **2021**, *24*, 1047–1060. [CrossRef]

land

MDPI

Article

Socio-Cultural Appropriateness of the Use of Historic Persian Gardens for Modern Urban Edible Gardens

Majid Amani-Beni [1], Gaodi Xie [2,3], Qingjuan Yang [1], Alessio Russo [4] and Mohammad Reza Khalilnezhad [5,*]

1 School of Architecture and Design, Southwest Jiaotong University, Chengdu 611756, China; majid@swjtu.edu.cn (M.A.-B.); Yqj@home.swjtu.edu.cn (Q.Y.)
2 Institute of Geographical Sciences and Natural Resources Research, Chinese Academy of Sciences, A11 Datun Road, Chaoyang District, Beijing 100101, China; xiegd@igsnrr.ac.cn
3 College of Resources and Environment, University of Chinese Academy of Sciences, 19 A Yuquan Road, No. 19, Shijingshan District, Beijing 100049, China
4 School of Arts, Francis Close Hall Campus, University of Gloucestershire, Swindon Road, Cheltenham GL50 4AZ, UK; arusso@glos.ac.uk
5 Faculty of Arts, University of Birjand, Birjand 9718854987, Iran
* Correspondence: smkhalilnejad@birjand.ac.ir

Abstract: Historic gardens have the ability to provide several ecosystem services in cities, including provisioning services (i.e., food production). The historic gardens in Iran (known as "Persian Gardens") have never been considered as places that could be used for food production. As a result, the purpose of this paper is to investigate whether the Iranian historic gardens' spatial and structural layout is suitable for modern urban food gardening. We conducted field studies in six recognized Persian gardens in four provinces of Iran via qualitative analysis according to socio-cultural guidelines drawn from a literature review. The results suggested that combining the elements of formal landscape design, non-edible decorative plants, and traditional artwork would increase the Persian gardens' attractiveness. Regarding encouraging users to become involved in urban gardening, we found that separating productive units containing edible plants from public units using a central meeting spot populated by aesthetic plants and items may attract ordinary visitors who are interested in gardening without disturbing anyone's activities. Furthermore, the Persian gardens' multifunctionality, aesthetic value, and health-promoting qualities constitute a considerable historic achievement in garden design, making the gardens a suitable model for edible urban gardening. The results of this study can enhance our understanding of the Persian gardens' spatial and structural design and provide practical implications for sustainable urban planning and landscape architecture.

Keywords: food gardening; edible green infrastructure; Persian garden; socio-cultural guidelines; cultural landscape; urban ecosystem services; landscape architecture

Citation: Amani-Beni, M.; Xie, G.; Yang, Q.; Russo, A.; Khalilnezhad, M.R. Socio-Cultural Appropriateness of the Use of Historic Persian Gardens for Modern Urban Edible Gardens. *Land* **2022**, *11*, 38. https://doi.org/10.3390/land11010038

Academic Editor: Thomas Panagopoulos

Received: 9 November 2021
Accepted: 21 December 2021
Published: 27 December 2021

1. Introduction

Urban agriculture as a tool to promote urban resilience and improve the health system of citizens has been well considered in the Global North [1–3]. In addition, countries such as China [4,5] and Cuba [6], among others have made remarkable progress in supporting and developing urban agriculture. Iran is a country that despite the rich history of urban agriculture and richness of native fruit species [7], today has not paid serious attention to urban agriculture. One of the most basic methods of preserving the vegetation of cities (both productive and decorative) is the laws that in the last 40 years many of them by legal authorities in Iran (such as the Islamic Consultative Assembly, Ministry of Agriculture, and the Supreme Council of Architecture and urban planning) has been approved [8]. Most of these are laws that seek to preserve the use of agricultural lands and gardens and the preservation and expansion of green space in cities, do not have a strong enforcement guarantee [8]. In general, despite the multiplicity of laws, there is not enough support for growing edible plants and horticultural activities in cities [9]. The issue of urban agriculture as a governance approach first entered the Iranian executive literature in 2021 through top-down governance by the Ministry of Roads and Urban Development (MRUD). MRUD announced the plan "Development of Urban Agriculture in Parks and Equipped Urban Spaces" (DUAP) carried out by Shahid Beheshti University to the executive areas of the subdivision. In this project, for the first time, the need to pay attention to the use of agricultural heritage and historical methods of urban agriculture in Iran in the planning and design of urban agricultural spaces was mentioned. Nevertheless, Iranian architecture and urban planning have traditionally had productive green spaces and agriculture [10]. Iranian yards and gardens that were built in or around cities have always had edible plants and ornamental plants [11]. Today, the remains of these historic green spaces, either in the form of Persian gardens or parks, are being protected by the Ministry of Cultural Heritage, Tourism and Handicrafts, or the municipalities.

Before the communiqué of DUAP, in many cities of Iran, productive green space was planted, but based on the interests of citizens or municipal experts, not based on a comprehensive study plan with a scientific approach. For example, the public green space of Shiraz, as one of the most important cities in southern Iran, has orange, palm and mulberry fruit trees. The streets of the northern cities of Iran are full of olive trees. In southwestern Iran, the cities of Bushehr and Qasr Shirin have palm trees. Some western cities of Iran, such as Tuyserkan, have walnuts and almonds. In Tehran, fruit trees have recently been planted by the Parks and Green Space Organization. For example, 36,000 fruit trees have been planted in Velayat Park and in 13th municipal district, the construction of fruit parks has been regarded as one of the urban green space projects [12]. In addition to these cities, many other cities in Iran pay more attention to fruit trees and productive green space. Study of [13] on the main strategies for cultivation of fruit trees and shrubs in Iran's urban landscapes showed using native fruit trees like almonds, grapes, pistachios, jujubes, olives, pomegranates, figs, and mulberries with low water requirements compared to conventional urban park design approach can be profitable by reducing the costs of the construction and maintenance, and also through increasing the incomes. The study of Hosseinpour et al. [14] provides a framework which helps landscape architects on a cost-benefit applying urban agriculture in sustainable park design in Iran through planting fruit trees, vegetables, and other productive plants in urban landscaping. For example, Golab cultivar of apple, have already been adapted to the local urban conditions of Iran, so landscape development through urban agriculture can benefit from these variations of plants [13].

But the disadvantage of many of these activities is that Iranian citizens do not play a significant role in planting and cultivating edible plants in cities. Therefore, to achieve a comprehensive and sustainable approach to urban agriculture in Iran, there is a need for pilot projects that test how public can participate in urban agriculture. Previous studies show that in order to find the best way to locate and involve citizens in urban agriculture and create an edible city, there are experimental sites as the live laboratories [15]. For example, in Switzerland the Agro-urban Park of Bernex is the first of its kind which is piloted by the canton of Geneva, aiming to propose a test plot for interaction between the urban population and the farming sector [16]. For the first time in Iran, such a park is to be built in Mashhad, the second largest city in Iran. Detailed documents regarding the approach of Mashhad Municipality to this project have not been published yet. But in general, little field studies have been done on how to integrate urban agriculture in Iranian cities. The lack of studies on urban agriculture and the lack of test plots to study the interaction between production and leisure, between farmers and the urban population made Khalilnezhad et al. [17] consider the historical gardens as one of the most suitable urban spaces to start urban agriculture. The study of Khalilnezhad et al. [18] on the historical gardens of Birjand shows that while many of these gardens are located in or on the outskirts of the city, agricultural production is still preserved in these gardens. Although some of these gardens are privately owned, many of the gardens today are managed by the municipality and the Ministry of Cultural Heritage, Tourism and Handicrafts (MCHTH). This means that these sites can be used to familiarize people with the subject of agriculture. But from about 50 years ago onwards, looking to the historic green spaces such as the Persian Garden was more of a conservation approach, and the garden was protected not as a living thing but as an antique object [19]. This has caused the historic gardens owned by the municipalities and MCHTH to gradually lose their agricultural capacity [20]. Therefore, despite the becoming publicly accessible of many gardens, little attention is paid to the issue of production and agricultural economics in these gardens. In fact, ecosystem services of the Persian gardens as the multifunctional landscape has fluctuated over time due to management and maintenance conditions and the way the garden is viewed. Therefore, while the provisioning services of the Persian gardens are declining, the cultural services in term of recreation and tourism have been upgraded due to the opening of the garden to the people [21]. While the Persian garden in the past had a great variety of agricultural products (different types of fruits, vegetables and medicinal plants) [22], today the variety of products has decreased sharply and from the old generation of edible landscape only a few fruit trees remain [23].

As getting citizens to commit to gardening is not always easy [24], understanding urbanites' motivations helps landscape architects design gardens that encourage citizens to actively experience ecosystem services and contribute to the management of urban green spaces [25,26]. Therefore, the design creates a landscape that could be interesting not only for food production, but also for leisure, ecology, and water conservation purposes [27,28].

This paper focuses on the Persian garden as edible green infrastructure [29,30] that is preserved as historic property in many Iranian cities. Despite great efforts to achieve recognition of the Persian garden, this type of green heritage's role has not yet been explored in urban agriculture development. Interestingly, edibility and planting fruit trees constitute a notable element of these gardens [31,32], and they encompass some of the remnant urban agricultural heritage in Iranian cities, where historical value closely links urban gardening to cultural heritage. At present, the historical Iranian urban gardens are, in many cases, publicly accessible open spaces where users can interact, which enhances social inclusion [33,34]. However, conservative approaches involving rules and regulations aimed at defining and protecting urban historical gardens as permanent land use [19] do not guarantee farming activity [17]. Even though land preservation is a necessary component of master plans and other planning tools, it will not guarantee long-term agricultural activity [35].

Based on Wright et al. [36], the edible landscape is valuable for its stunning scenic quality, but society requires that these gardens be planned and operated as more than recreational resources. Thus, it is necessary to evaluate historical Persian gardens in cultural and geographical contexts that suffer from urbanization and the widening inequalities that exist in many Iranian cities in order to assess the gardens' suitability as a model for contemporary urban gardening, with the aim of encouraging users to get involved in agricultural activities.

Hence, this paper addresses the following main question: What role can be designated for the Persian garden in urban agriculture development? More specifically, the authors seek to answer the following question: Is the Persian garden's spatial and structural design appropriate for accommodating urban food gardening? Therefore, the prime objective of this paper is to compare and assess the appropriateness of the Persian garden's design, structure, and heritage for the development of urban food gardening, based on criteria derived from related multidisciplinary studies.

2. Theoretical Framework

2.1. Urban Food Gardening

Among the different types of urban agriculture, this paper focuses on urban food gardening, which includes agricultural activities with relatively low financial dependence on material outputs, while using what is produced to achieve other—mainly social and cultural—goals [37]. In this regard, proper planning, design, and management are key to enhancing the well-being that urban gardens can deliver to society by maximizing accessibility and promoting fair harvesting [38]. While urban gardens provide social innovation [39–41] in addition to food, explorations of spatial quality and design guidelines in terms of garden architecture have received little attention [42]. Landscape architects are trained to design public spaces, but they often lack the experience to undertake the spatial design of a productive landscape [43]. An in-depth review of recently published scientific literature in the field of agriculture landscape revealed at least three major areas that determine the rules of edible landscape design, namely garden architecture health considerations [44–47], ecological considerations [48–52], and social and cultural considerations.

2.2. The Socio-Cultural Dimension of the Landscape Architecture of the Public Edible Garden

Urban food gardening provides opportunities to enhance community involvement, promote social interaction between communities, and catalyze community development [38]. Hence, designing urban agriculture greenery plays significant socio-cultural roles [53,54]. Morckel [55] proved that social functions and attractiveness direct design guidelines for community garden spaces. While the functionality aspect enables gardeners to conduct their activities, attractiveness ensures that garden and community space is aesthetically pleasing.

2.2.1. Functional Guidelines

Functional motivations such as the desire to produce fresh food, maintain personal health, and enjoy being outdoors have significant relationships with gardeners' intention to participate in community gardens [56]. Thus, well-designed edible community gardens might contribute to a secure food supply and provide necessary infrastructure that is designed with consideration to elements that promote social interaction and group gardening activities in order to enhance feelings of connectedness with nature, thereby reducing tension and stress [57]. Therefore, urban gardens' landscape architectural features affect community participation and activities at each site. Community orchards are generally managed with much less intensity than gardens and allotments; because of the comparatively minimal amount of landscape engineering, there may simply be less opportunity for physical activity due to their design. Community gardens and allotments combine designated edible gardening areas with common recreational areas and require higher levels of maintenance efforts. Therefore, incorporating multifunctionality and a greater degree of complexity in community garden design could significantly increase user participation [58]. Moreover, ensuring the presence of a sufficient number of well-appreciated and appropriately located specimens and species will ensure that public urban orchards have a positive impact [59].

According to Milburn and Vail's [60], Prové et al.'s [61], and Mack et al.'s [62] findings on the keys to community gardens' success included the following physical design considerations: proximity to users and accessibility, physical characteristics that support growing (solar gain, access to water and soil), a compact site (as opposed to long, linear sites), high visibility from the street and within the garden, and the inclusion of appropriate site elements for growing (including composting, storage, perimeter fencing, and a bulletin/message board). Although site access has been proposed as affecting the management of and sense of place associated with community-managed spaces [63], the data Dennis and James [58] have presented indicated that physical design could affect stakeholder involvement in urban gardening. Hou et al.'s [64] work on hybrid community garden and public space projects identified several important physical conditions for successfully designing production landscapes as public space, including improvements and adjustments made over time, the capability for incremental change, addressing user needs and sensitivity to the existing context, programs involving multi-use activities including non-gardening programs, and diversity and opportunities for artistic expressions.

Napawan [43] reviewed the literature specific to the development of urban agriculture projects, including communally-managed spaces, and classified the physical condition and spatial components into three predominant site criteria, namely site context, site perimeter, and site layout and design. Site context refers to the appropriate neighborhood context and pedestrian or transit site accessibility. Ease of site entry and visual connectivity are related to the site perimeter [3,35,65]. More importantly, the site layout and design criterion covers attributes such as flexible layout [17], within-site accessibility [33], site maintenance [45,66], flexible program opportunities [67,68], and the dual functions of urban farming and alternative recreational programs. This is done by separating production and public space functions [36,69], publicizing opportunities for nesting food production and public space programs to varying degrees [70,71], creating a central meeting spot [3,72], having an onsite gathering space [57], and cultivating hedgerow and installing raised planter beds [73,74].

Moreover, Francis and Griffith [75] introduced four design principles for designing farmers' markets in public space: wholeness, social life, flexibility, and design permanency. Regarding design permanency, landscape features (e.g., entry structures, gazebos, fountains, pavilions, and groves of mature trees) can establish garden permanency through design [54,76,77]. Flexibility refers to a resilient design that is adaptive and accommodating of work, allowing the space to adapt to seasonal variation and fluctuations [55,70,78]. In addition, the productive landscape must be simultaneously distinguished from and reflective of the adjacent urban context, demonstrating the importance of spatial consideration of the landscape's periphery [58]. Special design consideration should be given to supporting socialization by coordinating the promenade landscape with sedentary spaces; having adequate seating of both the fixed and movable variety; hosting a diversity of users in terms of age, gender, and social-cultural background; integrating children's play and activity needs into the design; offering social programs; and ensuring that the design process is participatory [15,34,40,54,68,79–86]. In this regard, Mangone et al. [87] have also suggested that urban gardens as natural open spaces that are viewed as highly flexible, multi-use spaces suitable for a diversity of activities.

Furthermore, access to diverse workspace types with different spatial qualities appears to be highly valued. This kind of structural complexity, which is associated with management intensity, offers a basis for greater volunteer involvement, well-being, and the generation of local biodiversity and associated ecosystem services [58]. Expanded management requirements also offer participation opportunities, and the resultant physical activity while engaging in horticultural and site maintenance promotes participants' health [88,89]. In turn, as Dennis and James [58] have demonstrated, site biodiversity grows in proportion to volunteer input, and given the primacy of horticultural activity in community-managed spaces, food provision productivity may also be a key gain of more intensively-managed communal spaces.

2.2.2. Attractive Features

While edibility has been mentioned as the primary function, recreation, connectedness to the natural environment, positive social interactions, education, and habitat value have also been discussed across different types of urban food systems with plants [68,90]. Thus, gardeners are not just driven by functional motivations; they are also motivated by the emotions they ascribe to the gardens and potentially impeded by conditional motivations [56], in which attractiveness plays a critical role [33]. Thus, the design side of urban food gardens aims to design attractive urban environments [27]. Additionally, Lee and Matarrita Cascante (2019) showed that emotional motivation with respect to psychological connectedness with and appreciation of a garden composition and configurations that satisfy gardeners' demands and specific goals increases gardeners' intention to participate in gardening. Therefore, both functional connectedness with the garden and gardeners' appreciation of the attractions help establish a positive affective bond between gardens and citizens. Additionally, for an edible landscape, a structure must be adopted based on the most influential gardening motives, among which aesthetics, shade, and deriving joy from the hobby [91] all demonstrate an attractive design's significance in promoting urban food gardening. More specifically, even minimizing the potential negative impacts of practical issues such as littering, maintenance, unequal sharing, and the potential presence of worms in fruit would affect whether public urban orchards flourish [59].

To provide insight into what can be done to improve green spaces in order to make them more attractive community assets year-round, the findings of previous researchers [20,23,32,33,35,55,92–95] revealed that the presence of each of the following features would be beneficial and effective: a focal point (such as a gazebo or an arch), fencing, plants arranged in rows, raised garden beds, formal landscaping (such as walkways), trees, decorative non-edible plants, and artwork. To increase gardeners' involvement, garden designers should promote gardeners' emotional attachment to their gardens, while decreasing the time required to manage them. In addition to offering a restorative environment in edible community gardens, Lee and Matarrita-Cascante (2019) argued that garden–user relationships could be improved by the establishment of shelter settings, walls of trees, or hedges between garden units to provide gardeners with quiet, cozy environments. Additionally, small zones would not only allow individual gardeners to leisurely cultivate vegetation and concentrate on their work, but also heighten gardeners' sense of personal responsibility to and ownership of their individual plot [96,97]. Thus, in the designs they produce, garden planners and designers must be capable of reflecting gardeners' motivations for sustained participation, while decreasing any obstacles to garden participation. Better garden designs, as Lee and Matarrita-Cascante [56] and April [53] have recommended, may bolster community gardens' long-term viability and contribute to the provision of more urban green spaces.

3. Materials and Methods

This study examines the existing physical features and functionality of Persian gardens to determine if these sites can play a role in providing utility to the urban environment. Persian gardens are historical examples of Iranian green landscapes that have established an intricate relationship with cities and become part of the public realm [98]. These gardens date back to the 6th century BC and were recognized as World Heritage Sites by the United Nations Educational, Scientific, and Cultural Organization (UNESCO) in 2011 [99].

The approach of this research is primarily descriptive and is based on the investigation of purposive case studies [100,101]. In this research, we use multiple cases to build a logical base for a comparison framework. Persian historic gardens have been individually selected as purposive samples because their similar characteristics and authentic situations enable the authors to draw conclusions and generalizations based on the gardens' landscape type rather than their statistical occurrence.

Among several types of purposive cases, the authors selected paradigmatic cases [102], which, in this research, means Persian gardens as historic Iranian landscape exemplars with prototypical value. These were selected because the researchers believe that they represent a generally relevant situation. The conclusions yielded from the analysis of these prototypical cases using a descriptive-qualitative approach will be relevant to urban garden design. These purposive case examples all share a common feature, in that they each facilitate the investigation and comparison of a particular type of relationship. Though a very long time has passed since the construction of the Persian garden of Pasargadae in 6th BC, the Persian garden has kept its architectural and geometrical principles during history. The common feature of Persian gardens encompasses a unique geometry, design, and architecture. In more detail, all the selected gardens bear a rectangular form consisting of several planting beds, streams and pathways, ponds and fountains, a central pavilion, and the walls that surround the gardens [103].

Case study comparisons were undertaken at the cross-case comparison level because the researchers sought to answer questions by comparing different cases using guidelines extracted from the mentioned literature, in addition to performing cross-case comparisons.

As a first step, inferences from the literature enabled the compilation of criteria associated with the physical patterns of food production landscapes, which are also public spaces. These data gave the authors insight into the characteristics of the productive landscape of the Persian garden, its ability to create productive public spaces, the design of the site, and relevant issues in the design of urban food gardening at the site level (Table 1).

Table 1. Landscape design criteria for productive urban gardens based on socio-cultural considerations (drafted based on [62] p. 149).

Design Guidelines	Guideline/Feature	Design Feature
Functional guidelines	Site context	Appropriate neighborhood context Pedestrian and/or transit accessibility to site
	Site perimeter	Ease of site entry Visual connectivity
	Site layout and design	Flexible layout and program opportunities Within-site accessibility Adaptive to alternative recreational programs Separating production from public space functions A central meeting spot An onsite gathering space Hedgerow and raised planter beds
	Permanency of design	Entry structures, bandstands, gazebos, fountains, market pavilions, groves of mature trees, socially interactive plazas, pedestrian-scaled lighting, and thematic gardens
	Flexibility	A resilient design Adaptability to seasonal variation Market patronage
	Wholeness	Physically: 1. Distinguished from the urban context 2. Reflective of the adjacent urban context 3. Spatial consideration of the periphery landscape Planning and maintenance: 1. Encourages communities to engage with the garden program
	Social life	More spontaneous public space Adequate seating Encourages visits from people who are diverse in terms of age, gender, and cultural background. Integrates children's play Social programs Participatory design process
Attractive features	Focal point	Includes a gazebo or arch
	Fencing	Clearly defined boundaries—e.g., a wall
	Plants arranged in rows	Planting edible plants based on a linear system
	Raised garden beds	Gardening availability for children, the elderly, and the disabled
	Formal landscape design	Walkways
	Integration of conifers	Planting a variety of pine and cypress trees
	Non-edible plants for decoration	Decorative plants
	Artwork	Historical/traditional decorations

In the second step, the information and guidelines gleaned from previous scholarly works helped us to orientate site observation work and to begin suggesting the how and why of the Persian garden landscape as a location for the development of urban gardening, rather than merely looking at the cultural heritage. All-important early research works were identified and helped us to develop an understanding of the landscape's socio-cultural appropriateness. Field studies were conducted in the summer of 2020. These involved qualitative analyses using a thematic case study methodology to assess the appropriateness of the structure of historic gardens according to the socio-cultural guidelines derived from our literature review.

The field study methods used (onsite mapping, observation, photography) were adopted in all the cases under investigation to identify the specific tangible landscape characteristics capable of creating collective landscape gardening. These components include circulation networks, spatial arrangements, buildings and structures, vegetation, boundaries, small-scale elements, and views and vistas. In the evaluation of the gardens based on the gathered criteria, three methods were used for evaluation. In the case of visual phenomena (such as plants), the presence/absence of the phenomenon was examined through field surveys. Regarding the structural dimensions of gardens (such as the circulation system) its functionality or non-functionality was evaluated. On non-structural and intangible dimensions (such as social life) both through garden visits and non-structural interviews with some gardens managers, the interaction of visitors with the gardens were assessed.

In the third step, we sought to fully understand the cultural landscape through its inherent tangible value and establish its significance as a guide for conservation and future sustainable use in the field of urban gardening. Theoretical collections, field surveys, and assessments of the heritage that created the landscape formed the basis for the understanding of its gardening value. The evaluation of the Persian gardens as a potential site for urban gardening development involved synthesizing information gleaned from the research and analysis of tangible data to develop a clear understanding of how and why these historic sites are appropriate locations for the future of urban agriculture in Iran. Through this endeavor, we produced a statement of the significance of each garden and an overall statement regarding the collective value of the studied gardens.

Case Studies

Table 2 shows the selected case studies—namely, the Akbarieh, Chehel Sotun, Dolat Abad, Fin, Pahlavanpur, and Shahzadeh gardens. These gardens are registered as World Heritage Sites by UNESCO and nested in different provinces of Iran. While Akbarieh is located in the central east of Iran (Birjand), Chehel Sotun (Isfahan), Dolat Abad (Yazd), Fin (Kashan), Pahlavanpur (Mehriz), and Shahzadeh (Mahan) are in the central provinces (Figure 1).

The six selected gardens, which together represent the outstanding features of the Persian garden, are managed under the supervision of the Iranian MCHTH. The existing management system considers the preservation and management of all the gardens while maintaining their authenticity and integrity and aims to preserve the outstanding overarching features of the Persian gardens.

From the agricultural point of view, the situation of these 6 gardens is different. Akbarieh, Pahlavanpur and Shahzadeh agricultural landscapes, mainly including native fruit trees (such as pistachios, pomegranates, berries, apricots and almonds) are preserved. But in Dolat Abad and Fin gardens, the agricultural landscape is being revived and doesn't yet bear a significant amount of fruits. Although Chehel Sotun Garden had an edible landscape in the past, today most of its greenery is decorative. The destinations of fruits produced in the mentioned gardens are different. In Akbarieh, Shahzadeh garden the edible products are distributed amongst or sold to the garden administration and workers. Pahlavanpur is a privately owned garden, thus the edible products is sold to the local market [18].

Table 2. Detailed information about the studied Persian gardens in Iran [65], (p. 4).

Ownership	Protective Designation	Area (ha)	Geographical Coordinates	Province/City	Number and Name of Garden
Endowment	1999	3.40	N: 32°51′10″ E: 59°13′40″	Southern Khorasan/Birjand	1—Akbariyeh
State property	1932	5.80	N: 32°39′27″ E: 51°40′51″	Isfahan/Isfahan	2—Chehel Sotun
Endowment	1967	8	N: 31°54′12.30″ E: 54°21′6.59″	Yazd/Yazd	3—Dolat Abad
State property	1935	7.60	N: 33°220′20.53″ E: 51°22′20.53″	Isfahan/Kashan	4—Fin
State property	2003	3.50	N: 31°54′12.30″ E: 54°21′6.59″	Yazd/Mehriz	5—Pahlavanpour
State property	1975	5.50	N: 30°01′30″ E: 57°16′59″	Kerman/Mahan	6—Shahzadeh

Figure 1. Location of the studied Persian gardens in Iran (1: Akbarieh; 2: Chehel Sotun; 3: Dolat Abad; 4: Fin; 5: Pahlavanpur; 6: Shahzadeh). Source of map: https://commons.wikimedia.org/wiki/Atlas_of_Iran#/media/File:Iran_2001_CIA_map.jpg (accessed on 8 November 2021), (the locations of the gardens were added by the authors).

4. Results

The existing elements of Persian gardens can be divided into two groups—namely, natural elements (soft landscape—e.g., trees and water) and artificial elements (hard landscape—e.g., pavilion, walls, paths. and entrances)—which together can play an important role in the creation of urban food gardens, both functionally and aesthetically. Accordingly, the relationship between each garden case's components and urban food gardening development can be described as follows (Table 3) based on the guidelines mentioned in Table 1. Based on the results of this study and the evaluation of the authors of this article, the studied gardens in terms of socio-cultural appropriateness for the development of urban agriculture, in each item had one of these three conditions:

- presently accommodated: currently, the garden has this criterion or element.
- cannot be accommodated: currently, the garden has not this criterion or element, and even in the future this item can not be executed in the garden due to the restrictions forced by laws or the limitation imposed by the garden structure and design.
- potential future accommodation: currently, the garden has not this criterion or element, but in the future this item can be executed in the garden.

Table 3. Socio-cultural appropriateness of the Persian Gardens for productive urban garden development (✓✓: presently accommodated; ×: cannot be accommodated; ✓: potential future accommodation).

	Appropriateness					
	Akbarieh	**Chehel Sotun**	**Dolat Abad**	**Fin**	**Pahlavanpur**	**Shahzadeh**
Site context	✓✓	✓✓	✓✓	✓✓	×	×
Site perimeter	×	✓	×	×	×	×
Site layout and design	✓✓	✓	✓✓	✓✓	✓	✓✓
Permanency of design	✓✓	✓✓	✓✓	✓✓	✓✓	✓✓
Flexibility	✓✓	✓✓	✓✓	✓✓	✓✓	✓✓
Wholeness	✓	✓	✓	✓	✓	✓✓
Social life	✓✓	✓✓	✓	✓✓	✓	✓
Focal point	✓✓	✓✓	✓✓	✓✓	✓✓	✓✓
Fencing	✓✓	✓✓	✓✓	✓✓	✓✓	✓✓
Plants arranged in rows	✓✓	✓✓	✓✓	✓✓	✓✓	✓✓
Raised garden beds	✓	✓	✓	✓	✓	✓
Formal landscape design	✓✓	✓✓	✓✓	✓✓	✓✓	✓✓
Integration of conifers	✓✓	✓✓	✓✓	✓✓	✓✓	✓✓
Decorative plants for embellishment	✓✓	✓✓	✓✓	✓✓	✓✓	✓✓
Works of art	✓✓	✓✓	✓✓	✓✓	✓✓	✓✓

4.1. Akbarieh Garden

It is nested adjacent to the urban neighborhood context in Birjand city. Indeed, according to several onsite observations and interviews with the garden manager, most visitors are attracted to the south side of the garden, which is the pleasure landscape. On the other hand, based on the basic garden layout, the garden's elongated north side is dedicated to agricultural production. Accordingly, the pleasure landscape and production landscape are separated from each other (Figure 2).

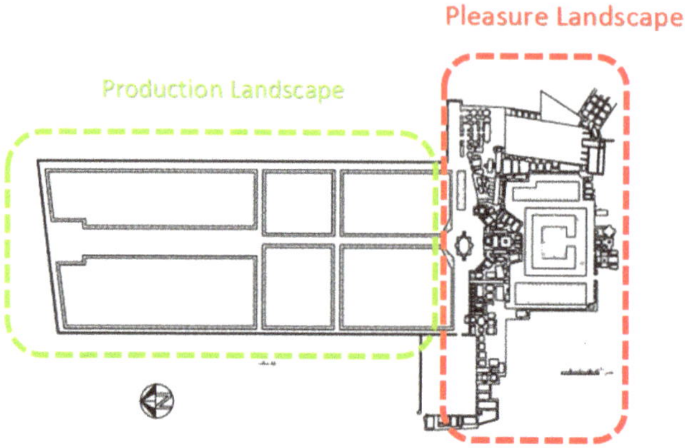

Figure 2. Separation of the production landscape from pleasure landscape in Akbarieh garden meets the functional appropriateness of this garden for development of urban food gardening.

The two-story historic monument is the garden's central focal point, as the recessed façade creates a welcoming atmosphere and encourages an appreciation of the building's glory, given that the main portico draws attention to the skyline. Behind this central building is a yard of about 3000 square meters where a horseshoe-shaped pool, surrounded by pine and mulberry trees, makes this area the recreational space of the garden.

As Figures 3 and 4 show, Akbarieh enjoys a wide variety of plant species, including fruit-bearing trees (Pistscio, apricot, pear, white mulberry, fig, pomegranate, barberry) alongside tall, old, evergreen coniferous decorative pines and cedars (Pinus eldarisa, Thuya orientalis, Cupressus Sempervirens). In the garden, the planting layout is such that pine tree rows run along the longitudinal axis in a north-to-south direction. On either side of this axis, there are six symmetrical planting plots allocated to food trees. At present, gardening affairs are the responsibility of the garden body management, which comprises part-time and full-time employees.

(a) (b) (c)

Figure 3. *Cont.*

(d) (e) (f)

Figure 3. Integration of fruit-bearing trees in Akbarieh garden as functional urban food gardening tools. (**a**) *Ficus carica* (Fig), (**b**) *Punica granatum* (Pomegranate), (**c**) *Prunus armeniaca* (Apricot), (**d**) *Pistacia vera* (Pistachio), (**e**) *Morus alba* (White mulberry), (**f**) *Berberis vulgaris* (Barberry).

Figure 4. Pistachio trees in Akbarieh garden.

Since the garden is situated in one of the driest areas of eastern Iran, the prime utilization of water here involves irrigating the garden's lush vegetation to nurture a lovely, refreshing environment.

4.2. Chehel Sotun Garden

This garden is located in the Isfahan urban context and is accessible to visitors via several peripheral pedestrian walkways. The garden is surrounded on all four sides by a brick lattice wall, which separates the grounds from the surrounding area to improve care and maintenance. Thus, from the peripheral streets and walkways, pedestrians enjoy visual connectivity with the garden scenery, but transit accessibility is only possible through a specific entrance gate.

At present, the Chehel Sotun garden reflects the attractiveness of Persian gardens, where vegetation, water, and architecture coexist to create a pleasant, relaxing historic garden. Its vegetation comprises many trees and shrubs, most of which are decorative trees (such as Pine, elm, maple, and sycamore). Unlike the other Persian gardens, this historic landscape does not accommodate fruit-bearing plants or an agricultural landscape (Table 3).

A grid-like access network facilitates access to the site via the main axes of the garden, which run east–west and along which plane trees are planted to connect the entrance and the central palace. The most important feature of the garden at Chehel Sotun is the long pool opposite the historic palace. The pool is closely related to the prestigious palace. Together, they serve as the central meeting spots, where many visitors gather. Social life flourishes around this pool due to the presence of shade casting-trees, a comfortable thermal environment, adequate seating, and visually delightful scenery, all of which encourage visits from people who are diverse in terms of age, gender, and cultural background.

4.3. Dolat Abad Garden

This garden is located in the urban fabric of one of central Iran's biggest cities, Yazd. In the garden of Dolat Abad, there are many trees, mainly pines, cedars, and fruit trees. The fruit trees include vines and pomegranates, which are planted in specific plots. The pines are in two rows along the main axis of the garden, which lies between two pavilions.

The oldest tree of the garden is an ancient mulberry tree that is located along the main entrance path. The garden consists of seven main beds separated by water streams; special food trees (pomegranate, fig, olive) are planted in each bed (Figure 5). In the western area, opposite the winter mansion, an exotic palm tree stands out among the other trees.

Figure 5. Edible landscape of Dolat Abad garden.

Other prominent features of the garden at Dolat Abad include the historic long pond and the several streams that flow through the garden area. The other notable element at Dolat Abad in terms of spatial structure is the strong east–west axis that separates the agricultural plots from the central promenade.

On the east side of the garden, the hexagonal porch is a traditional architectural element that attracts the attention of visitors. This two-storey building is located on the main axis of the garden and serves as a central meeting point, as it acts a reception and resting place for visitors. Considering that there are several buildings and water basins, as well as several streams, former stables, and servants' quarters, the garden reflects the permanent design and spatial elements that are prerequisites for the development of the urban garden. Table 3 briefly describes the suitability of the Dolat Abad World Heritage Garden for the development of productive urban gardening from the perspective of social and cultural guidelines.

4.4. Fin Garden

The Fin garden is located in the Kashan urban setting and is understood today as a tourist historic garden mainly by virtue of its plants, particularly its cedar trees, as well as its water circulation system, grid pedestrian network, and prestigious historic architectural monuments. The garden is enclosed by a high wall and the garden area is accessible via a specific entrance, which is the first structure that comes into view when approaching the garden from the peripheral street.

In addition, there are seventeen large planting areas in the Fin garden, the edges of which are lined with cedars, while insides are planted with trees that produce various types of fruit, such as figs, mulberries, pears, pomegranates, willows, quinces, greengage, and apricots (Figure 6). Therefore, unlike the Chehel Sotun garden, this historic landscape hosts urban gardening, albeit within the framework of the governmental agency of the MCHTH.

Figure 6. Maintenance of food trees in the Fin garden.

The presence and distribution of water at different levels and in all nooks and crannies of the garden not only enhance the visual scenery and increase the local humidity and coolness, especially in the warm seasons, but also support the thriving social activity taking place in the garden (Figure 7).

Considering the fact that the spatial structure of the Fin garden has two intersecting axes, which are considered to be important for the formation of the general space, and that the central pavilion serves as a focal point and gathering place, the layout and design are suitable for the development of the urban garden (Table 3).

The beautiful ornamental trees and shrubs in this garden contribute to the aesthetic value of the garden from the visitors' point of view, as the arrangement, configuration, order, and composition of the trees, flowers, and shrubs in the garden follow the principles of social farming.

By separating edible trees (planted in square plots) from non-edible and ornamental plants (planted in rows around walkways), the garden enjoys a high degree of legibility—a feature that is potentially important in the creation of attractive and functional urban garden spaces (Table 3).

Figure 7. Functional and attractive water features in Fin garden, especially as tools for supporting social life.

4.5. Pahlavanpur Garden

Pahlavanpur garden is a green complex comprising of a natural product garden and a promenade lined with and concealed by tall plane trees. Considering that a Qanat (an antiquated arrangement of underground channels developed to ship water from the inside of a slope to a town lying beneath it) goes through it and because of its moderate climate, the garden features rich vegetation, which draws in an enormous number of tourists from Mehriz and Yazd. Due to the garden's rural setting, the authors explore its inward properties to present this world heritage site as a model of urban edible gardens.

Pahlavanpur has a linear system of different trees, which are generally pines, cedars, and organic product trees. Organic product-bearing trees in the garden include fig and pomegranate trees planted in the agricultural areas (Figure 8). Plane trees are planted in two lines along a couple of streams; these create a healthy, attractive atmosphere. Thus, there is distinct separation between the utilitarian and recreational landscapes in this garden (Table 3).

Figure 8. Edible landscape of Pahlavanpur garden.

4.6. Shahzadeh Garden

While the garden is not part of the urban fabric and consequently does not meet the functional guidelines for urban gardening development since it is not situated in the appropriate neighborhood context, its layout and design satisfy the mentioned criteria (Table 3). Shahzadeh is an exemplar of terraced Persian gardens. Shahzadeh garden was designed to incorporate innovative irrigation measures that not only water the whole garden, but also integrate the emotional sensoria of water into the historic landscape. For example, the Baroque organizations of water cascades orchestrate a central water stream in the garden. In addition, blending nine fountains and nine cascades in this green space makes the Shahzadeh garden one of the most artistic manifestations of oasis gardens in Iran.

The linear spatial structure of the garden is emphasized by the plane and cypress trees planted along the central walkway and the stream running along it, which are formal landscape elements that facilitate walking, gathering, sitting, and photography (Figure 9).

The garden's vegetation follows a completely regular symmetrical design. The garden's tree planting design and appropriate plant selection are significant factors in creating shade and displaying the various colors of the different seasons, both of which contribute to a varied, visually interesting landscape. Planes and cypresses planted along the garden's central axis cast shadows on the walkways. Moreover, a wide variety of fruit trees, such as grape, apple, pear, apricot, pomegranate, quince, peach, and black plum trees, are planted in specific planting beds (Figures 10 and 11).

More importantly, as Figure 12 depicts, the productive landscape is separated from the central promenade, revealing that this landscape's layout and design can be considered as a prototype for designing future multifunctional edible urban gardens (Table 3).

4.7. A Comparison of the Cases

4.7.1. Site Context and Perimeter

With the exception of the Shahzadeh garden, which is located on a desert bed in the middle of an inter-city road, and the Pahlavanpur garden, which is located in a rural area, the other gardens are situated in an urban context where citizens live close to the garden. Even the Shahzadeh and Pahlavanpur gardens, despite their non-urban locations, have structures and spatial systems that, as described, are favorable for the development of civic gardening or can be regarded as a prototype for designing the new generation of urban edible gardens

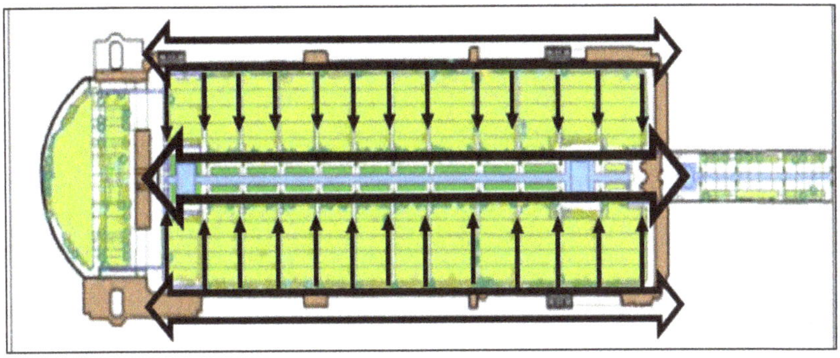

Figure 9. The functionality of the circulation system in the Shahzadeh garden as a means to make the whole garden accessible and attractive to public visitors.

Figure 10. New planted food trees in Shahzadeh garden.

Figure 11. Traditional framework for growing grapes in Shahzadeh garden.

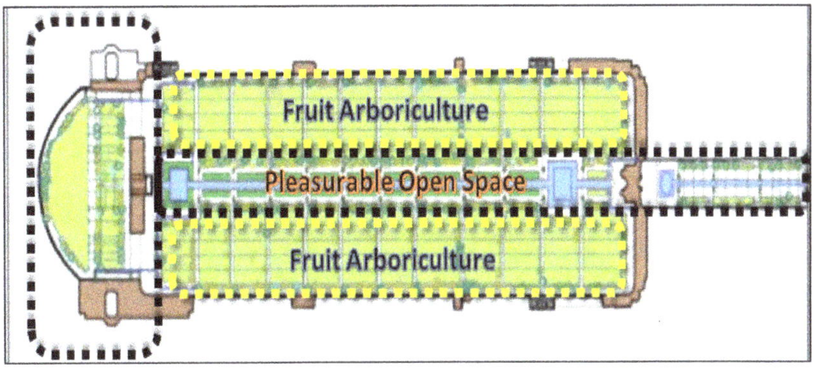

Figure 12. Separating production (yellow dashed line) from recreation (black dashed line) functions in the Shahzadeh garden as part of the basic site layout.

All the studied gardens have a perimeter wall; therefore, the only way to enter the garden is through a marked entrance that leads people into the garden. There are some differences between the studied gardens. For example, the Chehel Sotun garden has a lattice brick wall that allows people to see the garden's interior from the sidewalk. However, the other gardens have a solid brick wall that does not allow citizens to view the garden's interior from the outside. Although garden walls provide security and privacy, walls are problematic for urban gardening because they prevent people from visually connecting with the agricultural activities inside the garden. However it secures the garden's edible products.

The garden space's entrance structure catches visitors' attention with its geometrical features and traditional decorations. The Fin and Shahzadeh gardens have a ceremonial entrance as part of a a two-storey building, while the Pahlavanpur and Akbarieh gardens just have a simple gate. "Entry structures and ease of site entry are mentioned as requisite characteristics of a collective productive site" ([62], p. 159). Nevertheless, in some cases, such as in the Chehel Sotun and Fin gardens, visitors must pass through a series of hierarchically organized entrances to enter the main garden area. While the hierarchical access offered by Persian garden features may be appealing to visitors, it can be tedious for urban gardeners who must be constantly on the move. Some gardens, such as the Akbarieh garden, have side entrances that are closer to the productive landscape for the exclusive use of urban gardeners.

4.7.2. Site Layout and Permanency of Design

From the perspective of urban gardening, the most important structural feature is the separation of agricultural space from recreational space so that agricultural activities do not hinder recreational use and vice versa (Figures 4 and 10). Next, the existence of a fixed access network in the gardens, the separation of access paths to the main and secondary paths, and the existence of a central axis in each garden per formal landscape design allow people to access different parts of the garden and use different paths for a variety of purposes.

Moreover, the mansion has a significantly eye-catching frontage; thus, it can serve different basic functions in terms of envisioning urban gardening (Table 4). Generally, the mansion is surrounded by open spaces, making it the central meeting area as well as the gathering place for the community of gardeners.

Table 4. Persian garden's hard landscape elements' role in accommodating urban gardening [62], (p. 156).

Design Guidelines	Specific Guideline	Role of Architectural Heritage in Urban Gardening Development in the Persian Garden
Functional guidelines	Site perimeter	Ease of site entry due to a distinct entrance to the Persian garden.
	Site layout and design	Accessibility within the site via the permanent circulation system. Alternative recreational programs offered because of the existence of architectural heritage and non-agricultural spaces. Separation of production from public space functions. Pavilion as the central meeting spot. Onsite gathering space around the pavilion.
	Permanency of design	All the architectural features (pavilion, buildings, wall, entrance, walkways, canals, pools, and gazebos) in the Persian garden are permanent.
	Flexibility	Consideration of a resilient design, the regional climate, and adaptability to seasonal variation in the Iranian traditional landscape architectural heritage.
	Wholeness	The gardens' historical environment distinguishes them from the urban context.
	Social life	Seating facilities in the garden. The garden is notable among citizens as a site of cultural heritage. Architectural spaces (indoor and outdoor) for holding urban food gardening participatory events.
Attractivefeatures	Focal point	Mansion and pavilion as the focal points.
	Fencing	Persian garden wall system.
	Formal landscape design	Walkway on the main axis.
	Artwork	Traditional decorations on buildings, walkways, entrances, and outdoor furniture.

The next piece of architectural heritage in the studied gardens is the geometrical circular walkway, which provides wonderful opportunities for visitors to see all the sections of the garden (Figure 9). As the access network, the circular system plays an important role in introducing public visitors to historic urban gardens' agricultural identity and capabilities. Urban dwellers will feel an attachment to urban gardens if they can see and use them while engaged in gardening activities [35]. The Persian garden's access network is designed to welcome visitors without creating conflict with the garden's agricultural use (Figure 9). Beyond the main axes, a pre-existing agricultural road system provides the basic structure; this has been adapted to accommodate a variety of different uses [17]. Furthermore, special paths could be created to help guide visitors. The design of existing paths could also be transformed to provide adaptability to multifunctional uses, or new connections could be inserted into the existing circular system.

4.7.3. The Persian Gardens' Attractiveness

Among the attractive natural features of the garden landscape, water and plants can play vital roles in envisioning urban gardening in a Persian garden.

Based on the data presented in Table 5, water can play irreplaceable function in encouraging urban dwellers to get involved in the Persian gardens with the aim of urban food gardening. Firstly, the water features act as focal point and, in combination with the architectural monuments, enhance the attractiveness of the site. Secondly, the presence of ornamental non-edible plants beside the streams (in the Pahlavanpur, Fin, and Shahzadeh gardens) or other water surfaces (in the Dolat Abad and Chehel Sotun gardens) increases these historic gardens' palatability. Thirdly, the other aspect of water's potential to enhance attractiveness is the landscape architects' experimental use of water to showcase artwork. On the other hand, due to the variety of different visitors (in terms of age, gender, and cultural background) attracted to water-related features, the Persian gardens can be said to strengthen the social ties between people (Figure 7). "Thus, offering alternative recreational programs, having an onsite central meeting spot and gathering space, and integrating children's play are all conditions of the edible public landscape that can be furnished by the presence of water in traditional gardens, allowing us to reuse those gardens to enhance urban agricultural sites" ([62], p. 162)s.

Table 5. Significance of water features in envisioning urban gardening in the Persian garden (drafted based on [62] (p. 161)).

Design Guidelines	Specific Guideline/Feature	Role of Water Features in Urban Gardening Development in the Persian Garden
Functional guidelines	Site layout and design	Alternative recreational programs offered due to the various water features. A central meeting spot around the garden's main pool. An onsite gathering space around the water features.
	Permanency of design	Permanency of the fountains', streams', and pools' position and structure.
	Social life	The attractiveness of the water features facilitates the accommodation of visitors who are diverse in terms of age, gender, and cultural background. Integrating children's water play. Water-related social programs offered.
Attractive feature	Focal point	Pools, streams.
	Formal landscape design	Stream in the middle of the main axis. Water features around or in front of the mansion.
	Artwork	Cascading water in Shahzadeh garden. Water jets in the Fin garden.

But an important issue regarding the role of water in improving the production capacity of the studied gardens is the water crisis in Iran. It is interesting that the gardens of Fin, Shahzadeh and Pahlavanpur do not suffer of water shortage due to the abundant water of the aqueducts, so the strong and effective presence of water in these three gardens is quite tangible. But the gardens of Akbarieh and Dolat Abad are facing a water shortage crisis. In these gardens, the problem can be solved to some extent by upgrading traditional irrigation systems and replacing it with drip irrigation. But it should also be noted that the role of gardens as a bed of social food landscape is not just to increase productivity capacity. In fact, the involvement of urbanites into these gardens as urban gardeners does not necessarily mean an increase in planting volums, but citizens can participate in the maintenance of existing fruit trees or receive horticultural and agricultural training as discusses by [49,104,105]

The current plants in the studied Persian gardens can be categorized as conifers such as pines and cypresses; shading trees such as plane trees; decorative bushes and flowers such as Pyracantha, Juniperus, and Cotoneaster; and edible-bearing trees such as apple, apricot, quince, and peach (Figures 3 and 4), except in the Chehel Sotun garden, which cannot accommodate fructiferous vegetation.

From the perspective of the functional guidelines, as decorative vegetation is planted beside walkways in a manner that is compatible with the garden layout, the plants positively affect within-site accessibility. Additionally, positioning attractive plants alongside the garden's pedestrian networks and around the main building supports alternative recreational programs beyond agricultural activities. Moreover, the productive landscape is divided from the pleasure landscape, an aspect that has been pointed out by previous researchers [43,106,107] as a basic guideline for designing food gardens, and this strategy has been applied in all case studies (Table 6).

Table 6. Plants' functional and attractive role in realizing urban gardening in the Persian garden (drafted based on [62] (p. 168)).

Design Guidelines	Specific Guideline/Feature	Role of Green Heritage in Urban Gardening Development in the Persian Garden
Functional guidelines	Site layout and design	Separation of production from non-agricultural space. Guiding the visitor to the central spot by planting structural plants.
	Permanency of design	Groves of mature trees. Thematic gardens such as fruit, medicinal, and indigenous gardens.
	Flexibility	Implementing a resilient design by utilizing resistant indigenous species. Adapting to seasonal variation by utilizing the appropriate indigenous plants.
	Wholeness	Distinguished from the urban context due to agricultural scenery.
	Social life	Accommodating visitors who are diverse in terms of age, gender, and cultural background due to the production of a variety of fruits. Integrating children's play in agricultural ceremonies. Social programs related to gardening and production. Participatory process for collective gardening.
Attractivefeatures	Fencing	Huge vegetation on the garden's periphery. Dense planting along the axes.
	Plants arranged in rows	Ornamental planting in rows around the axis. Planting fruit-bearing trees in rows.
	Non-edible plants for decoration	Existence of a variety of flowers, ornamental bushes, and trees.
	Formal landscape design	Formal landscape design along the main axis, around the mansion, and in the private yards.
	Integration of conifers	Planting pine and cedar trees.

5. Discussion

This paper started with this main query that how can the components of the Persian gardens allow the development of urban food gardening in the located Iranian urban or peri-urban areas? We surveyed six UNESCO registered gardens in Iran to assess the adaptability of them with the guidelines of designing the urban food garden, both functionality and attractiveness. While the attractiveness of Persian garden as a beautiful landscape which can prepare the cultural ecosystem services was previously well discussed [11,33,100–110], the authors viewpoint of this paper to the garden aesthetics was focused on preparing a basement for implementing the public-involving strategy in development of urban agriculture [95], since aesthetic value often attract many citizens, creating additional benefits for food urbanism [95,111]. In this approach, the Persian garden is assessed to accommodate a diverse range of ecosystem services, both provisioning and cultural ecosystem services. Thus, despite the current perspective that defines the historic gardens as cultural landscapes where the architectural and vegetation monuments were embraced merely for the aim of attracting tourists [112,113], the garden attractive features also prepare the fundamental component of urban food gardens [114,115]. Many landscape features, such as channels, fountains, and planting systems, have been restored and preserved to make Persian gardens appropriate open spaces for offering cultural services [21]. However, beyond a beautiful landscape, the Persian gardens also accommodate a productive landscape that produces food. Such gardens (e.g., Rahimabad garden in Birjand city) can potentially produce about 4000 $kg/m^2/year$ of fruit and vegetable crops [18]. At present, some of such agricultural lands are vacant (in Akbarieh and Dolat Abad) or have been repurposed as beautiful landscapes (in Fin and Chelesotun). Thus, the development of urban gardening on these kinds of landscape heritage allows many people to recognize urban gardening as a meaningful way to engage with the urban food movement [81,116–119].

As mentioned previously Iran's MRUD announced the plan "DUAP" which emphasized the role of urban gardens in the development of urban agriculture in Iran. But this contrasts to the Iranian mission to protect historic gardens and landscapes as the antique objects, so maintaining or reintroducing the agricultural capacities of the historical gardens is not a priority. Since availability and access to productive land are crucial for urban agriculture [120], a program promoting the creation of productive open spaces in these gardens should be launched with the participation of educational farms, farmers' markets, and food activists under plan of DUAP.

As detailed characteristics of the Persian garden designs presented in this paper, drawing citizens to these types of edible landscapes will not disturb their current recreational and official functions, since the spatial separation between utilitarian agricultural space and pleasure landscape. Additionally, we can hypothesize and further investigate whether creating opportunities for public engagement will enhance the preservation of the other garden functions, as the garden will be moved from the margins into the heart of the urban living, work, education, training, and feeding environment.

Creating mutual respect between the different users of these gardens is another goal of food urbanism and means reducing disagreement between agricultural productivity, ornamental greenery functions, and human use of the garden [121]. So that the character of the heritage gardens goes beyond their historic meaning, since the garden has become a means of involving urban dwellers in production activities. The authors of this paper explained the potential of the Persian garden in creating opportunities for public engagement in urban food gardening. More than producing food stuffs, approaching the Iranian historic gardens as hub for the integration of different programs, whether organic gardening, commercial agriculture, educational activities, or therapeutic visiting, can be regarded a basic factor making the garden a successful place where people can have positive experiences experience [122]. From this perspective, the Iranian historic landscapes are regarded as a place for the integration of different urbanites, attracting people to view the aesthetic elements and attributes of these gardens. Based on this strategy, the traditional agricultural utilities of these gardens are enhanced by their historical elements and the production of agricultural goods through contacting urban citizens who can engage in co-production programs or simply consume products from a well-known landscape heritage.

We spoke of the opportunities accommodating in the Persian gardens for the aim of urban gardening. But there are still some important challenges which would act as the barrier of our hypothesis. Firstly, the viewpoint and MCHTH administration to the Persian garden as a cultural landscape demonstrated the authority concern about the conservation and protection of the gardens [20,34,123]. Therefore, as long as historic gardens are managed under such limited thinking, citizens will not have a role in re-planning gardens with an urban gardening development approach. The second challenge is the absence of the Persian garden brand in the food products of Iran's domestic and export markets. Many of the fruits produced in Iranian gardens are consumed by the staff and visitors of the garden or put up for auction by the authorities. Therefore, considering the importance of branding in the development of green businesses and agricultural economy, garden branding can strengthen the motivation to revive urban gardening [124–127]. The third challenge is how to maintain the traditional planting system in the process of developing social gardening in gardens. Historic gardens have certain plant species that are planted under the traditional planting system. However, urban agricultural activists may be interested in planting new plants and cultivars in historic gardens [128–131]. Of course, with the emergence of crises such as water scarcity and climate change [132], it is predictable that the Iran's MCHTH will have to replace some hydrophilic plant species with low-consumption cultivars. It should also be noted that traditional planting systems and native garden trees can be used as live classes, a platform for traditional gardening education for those interested urbanites [17]. Moreover, traditional planting systems and native plant cultivars are currently being regarded as a model for development of urban edible landscape such as development of urban agriculture in urban parks [13,14]. Thus, the Persian historic gardens can be regarded as a vanue for traning the landscape architects on the traditional system of edible landscaping.

6. Conclusions

In the Iranian urban context, a growing number of social, environmental, and economic problems are associated with great challenges such as food poverty, environmental injustice, and socio-cultural isolation. However, some areas of spatial heritage, such as the Persian gardens, can be planned and programmed with the view to developing new strategies promoting the multifunctional use of urban historical gardens. In this study, the authors' focus was on the following question: beyond accommodating historical elements architecturally and in terms of landscape architecture, does the Persian garden heritage have the potential to facilitate the development of urban gardening programs? From the perspective of urban gardening, citizens can be welcomed into historical gardens as food activists. On the one hand, the agricultural heritage features are still visible in the shape of Iranian gardens that are situated adjacent to residential neighborhoods. In addition, these historic landscapes encompass characteristic elements and traditional design features that could satisfy the guidelines that determine socio-cultural appropriateness for urban gardening. Astonishingly, in addition to having great cultivation potential and meeting the basic spatial requirements for developing urban gardens, the Persian garden's hard and soft landscapes are suitable for accommodating collective gardening activities. Therefore, from the socio-cultural perspective, the Persian garden is highly suitable for the development of urban food gardening.

Formal landscape design and the presence of non-edible decorative plants and traditional artwork increase Persian gardens' attractiveness in the context of the collective productive landscape. With their multifunctional spaces and tangible heritage, historical gardens are appropriate urban spaces for organizing multi-use programs. As a paradoxical fact, the wall frame impedes visual contact between the inside and outside of the garden while at the same time providing environmental security. Moreover, there are supportive and encouraging attributes in these gardens, including the separation of agriculture areas from pleasure landscapes and some public meeting spots, which enhance the potential of the Persian garden to attract diverse urbanites without disrupting gardening activities. Above all, the Persian garden's beauty, freshness, exceptional greenery, and health benefits are a testament to the historic achievement of designing a multifunctional landscape that can provide urban gardeners with an aesthetically pleasing experience.

If urban agriculture were implemented in Persian gardens, there would be an increase in ecosystem services such as provisioning and cultural services Possible contribution of urban agriculture in enhancement the provisioning services includes providing food, fibers and biomass, beyond enhancing pollination [30,133]. Moreover, investment of food activists in agricultural enterprises locating in historic landscape would contribute to the employment the labors as the current crises in Iran is high rate of unemployment. As in other parts of the world, local environmental associations, co-operatives, or non-governmental organizations (NGOs) could manage these historic gardens for the production of vegetables and fruit (for example like urban agriculture within the Archaeological Park of Pontecagnano in Italy that is managed by a local association [134]. Such organizations, in partnership with local authorities, might develop rules for access, as well as guidelines for implementing biodiversity-based urban agriculture to maximize ecosystem services.

One of the most important results of this research is that even if in the future managerial or conservation barriers hinder the full realization of urban agriculture in historic gardens, these gardens can be regarded as design model for landscape architects who seek to realize the edible landscape ideas in the Iranian context. Furthermore, citizens who are looking to build a productive personal garden can also embrace fundamental concepts of the Persian gardens. Thus, these gardens can also be regarded as a teaching resource and natural laboratory.

Additionally, since this is the first time such an approach to the Persian garden has been proposed, it is necessary to conduct additional research to assess the adaptability of intangible (maintenance and laws) dimensions of the Iranian garden for the development of urban agriculture. We recommend that future research investigate Iranian urban dwellers' preferences for and perceptions of participating in developing urban agriculture as urban food gardeners at specific Persian garden sites. Subsequent research can also explore government agencies' approaches to the development of urban gardening in the Persian garden context.

Author Contributions: Methodology, M.R.K.; original draft preparation, M.R.K., M.A.-B.; analysis, M.R.K., M.A.-B. and A.R.; writing—review and editing, M.R.K., M.A.-B., A.R., G.X. and Q.Y.; conceptualization, M.R.K.; validation, G.X. and Q.Y.; funding acquisition, G.X. and M.A.-B. All authors have read and agreed to the published version of the manuscript.

Funding: This research was funded by the Strategic Priority Research Program of Chinese Academy of Sciences, grant number XDA2002040203; National Key Research and Development Plan, grant number 2016YFC0503403 and 2016YFC0503706; Fundamental Research Funds for the Central Universities of China, grant number 2682019CX77; Natural Science Foundation of China, reference number 52150410414; and Interdisciplinary Research Projects of Southwest Jiaotong University (No. 2682021ZTPY085).

Institutional Review Board Statement: Not applicable.

Informed Consent Statement: Not applicable.

Data Availability Statement: Data available upon request.

Conflicts of Interest: The authors declare no conflict of interest.

References

1. Grewal, S.S.; Grewal, P.S. Can cities become self-reliant in food? *Cities* **2012**, *29*, 1–11. [CrossRef]
2. Artmann, M.; Sartison, K.; Vávra, J. The role of edible cities supporting sustainability transformation—A conceptual multi-dimensional framework tested on a case study in Germany. *J. Clean. Prod.* **2020**, *255*, 120220. [CrossRef]
3. Veen, E.J. *Community Gardens in Urban Areas: A Critical Reflection on the Extent to Which They Strengthen Social Cohesion and Provide Alternative Food*; Wageningen University: Wageningen, The Netherlands, 2015; ISBN 9789462573383.
4. Wang, X. Edible Landscapes within the Urban Area of Beijing. Ph.D. Thesis, Universität Stuttgart, Stuttgart, Germany, 2016.
5. Wang, N.; Zhu, L.; Bing, Y.; Chen, L.; Fei, S. Assessment of urban agriculture for evidence-based food planning: A case study in Chengdu, China. *Sustainability* **2021**, *13*, 3234. [CrossRef]
6. Spencer, L. Urban Agriculture in Cuba: Alternative Legal Structures, Crisis and Change. In *Balanced Urban Development: Options and Strategies for Liveable Cities*; Springer: Cham, Switzerland, 2016; pp. 343–354.
7. Maghrebi, M.; Noori, R.; Bhattarai, R.; Mundher Yaseen, Z.; Tang, Q.; Al-Ansari, N.; Danandeh Mehr, A.; Karbassi, A.; Omidvar, J.; Farnoush, H.; et al. Iran's Agriculture in the Anthropocene. *Earth's Future* **2020**, *8*, e2020EF001547. [CrossRef]
8. Khalilnezhad, S.M.R.; Tobias, K. The productive landscape in Persian gardens, foundations and features. *Bagh-e Nazar* **2016**, *133*, 3–16.
9. Emamian, A.; Rashki, A.; Kaskaoutis, D.G.; Gholami, A.; Opp, C.; Middleton, N. Assessing vegetation restoration potential under different land uses and climatic classes in northeast Iran. *Ecol. Indic.* **2021**, *122*, 107325. [CrossRef]
10. Fadaie, H.; Majid, S.; Shemirani, M. A Comparative Study on Gardens of Isfahan and Shiraz From Sustainability View (Case Studies: Gardens of Hashtbehesht and Jahannama). *Int. J. Archit. Urban Dev.* **2014**, *4*, 33–40.
11. Gharipour, M. *Persian Gardens and Pavilions*; Bloomsbury Publishing: London, UK, 2013.

12. Mojtahed Najafi, S.N. *More Resilient Economy, Healthier Environment by Planting Fruit Trees*; Thaqalayn: Qom, Iran, 2019.
13. Kazemi, F.; Abolhassani, L.; Rahmati, E.A.; Sayyad-Amin, P. Strategic planning for cultivation of fruit trees and shrubs in urban landscapes using the SWOT method: A case study for the city of Mashhad, Iran. *Land Use Policy* **2018**, *70*, 1–9. [CrossRef]
14. Hosseinpour, N.; Kazemi, F.; Mahdizadeh, H. A cost-benefit analysis of applying urban agriculture in sustainable park design. *Land Use Policy* **2022**, *112*, 105834. [CrossRef]
15. Säumel, I.; Reddy, S.E.; Wachtel, T. Edible city solutions-one step further to foster social resilience through enhanced socio-cultural ecosystem services in cities. *Sustainability* **2019**, *11*, 972. [CrossRef]
16. Verzone, C.; Woods, C. *Food Urbanism*; De Gruyter: Berlin, Germany, 2021; ISBN 9783035615678.
17. Khalilnezhad, S. Urban Agriculture as a Tool for City and Landscape Planning in Iran with Emphasis on the Role of Persian Garden. Ph.D. Thesis, Technical University of Kaiserslautern, Kaiserslautern, Germany, 2016.
18. Khalilnezhad, M.R.; Farzin, S.; Zohoriyan, M. Appropriateness of the Historic Gardens for Urban Agriculture Development in Birjand City (Iran). *Bagh-e Nazar* **2021**, *18*, 55–72. [CrossRef]
19. Mahdizadeh, S.; Rajendran, L.P. A renewed approach to conservation policy of historical gardens in Iran. *Landsc. Res.* **2019**, *1*, 48–61. [CrossRef]
20. Khalilnezhad, M.R. Misadventure of decorative management of the World Heritage's Persian gardens. *Manzar* **2019**, *11*, 44–51. [CrossRef]
21. Mahdizadeh, S. Shiraz's heritage gardens during the political turmoil in Twentieth-century Iran. *Int. J. Herit. Stud.* **2021**, *27*, 953–970. [CrossRef]
22. Shandiz, M.H. Retour aux sources pour une meilleure reconnaissance et valorisation du patrimoine paysager perse. *Stud. Hist. Gard. Des. Landsc.* **2012**, *32*, 164–181. [CrossRef]
23. Farzin, S.; Khalilnezhad, S.M.R.; Moradzadeh Mirzaei, S.; Zarei, A. Investigation on Recognition of the Type of Multifunctional Landscape in Persian Garden (Case Study: Akbariyeh World Heritage Garden). *MANZAR Sci. J. Landsc.* **2020**, *12*, 6–17.
24. Lewis, O.; Home, R.; Kizos, T. Digging for the roots of urban gardening behaviours. *Urban For. Urban Green.* **2018**, *34*, 105–113. [CrossRef]
25. Teuber, S.; Schmidt, K.; Kühn, P.; Scholten, T. Engaging with urban green spaces—A comparison of urban and rural allotment gardens in Southwestern Germany. *Urban For. Urban Green.* **2019**, *43*, 126381. [CrossRef]
26. Amani-Beni, M.; Chen, Y.; Vasileva, M.; Zhang, B.; Xie, G. Quantitative-spatial relationships between air and surface temperature, a proxy for microclimate studies in fine-scale intra-urban areas? *Sustain. Cities Soc.* **2022**, *77*, 103584. [CrossRef]
27. Roggema, R. Food in spatial planning and design. In *Food Roofs of Rio de Janeiro: The Pavao-Pavaozinho and Cantagalo Case Study*; Springer: Berlin/Heidelberg, Germany, 2017; ISBN 9783319567396.
28. Amani-Beni, M.; Zhang, B.; Xie, G.; Shi, Y. Impacts of urban green landscape patterns on land surface temperature: Evidence from the adjacent area of Olympic Forest Park of Beijing, China. *Sustainability* **2019**, *11*, 513. [CrossRef]
29. Russo, A.; Cirella, G.T. Edible urbanism 5.0. *Palgrave Commun.* **2019**, *5*, 163. [CrossRef]
30. Russo, A.; Escobedo, F.J.; Cirella, G.T.; Zerbe, S. Edible green infrastructure: An approach and review of provisioning ecosystem services and disservices in urban environments. *Agric. Ecosyst. Environ.* **2017**, *242*, 53–66. [CrossRef]
31. Gharipour, M.; Deshamudre, A. *Encyclopaedia of the History of Science, Technology, and Medicine in Non-Western Cultures*; Springer: Berlin/Heidelberg, Germany, 2008; pp. 1–6. [CrossRef]
32. Khalilnezhad, S.M.R. Distinctive features of productive landscapes in Persian gardens. In Proceedings of the VI International Conference on Landscape and Urban Horticulture, Athens, Greece, 20–25 June 2016; pp. 35–38. [CrossRef]
33. Bazrafshan, M.; Tabrizi, A.M.; Bauer, N.; Kienast, F. Place attachment through interaction with urban parks: A cross-cultural study. *Urban For. Urban Green.* **2021**, *61*, 127103. [CrossRef]
34. Rostami, R.R.; Lamit, H.; Khoshnava, S.M.; Rostami, R.R. Successful public places: A case study of historical Persian gardens. *Urban For. Urban Green.* **2016**, *15*, 211–224. [CrossRef]
35. Timpe, A.; Cieszewska, A.; Supuka, J.; Tóth, A. Urban Agriculture goes Green Infrastructure. In *Urban Agriculture Europe*; Lohrberg, F., Licka, L., Scazzosi, L., Timpe, A., Eds.; Jovis: Berlin, Germany, 2015; pp. 126–137, ISBN 978-3-86859-371-6.
36. Wright Wendel, H.E.; Zarger, R.K.; Mihelcic, J.R. Accessibility and usability: Green space preferences, perceptions, and barriers in a rapidly urbanizing city in Latin America. *Landsc. Urban Plan.* **2012**, *107*, 272–282. [CrossRef]
37. Lohrberg, F. Urban Agriculture Forms in Europe. In *Agrourbanism: Tools for Governance and Planning of Agrarian Landscape*; Gottero, E., Ed.; Springer International Publishing: Cham, Switzerland, 2019; pp. 133–147. ISBN 978-3-319-95576-6.
38. Salbitano, F.; Fini, A.; Borelli, S.; Konijnendijk, C.C. Editorial—Urban Food Forestry: Current state and future perspectives. *Urban For. Urban Green.* **2019**, *45*, 126482. [CrossRef]
39. Orsini, F. Innovation and sustainability in urban agriculture: The path forward. *J. Consum. Prot. Food Saf.* **2020**, *15*, 203–204. [CrossRef]
40. Rusciano, V.; Civero, G.; Scarpato, D. Social and Ecological High Influential Factors in Community Gardens Innovation: An Empirical Survey in Italy. *Sustainability* **2020**, *12*, 4651. [CrossRef]
41. Cattivelli, V. The Motivation of Urban Gardens in Mountain Areas. The Case of South Tyrol. *Sustainability* **2020**, *12*, 4304. [CrossRef]
42. Viljoen, A.; Bohn, K. *Second Nature Urban Agriculture*; Routledge: Oxfordshire, UK, 2014; ISBN 9781317674511.

43. Napawan, N.C. Production places: Evaluating communally-managed urban farms as public space. *Landsc. J.* **2015**, *34*, 37–55. [CrossRef]
44. Säumel, I.; Kotsyuk, I.; Hölscher, M.; Lenkereit, C.; Weber, F.; Kowarik, I. How healthy is urban horticulture in high traffic areas? Trace metal concentrations in vegetable crops from plantings within inner city neighbourhoods in Berlin, Germany. *Environ. Pollut.* **2012**, *165*, 124–132. [CrossRef]
45. Von Hoffen, L.P.; Säumel, I. Orchards for edible cities: Cadmium and lead content in nuts, berries, pome and stone fruits harvested within the inner city neighbourhoods in Berlin, Germany. *Ecotoxicol. Environ. Saf.* **2014**, *101*, 233–239. [CrossRef] [PubMed]
46. Wilschut, M.; Theuws, P.A.; Duchhart, I. Phytoremediative urban design: Transforming a derelict and polluted harbour area into a green and productive neighbourhood. *Environ. Pollut.* **2013**, *183*, 81–88. [CrossRef]
47. Mitchell, R.G.; Spliethoff, H.M.; Ribaudo, L.N.; Lopp, D.M.; Shayler, H.A.; Marquez-Bravo, L.G.; Lambert, V.T.; Ferenz, G.S.; Russell-Anelli, J.M.; Stone, E.B.; et al. Lead (Pb) and other metals in New York City community garden soils: Factors influencing contaminant distributions. *Environ. Pollut.* **2014**, *187*, 162–169. [CrossRef] [PubMed]
48. Moro, M.F.; Westerkamp, C.; De Araújo, F.S. How much importance is given to native plants in cities' treescape? A case study in Fortaleza, Brazil. *Urban For. Urban Green.* **2014**, *13*, 365–374. [CrossRef]
49. Fischer, L.K.; Brinkmeyer, D.; Karle, S.J.; Cremer, K.; Huttner, E.; Seebauer, M.; Nowikow, U.; Schütze, B.; Voigt, P.; Völker, S.; et al. Biodiverse edible schools: Linking healthy food, school gardens and local urban biodiversity. *Urban For. Urban Green.* **2019**, *40*, 35–43. [CrossRef]
50. Norfolk, O.; Eichhorn, M.P.; Gilbert, F. Traditional agricultural gardens conserve wild plants and functional richness in arid South Sinai. *Basic Appl. Ecol.* **2013**, *14*, 659–669. [CrossRef]
51. Amani-Beni, M.; Zhang, B.; Xie, G.-D.; Odgaard, A.J. Impacts of the Microclimate of a Large Urban Park on Its Surrounding Built Environment in the Summertime. *Remote Sens.* **2021**, *13*, 4703. [CrossRef]
52. Zhang, B.; Amani-Beni, M.; Shi, Y.; Xie, G.-D. The summer microclimate of green spaces in Beijing' Olympic park and their effects on human comfort index. *Ecol. Sci.* **2018**, *37*, 77–86. (In Chinese) [CrossRef]
53. April, P. *Designing Urban Agriculture: A Complete Guide to the Planning, Design, Construction, Maintenance and Management of Edible Landscapes*; Wiley: Hoboken, NJ, USA, 2013; ISBN 1118073835.
54. Auželienė, I.; Daubaras, L.; Eidimtienė, V.V. Urban gardening: Elements, social, cultural and recreational aspects. *Maz. Stud. Reg.* **2016**, 35–47. [CrossRef]
55. Morckel, V. Community gardens or vacant lots? Rethinking the attractiveness and seasonality of green land uses in distressed neighborhoods. *Urban For. Urban Green.* **2015**, *14*, 714–721. [CrossRef]
56. Lee, J.H.; Matarrita-Cascante, D. The influence of emotional and conditional motivations on gardeners' participation in community (allotment) gardens. *Urban For. Urban Green.* **2019**, *42*, 21–30. [CrossRef]
57. Shimpo, N.; Wesener, A.; McWilliam, W. How community gardens may contribute to community resilience following an earthquake. *Urban For. Urban Green.* **2019**, *38*, 124–132. [CrossRef]
58. Dennis, M.; James, P. User participation in urban green commons: Exploring the links between access, voluntarism, biodiversity and well being. *Urban For. Urban Green.* **2016**, *15*, 22–31. [CrossRef]
59. Colinas, J.; Bush, P.; Manaugh, K. The socio-environmental impacts of public urban fruit trees: A Montreal case-study. *Urban For. Urban Green.* **2019**, *45*, 126132. [CrossRef]
60. Milburn, L.A.S.; Vail, B.A. Sowing the seeds of success: Cultivating a future for community gardens. *Landsc. J.* **2010**, *29*, 71–89. [CrossRef]
61. Prové, C.; Dessein, J.; de Krom, M. Taking context into account in urban agriculture governance: Case studies of Warsaw (Poland) and Ghent (Belgium). *Land Use Policy* **2016**, *56*, 16–26. [CrossRef]
62. Mack, E.A.; Tong, D.; Credit, K. Gardening in the desert: A spatial optimization approach to locating gardens in rapidly expanding urban environments. *Int. J. Health Geogr.* **2017**, *16*, 37. [CrossRef]
63. Kurtz, H. Differentiating multiple meanings of garden and community. *Urban Geogr.* **2001**, *22*, 656–670. [CrossRef]
64. Hou, J.; Johnson, J.; Lawson, L.J. *Greening Cities, Growing Communities: Learning from Seattle's Urban Community Gardens*; Case Study in Land and Community Design; Landscape Architecture Foundation: Washington, DC, USA, 2009; ISBN 9780295989280.
65. Fletcher, M.; Rushlow, J.; Schwartz-Berky, J. Overcoming Barriers to Cultivating Urban Agriculture. *Real Estate Law J.* **2012**, *41*, 215–245.
66. Armstrong, D. A survey of community gardens in upstate New York: Implications for health promotion and community development. *Health Place* **2000**, *6*, 319–327. [CrossRef]
67. Koroļova, A.; Treija, S. Urban Gardening as a Multifunctional Tool to Increase Social Sustainability in the City. *Archit. Urban Plan.* **2018**, *14*, 91–95. [CrossRef]
68. Xie, M.; Li, M.; Li, Z.; Xu, M.; Chen, Y.; Wo, R.; Tong, D. Whom do urban agriculture parks provide landscape services to and how? A case study of Beijing, China. *Sustainability* **2020**, *12*, 4967. [CrossRef]
69. Khalilnezhad, S.M.R. Principles of integration of the agriculture and pleasure greeneries in Persian gardens. *Acta Hortic.* **2017**, *1189*, 39–42. [CrossRef]
70. Archdeacon, K.F. *Urban Agriculture Design for Resilient Cities*; The University of Melbourne: Parkville, Australia, 2015.
71. Lin, B.B.; Philpott, S.M.; Jha, S. *The Future of Urban Agriculture and Biodiversity-Ecosystem services: Challenges and Next Steps*; Elsevier GmbH: Amsterdam, The Netherlands, 2015; Volume 16, pp. 189–201.

72. Trendov, N.M. Comparative study on the motivations that drive urban community gardens in Central Eastern Europe. *Ann. Agrar. Sci.* **2018**, *16*, 85–89. [CrossRef]
73. Same, A.; Lee, E.A.L.; McNamara, B.; Rosenwax, L. The Value of a Gardening Service for the Frail Elderly and People With a Disability Living in the Community. *Home Health Care Manag. Pract.* **2016**, *28*, 256–261. [CrossRef]
74. Soga, M.; Gaston, K.J.; Yamaura, Y. Gardening is beneficial for health: A meta-analysis. *Prev. Med. Rep.* **2017**, *5*, 92–99. [CrossRef]
75. Francis, M.; Griffith, L. The meaning and design of farmers' markets as public space: An issue- based case study. *Landsc. J.* **2011**, *30*, 261–279. [CrossRef]
76. Horst, M.; Mcclintock, N.; Hoey, L. The Intersection of Planning, Urban Agriculture, and Food Justice: A Review of the Literature. *J. Am. Plan. Assoc.* **2017**, *83*, 277–295. [CrossRef]
77. Sousa, R.; Sales, D. Urban Agriculture: The Allotment Gardens as Structures of Urban Sustainability. In *Advances in Landscape Architecture*; InTech: London, UK, 2013.
78. Van Den Berg, L.M. Urban agriculture between allotment and market gardening: Contributions to the sustainability of African and Asian cities. *WIT Trans. Ecol. Environ.* **2002**, *14*, 945–959.
79. Laaksoharju, T.; Rappe, E.; Kaivola, T. Garden affordances for social learning, play, and for building nature–child relationship. *Urban For. Urban Green.* **2012**, *11*, 195–203. [CrossRef]
80. Riolo, F. The social and environmental value of public urban food forests: The case study of the Picasso Food Forest in Parma, Italy. *Urban For. Urban Green.* **2018**, *45*, 1–12. [CrossRef]
81. Dubbeling, M.; Bracalenti, L.; Lagorio, L. Participatory design of public spaces for urban agriculture, Rosario, Argentina. *Open House Int.* **2009**, *34*, 36–49. [CrossRef]
82. Gorgolewski, M.; Komisar, J.; Nasr, J. *Carrot City: Creating Places for Urban Agriculture*; The Monacelli Press: New York, NY, USA, 2011.
83. Veen, E.J.; Bock, B.B.; Van den Berg, W.; Visser, A.J.; Wiskerke, J.S.C. Community gardening and social cohesion: Different designs, different motivations. *Local Environ.* **2016**, *21*, 1271–1287. [CrossRef]
84. Atmakur-Javdekar, S. Childrens Play in Urban Areas. In *Play and Recreation, Health and Wellbeing*; Springer: Singapore, 2016; pp. 109–133.
85. Buchecker, M.; Hunziker, M.; Kienast, F. Participatory landscape development: Overcoming social barriers to public involvement. *Landsc. Urban Plan.* **2003**, *64*, 29–46. [CrossRef]
86. Hallett, S.; Hoagland, L.; Toner, E. Urban agriculture: Environmental, economic, and social perspectives. In *Horticultural Reviews*; Wiley Online Books: Hoboken, NJ, USA, 2016; pp. 65–120, ISBN 9781119281269.
87. Mangone, G.; Capaldi, C.A.; van Allen, Z.M.; Luscuere, P.G. Bringing nature to work: Preferences and perceptions of constructed indoor and natural outdoor workspaces. *Urban For. Urban Green.* **2017**, *23*, 1–12. [CrossRef]
88. Alaimo, K.; Packnett, E.; Miles, R.A.; Kruger, D.J. Fruit and Vegetable Intake among Urban Community Gardeners. *J. Nutr. Educ. Behav.* **2008**, *40*, 94–101. [CrossRef]
89. Bailey, S.; Hendrick, A.; Palmer, M. Eco-social Work in Action: A Place for Community Gardens. *Aust. Soc. Work* **2018**, *71*, 98–110. [CrossRef]
90. Park, H.; Kramer, M.; Rhemtulla, J.M.; Konijnendijk, C.C. Urban food systems that involve trees in Northern America and Europe: A scoping review. *Urban For. Urban Green.* **2019**, *45*, 126360. [CrossRef]
91. Al-Mayahi, A.; Al-Ismaily, S.; Gibreel, T.; Kacimov, A.; Al-Maktoumi, A. Home gardening in Muscat, Oman: Gardeners' practices, perceptions and motivations. *Urban For. Urban Green.* **2019**, *8*, 286–294. [CrossRef]
92. Askerlund, P.; Almers, E. Forest gardens—New opportunities for urban children to understand and develop relationships with other organisms. *Urban For. Urban Green.* **2016**, *20*, 187–197. [CrossRef]
93. Audrey, S.; Batista-Ferrer, H. Healthy urban environments for children and young people: A systematic review of intervention studies. *Health Place* **2015**, *36*, 97–117. [CrossRef] [PubMed]
94. Jansson, M.; Gunnarsson, A.; Mårtensson, F.; Andersson, S. Children's perspectives on vegetation establishment: Implications for school ground greening. *Urban For. Urban Green.* **2014**, *13*, 166–174. [CrossRef]
95. Branduini, P.N.; Laviscio, R.; Scazzosi, L.; Supuka, J.; Tóth, A.; Laviscio, R.; Supuka, J.; Toth, A. Urban agriculture and cultural heritage: An historical and spatial relationship. In *Urban Agriculture Europe*; Lohrberg, F., Licka, L., Scazzosi, L., Timpe, A., Eds.; Jovis: Berlin, Germany, 2016; pp. 138–147, ISBN 978-3-86859-371-6.
96. Spierings, B.; Van Liempt, I.; Maliepaard, E. Ownership and Membership: Practices and Experiences of Neighbourhood Residents in the Wijsgeren Community Garden in Amsterdam. *Tijdschr. Voor Econ. Soc. Geogr.* **2018**, *109*, 677–684. [CrossRef]
97. Van Holstein, E. *Community and Ownership: A Relational Study of Community Gardens*; University of Wollongong: Wollongong, Australia, 2017.
98. Tajaddini, L. Investigating the characteristics of Persian gardens: Taking a close look at Mahan Shah Zadeh garden. In *Proceedings of the Geo-Environment and Landscape Evolution III*; WIT Press: Southampton, UK, 2008; Volume I, pp. 211–218.
99. UNESCO. The Persian Garden, UNESCO World Heritage Centre. Available online: https://whc.unesco.org/en/list/1372/ (accessed on 2 April 2021).
100. Yin, R.K. Discovering the Future of the Case Study. Method in Evaluation Research. *Eval. Pract.* **1994**, *15*, 283–290. [CrossRef]
101. Groat, L.N.; Wang, D. *Architectural Research Methods*, 2nd ed.; John Wiley & Sons: Hoboken, NJ, USA, 2013; Volume 148, ISBN 978-0-470-90855-6.

102. Swaffield, S. Case Studies. In *Research in Landscape Architecture Methods and Methodology*; van den Brink, A., Bruns, D., Tobi, H., Bell, S., Eds.; Routledge: Oxfordshire, UK, 2017; pp. 188–211, ISBN 9781138020931.
103. Mahmoudi Farahani, L.; Motamed, B.; Jamei, E. Persian Gardens: Meanings, Symbolism, and Design. *Landsc. Online* **2016**, *46*, 1–19. [CrossRef]
104. Harvey, D.J.; Montgomery, L.N.; Harvey, H.; Hall, F.; Gange, A.C.; Watling, D. Psychological benefits of a biodiversity-focussed outdoor learning program for primary school children. *J. Environ. Psychol.* **2020**, *67*, 101381. [CrossRef]
105. Middle, I.; Dzidic, P.; Buckley, A.; Bennett, D.; Tye, M.; Jones, R. Integrating community gardens into public parks: An innovative approach for providing ecosystem services in urban areas. *Urban For. Urban Green.* **2014**, *13*, 638–645. [CrossRef]
106. Napawan, N.C.; Burke, E. Productive potential: Evaluating residential urban agriculture. *Landsc. Res.* **2016**, *41*, 773–779. [CrossRef]
107. Napawan, N.C.; Townsend, S.A. The landscape of urban agriculture in California's capital. *Landsc. Res.* **2016**, *41*, 780–794. [CrossRef]
108. Razavi, N. *Paradise Extended; Re-Examining the Cultural Anchors of Historic Pleasure Avenues*; Springer: Cham, Switzerland, 2020; ISBN 9783030227623.
109. Fallahi, E.; Fallahi, P.; Mahdavi, S. Ancient Urban Gardens of Persia: Concept, History, and Influence on Other World Gardens. *HortTechnology* **2020**, *30*, 6–12. [CrossRef]
110. Mansouri, M. Water as the Origin of Beauty in Persian Garden. *Manzar* **2019**, *11*, 32–43. [CrossRef]
111. Lovell, S.T. Multifunctional urban agriculture for sustainable land use planning in the United States. *Sustainability* **2010**, *2*, 2499–2522. [CrossRef]
112. Fekete, A.; Haidari, R. Special aspects of water use in Persian gardens. *Acta Univ. Sapientiae Agric. Environ.* **2015**, *7*, 82–88. [CrossRef]
113. Abbas, M.Y.; Nafisi, N.; Nafisi, S. Persian Garden, Cultural Sustainability and Environmental Design Case Study Shazdeh Garden. *Procedia Soc. Behav. Sci.* **2016**, *222*, 510–517. [CrossRef]
114. Lindemann-Matthies, P.; Brieger, H. Does urban gardening increase aesthetic quality of urban areas? A case study from Germany. *Urban For. Urban Green.* **2016**, *17*, 33–41. [CrossRef]
115. Van Herzele, A.; Wiedemann, T. A monitoring tool for the provision of accessible and attractive urban green spaces. *Landsc. Urban Plan.* **2003**, *63*, 109–126. [CrossRef]
116. Scharf, N.; Wachtel, T.; Reddy, S.E.; Säumel, I. Urban Commons for the Edible City-First Insights for Future Sustainable Urban Food Systems from Berlin, Germany. *Sustainability* **2019**, *11*, 966. [CrossRef]
117. Zhang, Y.; Min, Q.; Zhang, C.; He, L.; Zhang, S.; Yang, L.; Tian, M.; Xiong, Y. Traditional culture as an important power for maintaining agricultural landscapes in cultural heritage sites: A case study of the Hani terraces. *J. Cult. Herit.* **2017**, *25*, 170–176. [CrossRef]
118. McLain, R.; Poe, M.; Hurley, P.T.; Lecompte-Mastenbrook, J.; Emery, M.R. Producing edible landscapes in Seattle's urban forest. *Urban For. Urban Green.* **2012**, *11*, 187–194. [CrossRef]
119. Brown, J.; Hay-Edie, T. *Engaging Local Communities in Stewardship of World Heritage: A Methodology Based on the COMPACT Experience*; UNESCO: Paris, France, 2014; ISBN 9789231000546.
120. Kennard, N.J.; Bamford, R.H. *Urban Agriculture: Opportunities and Challenges for Sustainable Development—Zero Hunger*; Leal Filho, W., Azul, A.M., Brandli, L., Özuyar, P.G., Wall, T., Eds.; Springer International Publishing: Cham, Switzerland, 2019; pp. 1–14, ISBN 978-3-319-69626-3.
121. Martinho da Silva, I.; Oliveira Fernandes, C.; Castiglione, B.; Costa, L. Characteristics and motivations of potential users of urban allotment gardens: The case of Vila Nova de Gaia municipal network of urban allotment gardens. *Urban For. Urban Green.* **2016**, *20*, 56–64. [CrossRef]
122. Koopmans, M.E.; Keech, D.; Sovová, L.; Reed, M. Urban agriculture and place-making: Narratives about place and space in Ghent, Brno and Bristol. *Morav. Geogr. Rep.* **2017**, *25*, 154–165. [CrossRef]
123. Gheissari, A. Authorial Voices and the Sense of an Ending in Persian Diaries: Notes on E'temād al-Saltaneh and 'Alam. *Iran. Stud.* **2016**, *49*, 693–723. [CrossRef]
124. Grebitus, C.; Chenarides, L.; Muenich, R.; Mahalov, A. Consumers' Perception of Urban Farming—An Exploratory Study. *Front. Sustain. Food Syst.* **2020**, *4*, 79. [CrossRef]
125. Rehan, R.M. Urban branding as an effective sustainability tool in urban development. *HBRC J.* **2014**, *10*, 222–230. [CrossRef]
126. Van Tuijl, E.; Hospers, G.-J.; Van Den Berg, L. Opportunities and Challenges of Urban Agriculture for Sustainable City Development. *Eur. Spat. Res. Policy* **2018**, *25*, 5–22. [CrossRef]
127. Krikser, T.; Zasada, I.; Piorr, A. Socio-economic viability of urban agriculture—A comparative analysis of success factors in Germany. *Sustainability* **2019**, *11*, 1999. [CrossRef]
128. Qaim, M. Role of New Plant Breeding Technologies for Food Security and Sustainable Agricultural Development. *Appl. Econ. Perspect. Policy* **2020**, *42*, 129–150. [CrossRef]
129. Peschard, K.; Randeria, S. 'Keeping seeds in our hands': The rise of seed activism. *J. Peasant Stud.* **2020**, *47*, 613–647. [CrossRef]
130. Kuehne, G.; Llewellyn, R.; Pannell, D.J.; Wilkinson, R.; Dolling, P.; Ouzman, J.; Ewing, M. Predicting farmer uptake of new agricultural practices: A tool for research, extension and policy. *Agric. Syst.* **2017**, *156*, 115–125. [CrossRef]

131. Blakeney, M.; Krishnankutty, J.; Raju, R.K.; Siddique, K.H.M. Agricultural Innovation and the Protection of Traditional Rice Varieties: Kerala a Case Study. *Front. Sustain. Food Syst.* **2020**, *3*, 116. [CrossRef]
132. Amani-Beni, M.; Zhang, B.; Xie, G.-D.; Xu, J. Impact of urban park's tree, grass and waterbody on microclimate in hot summer days: A case study of Olympic Park in Beijing, China. *Urban For. Urban Green.* **2018**, *32*, 1–6. [CrossRef]
133. Clucas, B.; Parker, I.D.; Feldpausch-Parker, A.M. A systematic review of the relationship between urban agriculture and biodiversity. *Urban Ecosyst.* **2018**, *21*, 635–643. [CrossRef]
134. Russo, A.; Cirella, G.T. Edible Green Infrastructure for Urban Regeneration and Food Security: Case Studies from the Campania Region. *Agriculture* **2020**, *10*, 358. [CrossRef]

land

Review

Dynamic and Heterogeneity of Urban Heat Island: A Theoretical Framework in the Context of Urban Ecology

Zahra Mokhtari [1,†], Shahindokht Barghjelveh [1,†], Romina Sayahnia [1], Salman Qureshi [2] and Alessio Russo [3,*]

1 Department of Environmental Planning and Design, Environmental Sciences Research Institute, Shahid Beheshti University, Tehran 1983969411, Iran; z_mokhtari@sbu.ac.ir (Z.M.); s-barghjelveh@sbu.ac.ir (S.B.); r_sayahnia@sbu.ac.ir (R.S.)
2 Department of Geography, Humboldt University of Berlin, Rudower Chaussee 16, 12557 Berlin, Germany; salman.qureshi@geo.hu-berlin.de
3 School of Arts, University of Gloucestershire, Francis Close Hall Campus, Swindon Road, Cheltenham GL50 4AZ, UK
* Correspondence: arusso@glos.ac.uk
† These authors contributed equally to this work.

Abstract: The dynamic and heterogeneity of the urban heat island (UHI) is the result of the interactions between biotic, physical, social, and built components. Urban ecology as a transdisciplinary science can provide a context to understand the complex social–biophysical issues such as the thermal environment in cities. This study aimed at developing a theoretical framework to elucidate the interactions between the social–biophysical patterns and processes mediating UHI. To do it, we conducted a theoretical review to delineate UHI complexity using the concept of dynamic heterogeneity of pattern, process, and function in UHI phenomenon. Furthermore, a hypothetical heterogeneity spiral (i.e., driver-outcome spiral) related to the UHI was conceived as a model template. The adopted theoretical framework can provide a holistic vision of the UHI, contributing to a better understanding of UHI's spatial variations in long-term studies. Through the developed framework, we can devise appropriate methodological approaches (i.e., statistic-based techniques) to develop prediction models of UHI's spatial heterogeneity.

Keywords: process-based approach; transdisciplinary research; theoretical review; urban heat island mitigation; social–biophysical interaction; compositional and configurational heterogeneity

Citation: Mokhtari, Z.; Barghjelveh, S.; Sayahnia, R.; Qureshi, S.; Russo, A. Dynamic and Heterogeneity of Urban Heat Island: A Theoretical Framework in the Context of Urban Ecology. *Land* **2022**, *11*, 1155. https://doi.org/10.3390/land11081155

Academic Editor: Christine Fürst

Received: 7 June 2022
Accepted: 22 July 2022
Published: 26 July 2022

1. Introduction

Urban regions consist of human and natural components that constantly change due to complex interactions within and between biophysical and social systems [1–3]. Therefore, these changes lead to the formation of unique landscapes, characterized by an extraordinary variety of land uses [4,5], which affect the surface–atmosphere energy balance and urban thermal environment (UTE) [6,7]. Dense urban settings tend to be significantly warmer than the nearby rural area which is known as urban heat island (UHI) phenomenon [8,9]. The UHI phenomenon exerts impacts on human heat-related health and comfort, particularly during heat waves [10,11]; moreover, it affects energy consumption, water quality, carbon dioxide emissions, and air pollution [12–17]. Due to health and environmental concerns, the UHI effect has aroused widespread attention in recent decades, leading to a cumulative body of research aiming to explore its drivers, formation, and consequences [18–20].

The spatial pattern of UHI is commonly retrieved from thermal data of satellite images such as Landsat 8-thermal infrared sensor (TIRS) and Moderate Resolution Imaging Spectroradiometer (MODIS) thermal infrared data, which are known as land surface temperature (LST) [21]. To capture the temperature of the heterogeneous surface (i.e, various land uses) LST within an urban landscape, unmanned aerial vehicles or drones have been introduced to retrieve LST at sub-meter spatial resolutions. The drone's spatial and temporal resolutions are highly advantageous for evaluating the variability of LST at fine spatial and temporal scales in an urban heterogeneous system [22,23].

Due to a wide diversity of socio-economic and biophysical intertwining drivers and outcomes, the UHI is a complex issue to study [13,24–28]. In addition to the complex interactions, Cadenasso et al. [29] argued that the extreme complexity of urban issues arises from the spatial attribute (i.e., configuration and composition) of urban mosaic patches and their temporal changes. Similarly, the configuration and composition attributes of urban patches (i.e., green and built-up patches) affect the thermal environment [30–34].

The science of urban ecology deals with the complex social–biophysical issues of cities [35,36] and investigates the interactions between complex biological systems, built structures, and human actions [37–39]. Considering the principle of urban ecology that ecological (biophysical) patterns and processes affect ecosystem services [39], it gives an insight into how urban ecology would be beneficial in investigating the UHI. Urban ecology as a transdisciplinary science assists society in moving toward sustainability and resilience [40–42]. It focuses on spatial-temporal patterns of urbanization and how they affect social–ecological processes and functions, ecosystem services, human wellbeing, and urban sustainability [40,43,44]. McPhearson et al. [35] argued that urban ecology provides a robust and holistic approach to the study of cities, helping the decision-makers to understand the complex relationships among social, ecological, economic, and technological systems. Therefore, developing theoretical and empirical studies related to the different issues in the context of urban ecology is essential [35]. In urban ecology, the pattern of an urban area is considered to be spatially heterogeneous and to have an influence on ecological processes [45]. These processes can be investigated by considering three broad realms: the flow of material and energy, biotic performance, and human actions [46]. UHI, as a result of social–biophysical interactions, is a spatially heterogeneous and temporally dynamic phenomenon. Then, urban ecology can give a new insight into investigating UHI.

Landscape ecology, as a holistic transdisciplinary science [47,48], explicitly emphasizes spatial composition and configuration, and its consequences on biophysical processes like biogeochemical fluxes and socio-economic processes [49–51]. Recently, urban landscape ecology [50,52] as the invention of landscape ecology and urban ecology [52], provides an appropriate context for understanding the formation, the effects of spatial and dynamical heterogeneity, and the relationship between landscape patterns (i.e., land cover/land use composition and configuration) and biophysical and socio-economic processes in multiple scales of time and space [52]. According to the urban landscape ecology, the compositional and configurational attributes like connectivity, distance from green area, shape characteristics, density, and degree of aggregation of patches exert impacts on thermal processes and land surface temperature [32,53–55].

Integrating social and ecological knowledge and data is critical to promoting the modeling of an urban ecosystem [56–60]. Therefore, to study the integrated complex infrastructural–social–ecological issues in urban ecology, different approaches and frameworks like a human ecosystem, Metacity, ecological feedback model, pattern–process–function, and dynamic heterogeneity like patch dynamic, dynamic heterogeneity, pattern–process–function, urban–rural gradient, ecosystem service framework, ecosystem service integrity, human ecosystem framework have been developed in recent years [46,56,61–63].

To study the integrated complex infrastructural–social–ecological issues in urban ecology, different approaches and frameworks like a human ecosystem, pattern–process–function, and dynamic heterogeneity [46,64,65], which basically include similar concepts, have been developing. The concept of spatial heterogeneity can be applied to urban planning and management: social–biophysical processes mediate urban functions and sustainability [66]. In other words, urban ecology as transdisciplinary research integrates human actions and perceptions and policymaking with biophysical components [35,42,58,67–73]. For instance, residents may respond to the heat in different ways like tree planting or using air conditioners [68], meaning that decision-making and human actions affect biophysical processes that change the UHI. When exploring the mechanisms behind complex urban phenomena, the process-based 'dynamic heterogeneity' approach can help clarify the interactions between human and biophysical components [35,68].

A framework in urban ecology refers to intertwined mechanisms or processes which can be tested using various hypotheses [40,74]. Synthesizing conceptual frameworks is essential in advancing urban ecology towards a strong science of cities [35]. The magnitude of interactions is regulated by policy, governance, culture, and individual behavior of the urban system of the urban system [75]. Integrating human actions into the urban ecosystem is widely perceived at the conceptual level, but developing effective and integrative theories and their applications in an urban system study still remains a challenge [76,77].

A dynamic heterogeneity approach is a useful tool for enhancing ecological integration and exploring the interactions between social and biophysical patterns and processes in urban ecosystems [46]. Understanding the complex interactions between processes, identifying their driving factors, and, ultimately, predicting the behaviour of environmental systems are among the main objectives of environmental research [78]. It can be inferred that the concept of dynamic heterogeneity can be applied to long-term research, facilitating the integration of ecosystem components and the development of predictive models [46].

Although there are a large number of systematic reviews related to the different aspects of the UHI phenomenon [79–81], it has not been investigated using a "theoretical review" lying in urban ecology. For instance, Deilami et al. [79] organized a synthesis review to identify the spatio-temporal factors and their causal mechanisms or processes that mitigate the UHI effect. Considering all the above, we aimed to develop a theoretical framework for a better understanding of the social–biophysical mechanisms behind UHIs' heterogeneity through time. In this article, we conducted a theoretical review to illuminate how the social–biological–physical processes contribute to forming the UHI in an urban ecosystem and consequently cause the dynamic and heterogeneity of UHI. To achieve the goal, we made two main implementations: (1) the dynamic heterogeneity approach was adjusted to the UHIs' dynamics and heterogeneity and (2) a template model of the driver–outcome spiral (i.e., heterogeneity spiral) was conceived for the UHI phenomenon. The proposed conceptual framework offers a comprehensive perspective of the UHI phenomenon in the context of urban ecology, supporting the analysis of UHIs' spatial heterogeneity in long-term studies.

2. Method

In this article, we conducted a theoretical review [82,83] in regard to the concept of "dynamic heterogeneity" lying in the urban ecology context. A theoretical review consists of concepts, together with their definitions, and existing theories that were used for UHI study. A theoretical review is drawn based on the existing conceptual and empirical studies to provide a higher level of understanding of various concepts and relationships in the studied topic [83]. In this review, we attempted to demonstrate an understanding of theories and concepts that are relevant to the UHI. In fact, in this research, we saw the UHI phenomenon from the aspect of the dynamic heterogeneity framework which itself is an inclusive framework and includes many interrelated concepts.

Since the basic idea of the research originated from the "dynamic heterogeneity" approach, firstly, it was necessary to define the concept of dynamic and heterogeneity in urban ecosystems. In the first section of the paper, we reviewed the principles of urban ecology in order to elucidate the UHI. The second section reviewed papers that primarily consisted of specific variables related to the dynamics and heterogeneity of the UHI. The authors organized the papers according to whether they focused on the social, biological, or physical attributes that affect the dynamics and heterogeneity of the UHI. In the final section, a hypothetical spiral of dynamic heterogeneity of UHI was created based on empirical evidence of UHI studies. Following an initial search, the abstract and the content of the articles identified by the search engines were reviewed. The number of articles containing the keywords was extremely broad, and in many cases, the concepts that we sought were not recognizable only through keywords and titles. Therefore, the articles were screened and those which did not match our goals were excluded.

The literature review was based on searching peer-reviewed articles in search engines of ISI Web of Science and Scopus. To synthesize the literature, we used a broad range of keywords from diverse disciplines to identify papers related to our questions: What are the reciprocal interactions between physical and biological processes in the UHI phenomenon? What are the reciprocal interactions between physical and human processes in the UHI phenomenon? What are the reciprocal interactions between human and biological processes in the UHI phenomenon? For each process realm, we identified several variables, for instance, for the social process, we used human health, human comfort, energy consumption, household income, decision-making, and mitigation policy. These terms were chosen based on our knowledge acquired from basic literature. The keywords represented in Table 1 was related to the concepts of urban ecology and biological, physical, social, economic, and built variables that exert an influence on the dynamics and heterogeneity of the UHI.

Table 1. The basic query for paper selection by keywords in concepts of urban ecology and UHI.

Concepts of "Urban Ecology"	Keywords in UHI Literature
Urban ecosystem Social–biophysical (ecological) interaction Pattern–process Dynamic and Heterogeneity Spatial heterogeneity Social–ecological dynamic Biophysical dynamic Complexity Cause and effect	"Urbanization" OR "Urban development" OR "UHI" OR "cold spot and hot spot" OR "land surface temperature" OR "Spatial-temporal change" OR "Human intervention" OR "mitigation policy" OR "tree protection policy" OR "climate regulation" or "cooling effect" OR "decision-making" OR "artificial heat production" OR "human health" OR "energy consumption" OR "anthropogenic heat sources" OR "land architecture" OR "tree diversity" OR "tree attribute" OR "urban forest" OR "energy and water flow" OR "heat wave" OR "wind direction" OR "urban green space" OR "cooling effect"

3. Dynamic and Heterogeneity in Urban Ecosystems

In studying urban phenomena, understanding the causes and consequences of spatial heterogeneity of patterns, processes, and functions are considered critical issues [46,84]. Pickett et al. [46] developed the dynamic heterogeneity approach as an inclusive theory, which provides a framework to explore the mechanisms, outcomes, and drivers of spatial variability over time. In the urban scientific literature, the term 'dynamic' indicates how a patch or patch mosaic changes structurally and functionally through time [85], while 'heterogeneity' refers to the spatial variation of a property of interest across a landscape [86]. In particular, 'spatial heterogeneity' refers to the causal structure and spatial variability of a specified object [40,74].

However, Pickett et al. [46] argued that 'heterogeneity' is not just about the patterns, but also the social–biophysical processes which are spatially heterogeneous. It means that heterogeneity is an outcome of past social and biophysical processes, and can act as a driver of future social and biophysical processes (i.e., heterogeneity observed at a certain time is the result of prior conditions). Therefore, by analyzing heterogeneity within different time intervals, it is possible to conduct long-term research in the urban ecosystem [46,87].

The urban dynamic heterogeneity approach could help recognize the interactions between social and biophysical components. Human ecosystems consist of heterogeneous biological, physical, social, and infrastructural components—the heterogeneous layers interact with each other at different scales. Over time, these interactions create a new type of heterogeneity. Since there are potential interactions between all the components, the aim of the research determines which interactions should be investigated at a particular scale. The concept of the human ecosystem emphasizes how heterogeneity of human interventions influences heterogeneity of buildings and infrastructures; moreover, social and biophysical attributes and fluxes outside urban boundaries have been found to affect heterogeneity over time [46].

According to landscape ecology, patterns are defined as spatial attributes of a landscape: they encompass both the composition and configuration of patches and influence biophysical processes [84]. Therefore, pattern heterogeneity can be explained by both compositional and configurational heterogeneities [88]. In an urban ecosystem, the process refers to the transferring of energy, material, or organisms, flux, and cycling of elements within a city [65], which are inherently heterogeneous in space and occur in particular places on a landscape [89,90]. In an urban ecosystem, patterns and processes interact reciprocally and are theoretically inseparable (i.e., there is a coupling of patterns and processes) [65,79,91,92]. The function is the interaction between pattern and processes that supports delivering ecosystem services like climate regulation in urban areas [65]. In a time frame, pattern heterogeneity leads to process and functional heterogeneity [46]. Functional heterogeneity is defined as the spatial variation of the urban ecosystem's capacity to provide services [65].

Urban ecologists hypothesized that the interaction between social–biophysical patterns and processes can be observed in the form of surface cover or land use/land cover (LULC). LULC is regarded as an ecological indicator in urban studies. It affects ecological patterns and processes, causing broad environmental phenomena like the UHI. The new biophysical conditions such as UHI affects human attitudes which may lead to the establishment of new policies. These policies themselves change the LULC over time [2,93].

Pickett et al. [46] outlined the existence of three interactive processes that lead to the hybridization of biophysical, social, and built components of the human ecosystem. These processes include (1) flows of material and energy (e.g., heat fluxes); (2) biological potentials or biotic performances (e.g., spatial arrangement of organisms, their traits, and community dynamics); (3) human actions and interventions and decision-making in an urban ecosystem.

The vast realm of material and energy flow in the urban ecosystem refers to the transforming and transferring of food, goods, and fuel. In other words, they can be defined by the pathways as the input and output of water, food, air, fuel, and heat [94–96]. The resources that stream into cities shape and modify the structure of the urban biological system, empower, and drive urban capacities with an impact on common biological forms of cities, and in the long run, create yields that remain inside the boundary or are sent out beyond the boundary [97]. Biotic differentiation (biota differentiation) is defined as various biodiversity (fauna and flora) and species richness within an ecosystem [98,99]. Regarding the social or human-made process, it involves social–economic attributes like zoning regulation, lifestyle and livelihood arrangement, economic and political policy, neighborhood identity, housing price, the pattern of investment, access to the road and green area, house density, population distribution, the market economy, general patterns of income, and access to the service which make social–economic heterogeneity across the city [100–104]. Table 2 represents the main attributes of the urban ecosystem to illuminate dynamics and heterogeneity.

Table 2. The main attributes associated with h dynamic heterogeneity approach to elucidate UHI in an urban ecosystem (adapted from Pickett et al. [46]).

- Urban systems are extraordinarily heterogeneous.
- Heterogeneity encompasses space and time, patterns, and processes.
- There are different layers of biophysical, social, and infrastructural heterogeneity.
- The layers of heterogeneity interact with each other at different scales.
- Heterogeneity acts both as driver and outcome, so mediates between the social and biophysical components in the urban system.
- The interactions of different heterogeneous layers create new heterogeneities.
- Social and biophysical fluxes outside the urban boundary affect heterogeneity through time.
- Heterogeneity affects urban functions that lead to ecosystem services delivery, human wellbeing, and sustainability.
- Human beings' feedback amplify dynamic heterogeneity in urban systems.

4. Dynamic and Heterogeneity of UHI

In urban ecology, the human ecosystem consists of interacting biotic, physical, social, and built components that are temporally dynamic and spatially heterogeneous [46]. In association with UHI investigation, there are manifold types of biotic, physical, social, and built heterogeneities that mediate the spatial variation of UHI (Figure 1). There is a multitude of variables to study the biotic, physical, social, and built components that contribute to the UHI spatial heterogeneity. The arrows show the potential interactions between the heterogeneous components. The interactions between the components can be determined by the aim of particular research. Biotic heterogeneity can be defined as the heterogeneous distribution of natural and semi-natural patches (including forests, woodland, shrubs, green areas, and wetlands) across a city, which affect differentially the land surface temperature [55,105–109]. In particular, heterogeneity of vegetation distribution, abundance, and tree species can affect the temperature in various ways, such as providing shade, modifying the landscape's thermal properties (i.e., albedo and emissivity modification), altering the air movement, and heat exchange (i.e., wind blowing) through evapotranspiration [13,109–117]. The effect of biological differentiation on the thermal environment and the UHI phenomenon can be assessed using vegetation indices, like the greenness index and the normalized difference vegetation index (NDVI) [118–120].

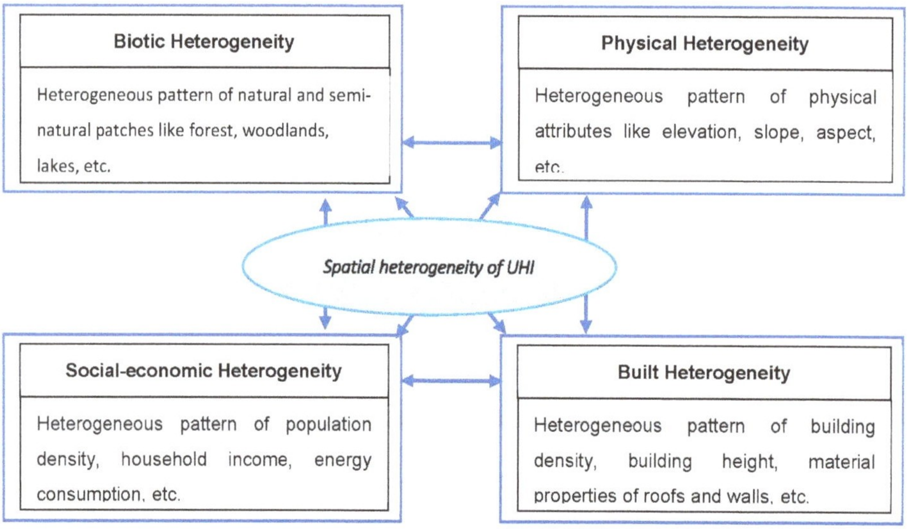

Figure 1. Four components of an urban ecosystem (adopted from Pickett et al. [46]) mediate the spatial heterogeneity of UHI.

Physical heterogeneity derives can be ascertained by topographic features (i.e., physical layers) like elevation, aspect, and slope. These features affect the thermal environment and control the UHI phenomenon [25,121,122]. Heterogeneous patterns of topographic attributes in an urban region alter the potential radiation and thermal loads (i.e., alter the energy flow process) [121].

In terms of the built component in the context of an urban ecosystem, it refers to a man-made built-up area characterized by infrastructural and technological components, changing through time due to human decision-making [46]. Notably, the characteristics of the built complex influence urban temperature and the formation of UHI. The height of buildings and their variability, as well as the spacing between buildings, affect air circulation, wind flow, and thermal energy absorption [18,24,123–125]. More importantly, the material properties of roofs and walls significantly affect both albedo and emissivity, leading to temperature alterations [13,126]. The sky view factor (SVF) is a parameter related to urban building and measures sky visibility. A reduction of the SVF leads to an increase in solar radiation absorption and a lowering of wind speed, ultimately amplifying the UHI effect [110,123,124,127–130]. Additionally, the normalized difference built-up index (NDBI), which reflects the amount of urban built-up areas, can be used to investigate the effect of the built-up surface on the intensity of the UHI phenomenon [119].

Social–economic heterogeneous patterns affect urban temperatures and support the occurrence of the UHI phenomenon [25,131,132]. For instance, heterogeneities in population density and household income influence the intensity of this phenomenon [25]. Furthermore, urban anthropogenic heat emission, derived from household energy consumption and vehicular traffic, is significantly related to socio-economic activities and is considered a key factor contributing to the formation of UHI [133–135]. In this context, human perception is considered an important process capable of altering the intensity of the UHI phenomenon. For instance, there can be a tendency to plant certain species (e.g., trees that provide more shading) in neighborhoods [136–138]; moreover, people living in the hot area usually apply strategies (e.g., using air conditioning or altering the neighborhood's biophysical structure through tree planting) to mitigate the UHI effect [68]. At the same time, policymaking outcomes (e.g., increasing vegetation, constructing living (green) roofs, and promoting light-coloured surfaces) effectively influence variations of the UHIs over time [139]. The application of policies targeting the alteration of urban structures (e.g., the placement and orientation of buildings) and the residents' lifestyles can also explain temperature variations across a city [140].

The ultimate result of the reciprocal biotic–physical–social–built interactions described above mediates a spatial heterogeneous mosaic of UHI (Figure 1). This mosaic affects the biophysical–social processes (i.e., evapotranspiration, heat exchange, and decision-making processes) in urban areas (Figure 2). Each of these three processes contributing to the UHI heterogeneity is itself a large topic. The researchers can focus on each of the three processes related to the others and study the feedbacks and interactions among them. For instance, how does the decision-making process change the vegetation surface, or how energy and material flow can affect human perception.

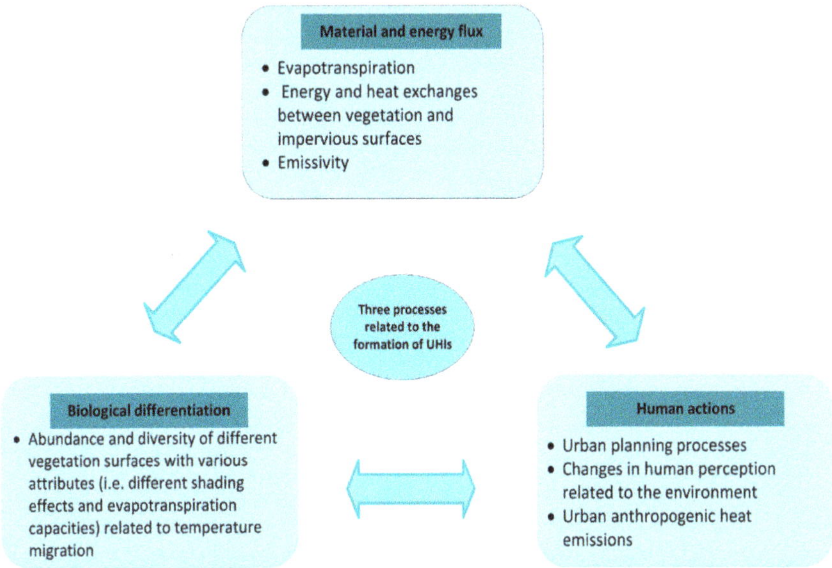

Figure 2. The reciprocal interaction between the processes of 'energy and material flows', 'biotic differentiation', and 'human action' (adapted from Pickett et al. [46]) contribute to the formation of UHI. For each process realm, several variables have been outlined.

Integrating Figures 1 and 2, we can adjust the dynamic heterogeneity for the case of UHI. As shown in Figure 3, the interactions between the above-mentioned coupling social–biophysical patterns and processes over time may lead to a new heterogonous UHI pattern. In other words, the interactions between the patterns and processes change the UHI heterogeneity over time which can be called "dynamic heterogeneity of UHIs". As shown in Figure 3, the interactions among a multitude of heterogeneous built–social–biophysical layers drive social–biophysical processes, and the process feedbacks change the pattern heterogeneity. The coupling interaction between heterogeneous patterns and processes can hence give birth to a new heterogeneous UHI pattern over time that is called "dynamic heterogeneity of UHI".

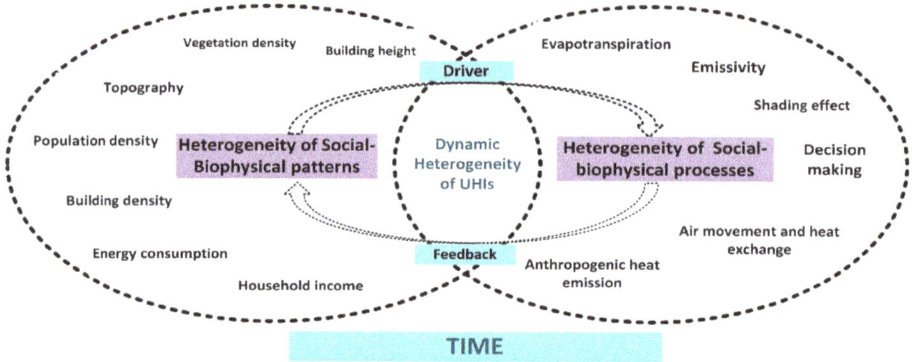

Figure 3. Dynamic heterogeneity approach in UHI: The interactions between four biotic–physical–social–built heterogeneous layers (variables in the left oval) make a heterogeneous pattern. The right oval shows different kinds of energy and material, biotic differentiation, and human action processes that interact reciprocally.

5. Driver–Outcome Spiral of UHI: Building a Model Template

The UHI is affected by numerous social–biophysical factors, as well as by the spatial arrangement of the LULC [25,141] and affects energy consumption, human health, water quality, and air pollution [14,142,143]. A template of heterogeneity or a spiral of dynamic heterogeneity [46] is a model template that indicates how a set of factors associated with a problem are potentially linked to each other. In addition, it represents the mechanisms, causes, effects, and interactions for a specific subject in a social–ecological system [68]. The above template, which was adopted from biological theories [64], follows the 'conditional statement' or 'if-then statements' (i.e., if A happens, then B is predicted: if a condition or relationship is verified, then certain results can be expected) [46,64].

In creating a driver–outcome spiral, due to the extremely wide diversity of the components, variables, and driver–outcome interactions involved, a myriad of templates can be developed to illustrate the causes and effects of the UHI. The choice of which template to build depends on the specific analytical goal: a large number of mechanistic spirals can be proposed by considering the various drivers and outcomes of the UHI. Figure 4 describes a simplified hypothetical driver–outcome spiral (i.e., a model template of the UHI dynamic heterogeneity), which was created based on a literature review. Here, heterogeneity is temporally dynamic and influences social–biophysical processes: physical, biological, and social–economic heterogeneities result from past interactions and are the drivers of future changes [46]. In this figure, the heterogeneous patterns of vegetation and impervious surfaces alter the land surface temperature pattern through biophysical processes (e.g., evapotranspiration and heat exchange) between time 1 and time 2. These, in turn, affect human comfort and health (between time 2 and time 3). Environmental concerns lead to the establishment of new policies for the mitigation of urban temperature. The decision-making policy process is expected to cause changes in land cover over time.

Notably, the occurrence of pulse events (i.e., regional events out of the urban boundary) at a given time may affect heterogeneity at a subsequent time. Note: the starting point of the driver–outcome spiral, which encompasses intrinsic physical attributes (e.g., topography and climatic zone) and corresponds to 0, is not shown in this figure.

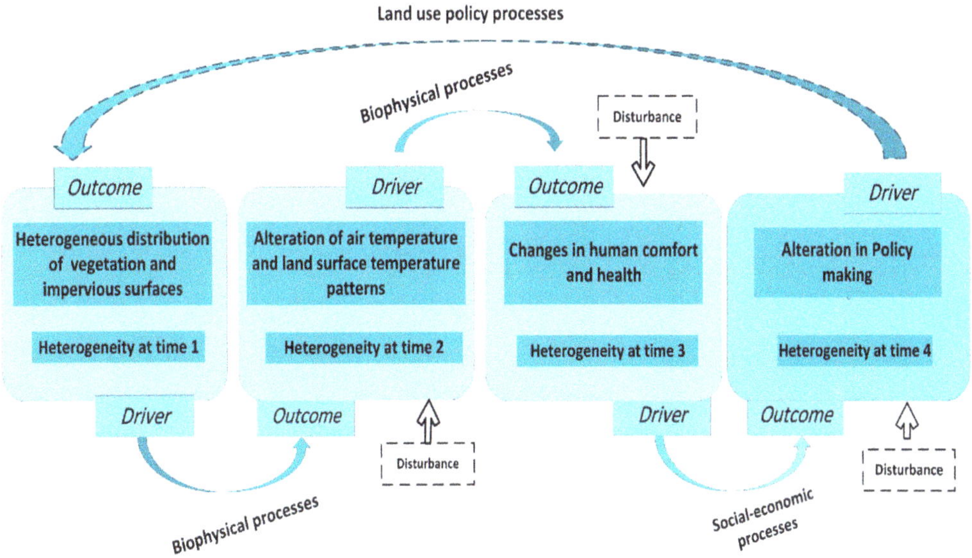

Figure 4. Model template: hypothetical driver–outcome spiral explaining the dynamics and heterogeneity of the UHI based on empirical evidence.

The assumptive spiral starts with a heterogeneous pattern of impervious and green patches, which are linked to changes in biophysical processes (e.g., evapotranspiration, shade, and heat exchange) through time (heterogeneity at time 1). The above heterogeneity led to a heterogeneous land surface temperature pattern (heterogeneity at time 2); in turn, temperature variations typically affect human thermal comfort and health (heterogeneity at time 3) [144–149]. In addition, high temperatures can trigger specific atmospheric chemistry procedures (e.g., increased ozone production, hydrocarbon, PM_{10}, and VOC concentrations), which lead to a worsening of air pollution [143,150]. Health and environmental issues deriving from high temperatures and air pollution may lead to changes in policies (heterogeneity at time 4), which would ultimately result in the alleviation of UHIs' effects [151]. Hence, policymaking processes would be the drivers of new land cover heterogeneities, starting a new turn of the spiral, which would continue to repeat through time [93]. Moreover, disturbances or pulsed events (e.g., heatwave) occurring outside urban regions (i.e., at a regional scale) [43,46] are expected to affect the UHI [152–154], giving rise to a new heterogeneity of the UHI in subsequent time.

6. Quantifying and Modeling the Interactions and Feedback among the Processes Mediating UHI

Due to the complexity and lacking direct measurement of different social–biological–physical processes and interactions in the context of urban ecology, various approaches and statistical models have been developed to quantify the specific goal [46]. The researcher can investigate the following interactions in UHI: how biotic differentiation (e.g., forest and woodland change) may influence physical processes like solar energy flux or the wind blowing; how physical processes like heat flux affect biotic performance; how the decision-making process and human perceptions can affect biotic differentiation; how human preferences and attitudes towards particular types of plants may affect the biodiversity of the urban area that consequently change UHI intensity; how biotic attributes can change human activities or how green space and plant diversity may influence human perceptions, leading to alteration of the urban temperature. Mechanistic models can describe a complex system by bringing the components together, providing a method to test the hypotheses in holistic ways. It also can describe a phenomenon through a hypothesized or assumed mechanism/process [155,156]. This model can be applied in studying the complex issue of UHI. In addition, a Bayesian Network model is a useful tool to deal with various social–ecological processes in a specific phenomenon [157,158] and can be used in evaluating probable outcomes in complex ecological systems [159]. This approach allows for a combination of different types of data like quantitative data, expert or local knowledge, and outputs from scenario building, and can deal with lacking data, hence they are useful in areas such as ecology or social science [157]. To analyze the relationship between the unobservable variables and the observed measurement, the state-space model can also be used [160] as a flexible approach [161]. In the case of UHI, for instance, there is some unobservable variable like human perception. In this case, the researcher's knowledge from the past is needed to estimate the future change of each variable [162].

7. Implications of the Theoretical Framework for Urban Planning toward UHI Mitigation

The urban ecology defines the cities as complex mosaics [66], engaging numerous social, ecological, and economic issues and strategies. In addition, landscape ecology as a science for dynamic and heterogeneity study focuses on spatial patchiness [163]. A city can be planned in a way to mitigate the UHI based on the transdisciplinary science of urban ecology and landscape ecology. In urban ecology, the urban heterogeneity comprises spatial variation within the physical, natural, and technological structures [40,46]. Urban planners consider how heterogeneity changes over time as the fundamental aspect of an urban ecosystem [66]. In addition, the compositional and configurational heterogeneity also affects the UHI intensity.

Therefore, the mitigation measures lying in this theoretical framework not only focus on the biotic components but also consider a hybrid of social–biological–physical–built components. Further, it emphasizes the pattern–process–function, considering how the composition and configuration of different patches within an urban landscape change the processes and functions and consequently, alleviate urban temperature.

8. Conclusions

Urban ecosystems are considered thermally heterogeneous because they typically comprise many small hot and cold spots which form a spatially heterogeneous pattern [164]. When dealing with this complexity, it is hence essential to recognize the mechanisms, components, and interactions between the social–biophysical components that contribute to the creation of UHI. In this context, the holistic science of urban ecology can be appropriate for investigating urban complex issues. Urban ecology studies are generally based on custodial frameworks, which enable the integration of biophysical and social components [68]. The concepts and tools introduced by transdisciplinary urban ecology have opened new pathways to tackle urban environmental concerns and ultimately improve related planning and management activities [66].

In this study, conceptualization and delineation of the causes and effects of spatial heterogeneity are essential in urban development [46]. In the case of UHI, the literature review indicated that pattern–process–function is heterogeneous and dynamic within an urban landscape (see the previous sections). In this study, by implying dynamic heterogeneity as an underlying approach in urban ecology, we developed a theoretical framework to understand the mechanisms behind the formation of UHI. In other words, the concept of dynamic heterogeneity was adopted to UHI: the interaction between social–biophysical patterns and processes over time leads to a new heterogeneous thermal environment. Furthermore, a hypothetical 'driver–outcome' spiral (i.e., heterogeneity spiral) was set up to better understand the UHI. In creating a driver–outcome spiral, due to an excessive diversity of components, variables, and driver–outcome interactions, a myriad of templates can be developed to illustrate a spiral of heterogeneity. Building a template depends on a specific analytical goal. Pickett and colleagues outlined that an "if-then" statement or "conditional statement" (i.e., if A happens then B is predicted, can support setting up a driver–outcome hypothesis. The synthesis of the literature review in this research demonstrated that UHI, as a specific subject that lies in a human ecosystem, can be defined through the dynamic heterogeneity approach. It enables us to integrate biophysical and social processes and patterns contributing to the UHI.

However, there are limitations to responding to all the questions related to the interaction between social–biophysical processes and their impact on UHI. As many variables and their effects are not directly observable, it means that the social–ecological feedback is not well understood. So, computer programs, simulation models, and special statistical models facilitate quantitative analysis of long-term data. Further, because of excessive diversity of components and driver–outcome interactions, a myriad of templates to illustrate the dynamic heterogeneity spiral can be developed. Building the model template is dependent on the specific analytical goal.

Overall, the theoretical framework in this paper allowed the examination of UHI from an ecological point of view, demonstrating that the concept of dynamic heterogeneity can describe UHI complexity. However, there are limitations to responding to all these questions related to the interaction between processes and their impact on the social–built–biological–physical components. As many of the variables and their effects are not directly observable, then social and biophysical complexes, their feedback, and interaction are not well understood. So, computer programs, simulations, and statistical models should be used to facilitate the quantitative analysis of long-term data for sustainable urban planning. The conceptual framework can be insightful in heterogeneity management of an urban system in a way to achieve temperature mitigation and an increase of climate regulation services. According to the transdisciplinary urban ecology, in future studies, ecologists and landscape architects are urged to collaborate with city residents to mitigate the UHI effects. Moreover, potentially, the developed framework can give the insight to understand the complexity of social–biophysical phenomena like air pollution, water flow and pollution, and soil pollution toward urban sustainability.

Author Contributions: Conceptualization, Z.M., S.B., S.Q., and A.R.; writing—original draft preparation, Z.M.; writing—review and editing, Z.M., S.B., A.R., R.S., and S.Q.; visualization, Z.M.; supervision, A.R. and S.B.; project administration, S.B.; funding acquisition, A.R. All authors have read and agreed to the published version of the manuscript.

Funding: This research received no external funding.

Institutional Review Board Statement: Not applicable.

Informed Consent Statement: Not applicable.

Data Availability Statement: Not applicable.

Conflicts of Interest: The authors declare no conflict of interest.

References

1. McDonnell, M. Editorial: Linking and promoting research and practice in the evolving discipline of urban ecology. *J. Urban. Ecol.* **2015**, *1*, 6. [CrossRef]
2. Pauleit, S.; Breuste, J.; Qureshi, S.; Sauerwein, M. Transformation of rural-urban cultural landscapes in Europe: Integrating approaches from ecological, socio-economic and planning perspectives. *Landsc. Online* **2010**, *20*, 1–10. [CrossRef]
3. Qureshi, S.; Haase, D.; Coles, R. The theorized urban gradient (TUG) method—A conceptual framework for socio-ecological sampling in complex urban agglomerations. *Ecol. Indic.* **2014**, *36*, 100–110. [CrossRef]
4. Elmqvist, T.; Colding, J.; Barthel, S.; Borgström, S.; Duit, A.; Lundberg, J.; Andersson, E.; Ahrné, K.; Ernstson, H.; Folke, C.; et al. The dynamics of Social-Ecological systems in urban landscapes: Stockholm and the national urban park, Sweden. *Ann. N. Y. Acad. Sci.* **2004**, *1023*, 308–322. [CrossRef]
5. Qureshi, S.; Haase, D. Compact, eco-, hybrid or teleconnected? Novel aspects of urban ecological research seeking compatible solutions to socio-ecological complexities. *Ecol. Indic.* **2014**, *42*, 1–5. [CrossRef]

6. Voogt, J.A.; Oke, T.R. Thermal remote sensing of urban climates. *Remote Sens. Environ.* **2003**, *86*, 370–384. [CrossRef]
7. Weng, Q. Thermal infrared remote sensing for urban climate and environmental studies: Methods, applications, and trends. *ISPRS J. Photogramm. Remote Sens.* **2009**, *64*, 335–344. [CrossRef]
8. Akbari, H.; Pomerantz, M.; Taha, H. Cool surfaces and shade trees to reduce energy use and improve air quality in urban areas. *Sol. Energy* **2001**, *70*, 295–310. [CrossRef]
9. Huang, G.; Zhou, W.; Cadenasso, M. Is everyone hot in the city? Spatial pattern of land surface temperatures, land cover and neighborhood socioeconomic characteristics in Baltimore, MD. *J. Environ. Manag.* **2011**, *92*, 1753–1759. [CrossRef] [PubMed]
10. Heaviside, C.; Macintyre, H.; Vardoulakis, S. The urban heat island: Implications for health in a changing environment. *Curr. Environ. Health Rep.* **2017**, *4*, 296–305. [CrossRef] [PubMed]
11. Steeneveld, G.-J.; Koopmans, S.; Heusinkveld, B.; Van Hove, L.; Holtslag, A. Quantifying urban heat island effects and human comfort for cities of variable size and urban morphology in the Netherlands. *J. Geophys. Res. Atmos.* **2011**, *116*, D20129. [CrossRef]
12. Sun, R.; Wang, Y.; Chen, L. A distributed model for quantifying temporal-spatial patterns of anthropogenic heat based on energy consumption. *J. Clean. Product.* **2018**, *170*, 601–609. [CrossRef]
13. Phelan, P.E.; Kaloush, K.; Miner, M.; Golden, J.; Phelan, B.; Silva, H., III; Taylor, R.A. Urban heat island: Mechanisms, implications, and possible remedies. *Annu. Rev. Environ. Resour.* **2015**, *40*, 285–307. [CrossRef]
14. Ulpiani, G. On the linkage between urban heat island and urban pollution island: Three-decade literature review towards a conceptual framework. *Sci. Total Environ.* **2021**, *751*, 141727. [CrossRef]
15. Romano, P.; Prataviera, E.; Carnieletto, L.; Vivian, J.; Zinzi, M.; Zarrella, A. Assessment of the urban heat island impact on building energy performance at district level with the eureca platform. *Climate* **2021**, *9*, 48. [CrossRef]
16. Li, X.; Zhou, Y.; Yu, S.; Jia, G.; Li, H.; Li, W. Urban heat island impacts on building energy consumption: A review of approaches and findings. *Energy* **2019**, *174*, 407–419. [CrossRef]
17. Narumi, D.; Levinson, R.; Shimoda, Y. Effect of urban heat island and global warming countermeasures on heat release and carbon dioxide emissions from a detached house. *Atmosphere* **2021**, *12*, 572. [CrossRef]
18. Khamchiangta, D.; Dhakal, S. Physical and non-physical factors driving urban heat island: Case of Bangkok Metropolitan Administration, Thailand. *J. Environ. Manag.* **2019**, *248*, 109285. [CrossRef] [PubMed]
19. Constantinescu, D.; Cheval, S.; Caracaş, G.; Dumitrescu, A. Effective monitoring and warning of Urban Heat Island effect on the indoor thermal risk in Bucharest (Romania). *Energy Build.* **2016**, *127*, 452–468. [CrossRef]
20. Zhang, B.; Amani-Beni, M.; Shi, Y.; Xie, G. The summer microclimate of green spaces in Beijing'Olympic park and their effects on human comfort index. *Ecol. Sci.* **2018**, *37*, 77–86.
21. Avdan, U.; Jovanovska, G. Algorithm for automated mapping of land surface temperature using LANDSAT 8 satellite data. *J. Sens.* **2016**, *2016*, 1480307. [CrossRef]
22. Gaitani, N.; Burud, I.; Thiis, T.; Santamouris, M. High-resolution spectral mapping of urban thermal properties with Unmanned Aerial Vehicles. *Build. Environ.* **2017**, *121*, 215–224. [CrossRef]
23. Naughton, J.; McDonald, W. Evaluating the variability of urban land surface temperatures using drone observations. *Remote Sens.* **2019**, *11*, 1722. [CrossRef]
24. Alavipanah, S.; Schreyer, J.; Haase, D.; Lakes, T.; Qureshi, S. The effect of multi-dimensional indicators on urban thermal conditions. *J. Clean. Product.* **2018**, *177*, 115–123. [CrossRef]
25. Buyantuyev, A.; Wu, J. Urbanization diversifies land surface phenology in arid environments: Interactions among vegetation, climatic variation, and land use pattern in the Phoenix metropolitan region, USA. *Landsc. Urban Plan.* **2012**, *105*, 149–159. [CrossRef]
26. Mirzaei, P.A. Recent challenges in modeling of urban heat island. *Sustain. Cities Soc.* **2015**, *19*, 200–206. [CrossRef]
27. Deng, C.; Wu, C. Examining the impacts of urban biophysical compositions on surface urban heat island: A spectral unmixing and thermal mixing approach. *Remote Sens. Environ.* **2013**, *131*, 262–274. [CrossRef]
28. Nuruzzaman, M. Urban heat island: Causes, effects and mitigation measures-a review. *Int. J. Environ. Monit. Anal.* **2015**, *3*, 67–73. [CrossRef]
29. Cadenasso, M.; Pickett, S.; Grove, J. Dimensions of ecosystem complexity: Heterogeneity, connectivity, and history. *Ecol. Complex.* **2006**, *3*, 1–12. [CrossRef]
30. Connors, J.P.; Galletti, C.S.; Chow, W.T. Landscape configuration and urban heat island effects: Assessing the relationship between landscape characteristics and land surface temperature in Phoenix, Arizona. *Landsc. Ecol.* **2013**, *28*, 271–283. [CrossRef]
31. Xie, M.; Wang, Y.; Chang, Q.; Fu, M.; Ye, M. Assessment of landscape patterns affecting land surface temperature in different biophysical gradients in Shenzhen, China. *Urban Ecosyst.* **2013**, *16*, 871–886. [CrossRef]
32. Masoudi, M.; Tan, P.Y. Multi-year comparison of the effects of spatial pattern of urban green spaces on urban land surface temperature. *Landsc. Urban Plan.* **2019**, *184*, 44–58. [CrossRef]
33. Adulkongkaew, T.; Satapanajaru, T.; Charoenhirunyingyos, S.; Singhirunnusorn, W. Effect of land cover composition and building configuration on land surface temperature in an urban-sprawl city, case study in Bangkok Metropolitan Area, Thailand. *Heliyon* **2020**, *6*, e04485. [CrossRef] [PubMed]
34. Aram, F.; García, E.H.; Solgi, E.; Mansournia, S. Urban green space cooling effect in cities. *Heliyon* **2019**, *5*, e01339. [CrossRef] [PubMed]

35. McPhearson, T.; Pickett, S.T.; Grimm, N.B.; Niemelä, J.; Alberti, M.; Elmqvist, T.; Weber, C.; Haase, D.; Breuste, J.; Qureshi, S. Advancing urban ecology toward a science of cities. *BioScience* **2016**, *66*, 198–212. [CrossRef]
36. McIntyre, N.E.; Knowles-Yánez, K.; Hope, D. Urban ecology as an interdisciplinary field: Differences in the use of "urban" between the social and natural sciences. In *Urban Ecology*; Springer: Berlin/Heidelberg, Germany, 2008; pp. 49–65.
37. Barot, S.; Abbadie, L.; Auclerc, A.; Barthelemy, C.; Bérille, E.; Billet, P.; Clergeau, P.; Consalès, J.-N.; Deschamp-Cottin, M.; David, A. Urban ecology, stakeholders and the future of ecology. *Sci. Total Environ.* **2019**, *667*, 475–484. [CrossRef]
38. Inostroza, L.; Hamstead, Z.; Spyra, M.; Qureshi, S. Beyond urban–rural dichotomies: Measuring urbanisation degrees in central European landscapes using the technomass as an explicit indicator. *Ecol. Indic.* **2019**, *96*, 466–476. [CrossRef]
39. Forman, R.T. Urban ecology principles: Are urban ecology and natural area ecology really different? *Landsc. Ecol.* **2016**, *31*, 1653–1662. [CrossRef]
40. Wu, J. Landscape sustainability science: Ecosystem services and human well-being in changing landscapes. *Landsc. Ecol.* **2013**, *28*, 999–1023. [CrossRef]
41. Zhou, W.; Pickett, S.; McPhearson, T. Conceptual frameworks facilitate integration for transdisciplinary urban science. *NJP Urban Sustain.* **2021**, *1*, 1–11. [CrossRef]
42. Pickett, S.T.; Cadenasso, M.L.; Childers, D.L.; McDonnell, M.J.; Zhou, W. Evolution and future of urban ecological science: Ecology in, of, and for the city. *Ecosyst. Health Sustain.* **2016**, *2*, e01229. [CrossRef]
43. Grimm, N.B.; Pickett, S.T.; Hale, R.L.; Cadenasso, M.L. Does the ecological concept of disturbance have utility in urban social–ecological–technological systems? *Ecosyst. Health Sustain.* **2017**, *3*, e01255. [CrossRef]
44. Niemelä, J. Ecology of urban green spaces: The way forward in answering major research questions. *Landsc. Urban Plan.* **2014**, *125*, 298–303. [CrossRef]
45. Li, Y.; Li, Y.; Qureshi, S.; Kappas, M.; Hubacek, K. On the relationship between landscape ecological patterns and water quality across gradient zones of rapid urbanization in coastal China. *Ecol. Model.* **2015**, *318*, 100–108. [CrossRef]
46. Pickett, S.; Cadenasso, M.; Rosi-Marshall, E.J.; Belt, K.T.; Groffman, P.M.; Grove, J.M.; Irwin, E.G.; Kaushal, S.S.; LaDeau, S.L.; Nilon, C.H. Dynamic heterogeneity: A framework to promote ecological integration and hypothesis generation in urban systems. *Urban Ecosyst.* **2017**, *20*, 1–14. [CrossRef]
47. Naveh, Z. Ecosystem and landscapes–a critical comparative appraisal. *J. Landsc. Ecol.* **2010**, *3*, 64–81. [CrossRef]
48. Ahern, J. Urban landscape sustainability and resilience: The promise and challenges of integrating ecology with urban planning and design. *Landsc. Ecol.* **2013**, *28*, 1203–1212. [CrossRef]
49. Wu, J.; Hobbs, R. Key issues and research priorities in landscape ecology: An idiosyncratic synthesis. *Landsc. Ecol.* **2002**, *17*, 355–365. [CrossRef]
50. Francis, R.A.; Millington, J.D.; Chadwick, M.A. *Urban Landscape Ecology: Science, Policy and Practice*; Routledge: Oxfordshire, UK, 2016.
51. Darvishi, A.; Yousefi, M.; Dinan, N.M.; Angelstam, P. Assessing levels, trade-offs and synergies of landscape services in the Iranian province of Qazvin: Towards sustainable landscapes. *Landsc. Ecol.* **2022**, *37*, 305–327. [CrossRef]
52. Wu, J.; He, C.; Huang, G.; Yu, D. Urban landscape ecology: Past, present, and future. In *Landscape Ecology for Sustainable Environment and Culture*; Springer: Berlin/Heidelberg, Germany, 2013; pp. 37–53.
53. Asgarian, A.; Amiri, B.J.; Sakieh, Y. Assessing the effect of green cover spatial patterns on urban land surface temperature using landscape metrics approach. *Urban Ecosyst.* **2015**, *18*, 209–222. [CrossRef]
54. Mokhtari, Z.; Barghjelveh, S.; Sayahnia, R.; Karami, P.; Qureshi, S.; Russo, A. Spatial pattern of the green heat sink using patch-and network-based analysis: Implication for urban temperature alleviation. *Sustain. Cities Soc.* **2022**, *83*, 103964. [CrossRef]
55. Estoque, R.C.; Murayama, Y.; Myint, S.W. Effects of landscape composition and pattern on land surface temperature: An urban heat island study in the megacities of Southeast Asia. *Sci. Total Environ.* **2017**, *577*, 349–359. [CrossRef] [PubMed]
56. Stoll, S.; Frenzel, M.; Burkhard, B.; Adamescu, M.; Augustaitis, A.; Baeßler, C.; Bonet, F.J.; Carranza, M.L.; Cazacu, C.; Cosor, G.L. Assessment of ecosystem integrity and service gradients across Europe using the LTER Europe network. *Ecol. Model.* **2015**, *295*, 75–87. [CrossRef]
57. Yli-Pelkonen, V.; Niemelä, J. Conservation, Linking ecological and social systems in cities: Urban planning in Finland as a case. *Biodivers. Conserv.* **2005**, *14*, 1947–1967. [CrossRef]
58. Pickett, S.T.; Cadenasso, M.L.; Grove, J.M.; Nilon, C.H.; Pouyat, R.V.; Zipperer, W.C.; Costanza, R. Urban ecological systems: Linking terrestrial ecological, physical, and socioeconomic components of metropolitan areas. In *Urban Ecology*; Springer: Berlin/Heidelberg, Germany, 2008; pp. 99–122.
59. Anderson, P.; Elmqvist, T. Urban ecological and social-ecological research in the City of Cape Town: Insights emerging from an urban ecology CityLab. *Ecol. Soc.* **2012**, *17*, 23. [CrossRef]
60. Alberti, M.; Marzluff, J.M.; Shulenberger, E.; Bradley, G.; Ryan, C.; Zumbrunnen, C. Integrating humans into ecology: Opportunities and challenges for studying urban ecosystems. *BioScience* **2003**, *53*, 1169–1179. [CrossRef]
61. Wu, J.; Levin, S.A. A patch-based spatial modeling approach: Conceptual framework and simulation scheme. *Ecol. Model.* **1997**, *101*, 325–346. [CrossRef]
62. Müller, F.; Hoffmann-Kroll, R.; Wiggering, H. Indicating ecosystem integrity—Theoretical concepts and environmental requirements. *Ecol. Model.* **2000**, *130*, 13–23. [CrossRef]
63. McGrath, B.; Pickett, S.T. The metacity: A conceptual framework for integrating ecology and urban design. *Challenges* **2011**, *2*, 55–72. [CrossRef]

64. Pickett, S.T.; Kolasa, J.; Jones, C.G. *Ecological Understanding: The Nature of Theory and the Theory of Nature*; Elsevier: Amsterdam, The Netherlands, 2010.
65. Alberti, M.J. Maintaining ecological integrity and sustaining ecosystem function in urban areas. *Curr. Opin. Environ. Sustain.* **2010**, *2*, 178–184. [CrossRef]
66. Zhou, W.; Pickett, S.T.; Cadenasso, M.L. Shifting concepts of urban spatial heterogeneity and their implications for sustainability. *Landsc. Ecol.* **2017**, *32*, 15–30. [CrossRef]
67. Childers, D.L.; Cadenasso, M.L.; Grove, J.M.; Marshall, V.; McGrath, B.; Pickett, S.T. An ecology for cities: A transformational nexus of design and ecology to advance climate change resilience and urban sustainability. *Sustainability* **2015**, *7*, 3774–3791. [CrossRef]
68. Rademacher, A.; Cadenasso, M.L.; Pickett, S.T. From feedbacks to coproduction: Toward an integrated conceptual framework for urban ecosystems. *Urban Ecosyst.* **2019**, *22*, 65–76. [CrossRef]
69. Guerrero, A.M.; Bennett, N.J.; Wilson, K.A.; Carter, N.; Gill, D.; Mills, M.; Ives, C.D.; Selinske, M.J.; Larrosa, C.; Bekessy, S. Achieving the promise of integration in social-ecological research. *Ecol. Soc.* **2018**, *23*, 28. [CrossRef]
70. McPhearson, T.; Haase, D.; Kabisch, N.; Gren, Å. Advancing Understanding of the Complex Nature of Urban Systems. *Ecol. Indic.* **2016**, *70*, 566–573. [CrossRef]
71. Alberti, M. *Advances in Urban Ecology: Integrating Humans and Ecological Processes in Urban Ecosystems;* Springer: Berlin/Heidelberg, Germany, 2008.
72. Andersson, E.; Haase, D.; Anderson, P.; Cortinovis, C.; Goodness, J.; Kendal, D.; Lausch, A.; McPhearson, T.; Sikorska, D.; Wellmann, T. What are the traits of a social-ecological system: Towards a framework in support of urban sustainability. *NPJ Urban Sustain.* **2021**, *1*, 1–8. [CrossRef]
73. Tress, G.; Tress, B.; Fry, G. Clarifying integrative research concepts in landscape ecology. *Landsc. Ecol.* **2005**, *20*, 479–493. [CrossRef]
74. Pickett, S.T.; Cadenasso, M.L.; Baker, M.E.; Band, L.E.; Boone, C.G.; Buckley, G.L.; Groffman, P.M.; Grove, J.M.; Irwin, E.G.; Kaushal, S.S. Theoretical perspectives of the baltimore ecosystem study: Conceptual evolution in a social-ecological research project. *BioScience* **2020**, *70*, 297–314. [CrossRef] [PubMed]
75. Bai, X.; Surveyer, A.; Elmqvist, T.; Gatzweiler, F.W.; Güneralp, B.; Parnell, S.; Prieur-Richard, A.-H.; Shrivastava, P.; Siri, J.G.; Stafford-Smith, M. Defining and advancing a systems approach for sustainable cities. *Curr. Opin. Environ. Sustain.* **2016**, *23*, 69–78. [CrossRef]
76. Collins, J.P.; Kinzig, A.; Grimm, N.B.; Fagan, W.F.; Hope, D.; Wu, J.; Borer, E.T. A new urban ecology: Modeling human communities as integral parts of ecosystems poses special problems for the development and testing of ecological theory. *Am. Sci.* **2000**, *88*, 416–425.
77. Coelho, D.; Ruth, M. Seeking a unified urban systems theory. *WIT Trans. Ecol. Environ.* **2006**, *93*, 179–188.
78. Schröder, B. Pattern, process, and function in landscape ecology and catchment hydrology—How can quantitative landscape ecology support predictions in ungauged basins? *Hydrol. Earth Syst. Sci.* **2006**, *10*, 967–979. [CrossRef]
79. Deilami, K.; Kamruzzaman, M.; Liu, Y. Urban heat island effect: A systematic review of spatio-temporal factors, data, methods, and mitigation measures. *Int. J. Appl. Earth Observ. Geoinform.* **2018**, *67*, 30–42. [CrossRef]
80. Jamei, E.; Rajagopalan, P.; Seyedmahmoudian, M.; Jamei, Y. Review on the impact of urban geometry and pedestrian level greening on outdoor thermal comfort. *Renew. Sustain. Energy Rev.* **2016**, *54*, 1002–1017. [CrossRef]
81. Santamouris, M. Cooling the cities—A review of reflective and green roof mitigation technologies to fight heat island and improve comfort in urban environments. *Sol. Energy* **2014**, *103*, 682–703. [CrossRef]
82. Kennedy, M.M. Defining a literature. *Educ. Res.* **2007**, *36*, 139–147. [CrossRef]
83. Paré, G.; Trudel, M.-C.; Jaana, M.; Kitsiou, S. Synthesizing information systems knowledge: A typology of literature reviews. *Inf. Manag.* **2015**, *52*, 183–199. [CrossRef]
84. Turner, M.G.; Chapin, F.S. Causes and consequences of spatial heterogeneity in ecosystem function. In *Ecosystem Function in Heterogeneous Landscapes*; Springer: Berlin/Heidelberg, Germany, 2005; pp. 9–30.
85. Zipperer, W.C.; Wu, J.; Pouyat, R.V.; Pickett, S.T. The application of ecological principles to urban and urbanizing landscapes. *Ecol. Appl.* **2000**, *10*, 685–688. [CrossRef]
86. Lovett, G.M.; Jones, C.G.; Turner, M.G.; Weathers, K.C. Ecosystem function in heterogeneous landscapes. In *Ecosystem Function in Heterogeneous Landscapes*; Springer: Berlin/Heidelberg, Germany, 2005; pp. 1–4.
87. Carpenter, S.R. The need for large-scale experiments to assess and predict the response of ecosystems to perturbation. In *Successes, Limitations, and Frontiers in Ecosystem Science*; Springer: Berlin/Heidelberg, Germany, 1998; pp. 287–312.
88. Fahrig, L.; Baudry, J.; Brotons, L.; Burel, F.G.; Crist, T.O.; Fuller, R.J.; Sirami, C.; Siriwardena, G.M.; Martin, J.L. Functional landscape heterogeneity and animal biodiversity in agricultural landscapes. *Ecol. Lett.* **2011**, *14*, 101–112. [CrossRef]
89. Turner, M.G.; Gardner, R.H.; O'neill, R.V.; O'Neill, R.V. *Landscape Ecology in Theory and Practice*; Springer: Berlin/Heidelberg, Germany, 2001; Volume 401.
90. Brown, D.G.; Aspinall, R.; Bennett, D.A. Landscape models and explanation in landscape ecology—A space for generative landscape science? *Prof. Geogr.* **2006**, *58*, 369–382. [CrossRef]
91. Wiens, J.A.; Chr, N.; Van Horne, B.; Ims, R.A. Ecological mechanisms and landscape ecology. *Oikos* **1993**, *66*, 369–380. [CrossRef]
92. Schröder, B.; Seppelt, R. Analysis of pattern–process interactions based on landscape models—Overview, general concepts, and methodological issues. *Ecol. Model.* **2006**, *199*, 505–516. [CrossRef]

93. Pauleit, S.; Breuste, J.H. Land use and surface cover as urban ecological indicators. In *Handbook of Urban Ecology*; Oxford University Press: Oxford, UK, 2011.
94. Decker, E.H.; Elliott, S.; Smith, F.A.; Blake, D.R.; Rowland, F.S. Energy and material flow through the urban ecosystem. *Annu. Rev. Energy Environ.* **2000**, *25*, 685–740. [CrossRef]
95. Zhao, W. Analysis on the characteristic of energy flow in urban ecological economic system—A case of Xiamen city. *Procedia Environ. Sci.* **2012**, *13*, 2274–2279. [CrossRef]
96. Barles, S. Society, energy and materials: The contribution of urban metabolism studies to sustainable urban development issues. *J. Environ. Plan. Manag.* **2010**, *53*, 439–455. [CrossRef]
97. Bai, X. Eight energy and material flow characteristics of urban ecosystems. *Ambio* **2016**, *45*, 819–830. [CrossRef]
98. Qian, S.; Qin, D.; Wu, X.; Hu, S.; Hu, L.; Lin, D.; Zhao, L.; Shang, K.; Song, K.; Yang, Y. Urban growth and topographical factors shape patterns of spontaneous plant community diversity in a mountainous city in southwest China. *Urban For. Urban Green.* **2020**, *55*, 126814. [CrossRef]
99. Farinha-Marques, P.; Lameiras, J.; Fernandes, C.; Silva, S.; Guilherme, F. Urban biodiversity: A review of current concepts and contributions to multidisciplinary approaches. *Innov. Eur. J. Soc. Sci. Res.* **2011**, *24*, 247–271. [CrossRef]
100. Boone, C.G.; Fragkias, M. *Urbanization and Sustainability: Linking Urban Ecology, Environmental Justice and Global Environmental Change*; Springer Science & Business Media: Berlin/Heidelberg, Germany, 2012; Volume 3.
101. Brelsford, C.; Dumas, M.; Schlager, E.; Dermody, B.J.; Aiuvalasit, M.; Allen-Dumas, M.R.; Beecher, J.; Bhatia, U.; D'odorico, P.; Garcia, M. Developing a sustainability science approach for water systems. *Ecol. Soc.* **2020**, *25*, 1–6. [CrossRef]
102. Irwin, E.G. New directions for urban economic models of land use change: Incorporating spatial dynamics and heterogeneity. *J. Reg. Sci.* **2010**, *50*, 65–91. [CrossRef]
103. Feng, J.; Wu, F.; Logan, J. From homogenous to heterogeneous: The transformation of Beijing's socio-spatial structure. *Built Environ.* **2008**, *34*, 482–498. [CrossRef]
104. Tang, M.; Li, Z.; Hu, F.; Wu, B. How does land urbanization promote urban eco-efficiency? The mediating effect of industrial structure advancement. *J. Clean. Prod.* **2020**, *272*, 122798. [CrossRef]
105. Chun-ye, W.; Wei-ping, Z. Analysis of the impact of urban wetland on urban temperature based on remote sensing technology. *Procedia Environ. Sci.* **2011**, *10*, 1546–1552. [CrossRef]
106. Shashua-Bar, L.; Pearlmutter, D.; Erell, E. The cooling efficiency of urban landscape strategies in a hot dry climate. *Landsc. Urban Plan.* **2009**, *92*, 179–186. [CrossRef]
107. Sung, C.Y. Mitigating surface urban heat island by a tree protection policy: A case study of The Woodland, Texas, USA. *Urban For. Urban Green.* **2013**, *12*, 474–480. [CrossRef]
108. Wang, W.; Zhang, B.; Zhou, W.; Lv, H.; Xiao, L.; Wang, H.; Du, H.; He, X. The effect of urbanization gradients and forest types on microclimatic regulation by trees, in association with climate, tree sizes and species compositions in Harbin city, northeastern China. *Urban Ecosyst.* **2019**, *22*, 367–384. [CrossRef]
109. Wang, X.; Dallimer, M.; Scott, C.E.; Shi, W.; Gao, J. Tree species richness and diversity predicts the magnitude of urban heat island mitigation effects of greenspaces. *Sci. Total. Environ.* **2021**, *770*, 145211. [CrossRef]
110. Grimmond, C.S.B.; Oke, T.R. An evapotranspiration-interception model for urban areas. *Water Resour. Res.* **1991**, *27*, 1739–1755. [CrossRef]
111. Kjelgren, R.; Montague, T. Urban tree transpiration over turf and asphalt surfaces. *Atmos. Environ.* **1998**, *32*, 35–41. [CrossRef]
112. Qiu, G.-Y.; LI, H.-Y.; Zhang, Q.-T.; Wan, C.; Liang, X.-J.; Li, X.-Z. Effects of evapotranspiration on mitigation of urban temperature by vegetation and urban agriculture. *J. Integr. Agric.* **2013**, *12*, 1307–1315. [CrossRef]
113. Taha, H.; Akbari, H.; Rosenfeld, A.; Huang, J. Residential cooling loads and the urban heat island—The effects of albedo. *Build. Environ.* **1988**, *23*, 271–283. [CrossRef]
114. Zardo, L.; Geneletti, D.; Pérez-Soba, M.; Van Eupen, M. Estimating the cooling capacity of green infrastructures to support urban planning. *Ecosyst. Serv.* **2017**, *26*, 225–235. [CrossRef]
115. Sen, S.; Roesler, J. Wind direction and cool surface strategies on microscale urban heat island. *Urban Clim.* **2020**, *31*, 100548. [CrossRef]
116. Schwaab, J.; Meier, R.; Mussetti, G.; Seneviratne, S.; Bürgi, C.; Davin, E.L. The role of urban trees in reducing land surface temperatures in European cities. *Nat. Commun.* **2021**, *12*, 1–11. [CrossRef]
117. Barbierato, E.; Bernetti, I.; Capecchi, I.; Saragosa, C. Quantifying the impact of trees on land surface temperature: A downscaling algorithm at city-scale. *Eur. J. Remote Sens.* **2019**, *52* (Suppl. 4), 74–83. [CrossRef]
118. Firozjaei, M.K.; Alavipanah, S.K.; Liu, H.; Sedighi, A.; Mijani, N.; Kiavarz, M.; Weng, Q. A PCA–OLS model for assessing the impact of surface biophysical parameters on land surface temperature variations. *Remote Sens.* **2019**, *11*, 2094. [CrossRef]
119. Guha, S.; Govil, H.; Dey, A.; Gill, N. Analytical study of land surface temperature with NDVI and NDBI using Landsat 8 OLI and TIRS data in Florence and Naples city, Italy. *Eur. J. Remote Sens.* **2018**, *51*, 667–678. [CrossRef]
120. Carrillo-Niquete, G.A.; Andrade, J.L.; Valdez-Lazalde, J.R.; Reyes-García, C.; Hernández-Stefanoni, J.L. Characterizing spatial and temporal deforestation and its effects on surface urban heat islands in a tropical city using Landsat time series. *Landsc. Urban Plan.* **2022**, *217*, 104280. [CrossRef]
121. Peng, X.; Wu, W.; Zheng, Y.; Sun, J.; Hu, T.; Wang, P. Correlation analysis of land surface temperature and topographic elements in Hangzhou, China. *Sci. Rep.* **2020**, *10*, 1–16. [CrossRef] [PubMed]

122. Serrano, S.V.; Prats, J.C.; Sánchez, M. Topography and Vegetation Cover Influence on Urban Heat Island of Zaragoza (Spain). In Proceedings of the 5th international conference on urban climate, Łódź, Poland, 1–5 September 2003.
123. Chun, B.; Guldmann, J.-M. Spatial statistical analysis and simulation of the urban heat island in high-density central cities. *Landsc. Urban Plan.* **2014**, *125*, 76–88. [CrossRef]
124. Kuang, W.; Liu, Y.; Dou, Y.; Chi, W.; Chen, G.; Gao, C.; Yang, T.; Liu, J.; Zhang, R. What are hot and what are not in an urban landscape: Quantifying and explaining the land surface temperature pattern in Beijing, China. *Landsc. Ecol.* **2015**, *30*, 357–373. [CrossRef]
125. Nassar, A.K.; Blackburn, G.A.; Whyatt, J.D. Dynamics and controls of urban heat sink and island phenomena in a desert city: Development of a local climate zone scheme using remotely-sensed inputs. *Int. J. Appl. Earth Observ. Geoinform.* **2016**, *51*, 76–90. [CrossRef]
126. Alchapar, N.L.; Correa, E.N.; Cantón, M.A. Classification of building materials used in the urban envelopes according to their capacity for mitigation of the urban heat island in semiarid zones. *Energy Build.* **2014**, *69*, 22–32. [CrossRef]
127. Arnfield, A.J. Two decades of urban climate research: A review of turbulence, exchanges of energy and water, and the urban heat island. *Int. J. Climatol. J. R. Meteorol. Soc.* **2003**, *23*, 1–26. [CrossRef]
128. Arnold, C.L., Jr.; Gibbons, C.J. Impervious surface coverage: The emergence of a key environmental indicator. *J. Am. Plan. Assoc.* **1996**, *62*, 243–258. [CrossRef]
129. Gill, S.E.; Handley, J.F.; Ennos, A.R.; Pauleit, S. Adapting cities for climate change: The role of the green infrastructure. *Built Environ.* **2007**, *33*, 115–133. [CrossRef]
130. Gunawardena, K.R.; Wells, M.J.; Kershaw, T. Utilising green and bluespace to mitigate urban heat island intensity. *Sci. Total Environ.* **2017**, *584*, 1040–1055. [CrossRef] [PubMed]
131. Jenerette, G.D.; Harlan, S.L.; Brazel, A.; Jones, N.; Larsen, L.; Stefanov, W.L. Regional relationships between surface temperature, vegetation, and human settlement in a rapidly urbanizing ecosystem. *Landsc. Ecol.* **2007**, *22*, 353–365. [CrossRef]
132. Li, Y.; Sun, Y.; Li, J.; Gao, C. Socioeconomic drivers of urban heat island effect: Empirical evidence from major Chinese cities. *Sustain. Cities Soc.* **2020**, *63*, 102425. [CrossRef]
133. Boehme, P.; Berger, M.; Massier, T. Estimating the building based energy consumption as an anthropogenic contribution to urban heat islands. *Sustain. Cities Soc.* **2015**, *19*, 373–384. [CrossRef]
134. Dong, Y.; Varquez, A.; Kanda, M. Global anthropogenic heat flux database with high spatial resolution. *Atmos. Environ.* **2017**, *150*, 276–294. [CrossRef]
135. Zhu, R.; Wong, M.S.; Guilbert, É.; Chan, P.-W. Understanding heat patterns produced by vehicular flows in urban areas. *Sci. Rep.* **2017**, *7*, 1–14. [CrossRef] [PubMed]
136. Avolio, M.; Pataki, D.E.; Gillespie, T.; Jenerette, G.D.; McCarthy, H.R.; Pincetl, S.; Weller-Clarke, L. Tree diversity in southern California's urban forest: The interacting roles of social and environmental variables. *Front. Ecol. Evol.* **2015**, *3*, 73. [CrossRef]
137. Avolio, M.L.; Pataki, D.E.; Pincetl, S.; Gillespie, T.W.; Jenerette, G.D.; McCarthy, H.R. Understanding preferences for tree attributes: The relative effects of socio-economic and local environmental factors. *Urban Ecosyst.* **2015**, *18*, 73–86. [CrossRef]
138. Morakinyo, T.E.; Ouyang, W.; Lau, K.K.-L.; Ren, C.; Ng, E. Right tree, right place (urban canyon): Tree species selection approach for optimum urban heat mitigation-development and evaluation. *Sci. Total Environ.* **2020**, *719*, 137461. [CrossRef] [PubMed]
139. Rosenzweig, C.; Solecki, W.; Slosberg, R. Mitigating New York City's Heat Island with Urban Forestry, Living Roofs, and Light Surfaces. *Rep. N. Y. State Energy Res. Dev. Auth.* **2006**.
140. Yamamoto, Y. *Measures to Mitigate Urban Heat Islands*; NISTEP Science & Technology Foresight Center: Tokyo, Japan, 2006.
141. Myint, S.W.; Brazel, A.; Okin, G.; Buyantuyev, A.J.G.; Sensing, R. Combined effects of impervious surface and vegetation cover on air temperature variations in a rapidly expanding desert city. *GISci. Remote Sens.* **2010**, *47*, 301–320. [CrossRef]
142. Sarrat, C.; Lemonsu, A.; Masson, V.; Guédalia, D. Impact of urban heat island on regional atmospheric pollution. *Atmos. Environ.* **2006**, *40*, 1743–1758. [CrossRef]
143. Li, H.; Meier, F.; Lee, X.; Chakraborty, T.; Liu, J.; Schaap, M.; Sodoudi, S. Interaction between urban heat island and urban pollution island during summer in Berlin. *Sci. Total. Environ.* **2018**, *636*, 818–828. [CrossRef]
144. Patz, J.A.; Campbell-Lendrum, D.; Holloway, T.; Foley, J.A. Impact of regional climate change on human health. *Nature* **2005**, *438*, 310–317. [CrossRef]
145. Vargo, J.; Stone, B.; Habeeb, D.; Liu, P.; Russell, A. The social and spatial distribution of temperature-related health impacts from urban heat island reduction policies. *Environ. Sci. Policy* **2016**, *66*, 366–374. [CrossRef]
146. Venter, Z.S.; Krog, N.H.; Barton, D.N. Linking green infrastructure to urban heat and human health risk mitigation in Oslo, Norway. *Sci. Total. Environ.* **2020**, *709*, 136193. [CrossRef] [PubMed]
147. Van Hove, L.; Jacobs, C.; Heusinkveld, B.; Elbers, J.; Van Driel, B.; Holtslag, A. Temporal and spatial variability of urban heat island and thermal comfort within the Rotterdam agglomeration. *Build. Environ.* **2015**, *83*, 91–103. [CrossRef]
148. Hiemstra, J.A.; Saaroni, H.; Amorim, J.H. The urban heat Island: Thermal comfort and the role of urban greening. In *The Urban Forest*; Springer: Berlin/Heidelberg, Germany, 2017; pp. 7–19.
149. Alvarez, I.; Quesada-Ganuza, L.; Briz, E.; Garmendia, L. Urban heat islands and thermal comfort: A case study of Zorrotzaurre island in Bilbao. *Sustainability* **2021**, *13*, 6106. [CrossRef]
150. Elsayed, I. Mitigation of the urban heat island of the City of Kuala Lumpur. *Malaysia* **2012**, *11*, 1602–1613.

151. Bush, J.J.E.I.; Transitions, S. The role of local government greening policies in the transition towards nature-based cities. *Environ. Innov. Soc. Trans.* **2020**, *35*, 35–44. [CrossRef]
152. Founda, D.; Santamouris, M. Synergies between Urban Heat Island and Heat Waves in Athens (Greece), during an extremely hot summer (2012). *Sci. Rep.* **2017**, *7*, 1–11. [CrossRef] [PubMed]
153. Jiang, S.; Lee, X.; Wang, J.; Wang, K. Amplified urban heat islands during heat wave periods. *J. Geophys. Res. Atmos.* **2019**, *124*, 7797–7812. [CrossRef]
154. Zhao, L.; Oppenheimer, M.; Zhu, Q.; Baldwin, J.W.; Ebi, K.L.; Bou-Zeid, E.; Guan, K.; Liu, X. Interactions between urban heat islands and heat waves. *Environ. Res. Lett.* **2018**, *13*, 034003. [CrossRef]
155. Kelly, J.M. Mechanistic Models: An Underutilized Tool in Soil Science Research. *Soil Horiz.* **2014**, *55*, 1–2. [CrossRef]
156. Cabral, J.S.; Valente, L.; Hartig, F. Mechanistic simulation models in macroecology and biogeography: State-of-art and prospects. *Ecography* **2017**, *40*, 267–280. [CrossRef]
157. Kragt, M.E. *A Beginners Guide to Bayesian Network Modelling for Integrated Catchment Management*; Landscape Logic: Charlevoix, MI, USA, 2009.
158. Pollino, C.; Hart, B.T. Developing Bayesian Network Models within a Risk Assessment Framework. In Proceedings of the Integrating Sciences and Information Technologyfor Environmental Assessment and Decision Making, Barcelona, Spain, 7–10 July 2008.
159. Maxwell, P.S.; Pitt, K.A.; Olds, A.D.; Rissik, D.; Connolly, R.M. Identifying habitats at risk: Simple models can reveal complex ecosystem dynamics. *Ecol. Appl.* **2015**, *25*, 573–587. [CrossRef]
160. Koller, D.; Friedman, N. *Probabilistic Graphical Models: Principles and Techniques*; MIT Press: Cambridge, MA, USA, 2009.
161. Auger-Méthé, M.; Newman, K.; Cole, D.; Empacher, F.; Gryba, R.; King, A.A.; Leos-Barajas, V.; Mills Flemming, J.; Nielsen, A.; Petris, G. A guide to state–space modeling of ecological time series. *Ecol. Monogr.* **2021**, *91*, e01470. [CrossRef]
162. Gelman, A.; Carlin, J.B.; Stern, H.S.; Rubin, D.B. *Bayesian Data Analysis*; Chapman and Hall/CRC: Boca Raton, FL, USA, 1995.
163. Forman, R.T. Interaction among landscape elements: A core of landscape ecology. *Perspect. Landsc. Ecol.* **1981**, 35–48.
164. Mokhtari, Z.; Barghjelveh, S.; Sayahnia, R. Heterogeneity of the thermal environment and its ecological evaluation in the urban region of Karaj. *Geogr. Environ. Sustain.* **2021**, *11*, 37–58.

MDPI
St. Alban-Anlage 66
4052 Basel
Switzerland
www.mdpi.com

Land Editorial Office
E-mail: land@mdpi.com
www.mdpi.com/journal/land

www.ingramcontent.com/pod-product-compliance
Lightning Source LLC
Chambersburg PA
CBHW040834250926
51850CB00007BA/1313
9783036587479